Lecture Notes in Artificial Intelligence 4884

Edited by R. Goebel, J. Siekmann, and W. Wahlster

Subseries of Lecture Notes in Computer Science

T0223171

Pompeu Casanovas   Giovanni Sartor
Núria Casellas   Rossella Rubino (Eds.)

Computable Models of the Law

Languages, Dialogues, Games, Ontologies

 Springer

Series Editors

Randy Goebel, University of Alberta, Edmonton, Canada
Jörg Siekmann, University of Saarland, Saarbrücken, Germany
Wolfgang Wahlster, DFKI and University of Saarland, Saarbrücken, Germany

Volume Editors

Pompeu Casanovas
Núria Casellas
Universitat Autònoma de Barcelona, Institute of Law and Technology
08193 Bellaterra, Barcelona, Spain
E-mail: {pompeu.casanovas, nuria.casellas}@uab.es

Giovanni Sartor
European University Institute, Badia Fiesolana
Via dei Roccettini 9, 50016 San Domenico di Fiesole, Florence, Italy
E-mail: giovanni.sartor@eui.eu

Rossella Rubino
Università di Bologna, CIRSFID
Via Galliera 3, 40121 Bologna Italy
E-mail: rossella.rubino@unibo.it

Library of Congress Control Number: Applied for

CR Subject Classification (1998): I.2, H.4, H.5, J.1, K.4.1-2

LNCS Sublibrary: SL 7 – Artificial Intelligence

ISSN   0302-9743
ISBN-10   3-540-85568-8 Springer Berlin Heidelberg New York
ISBN-13   978-3-540-85568-2 Springer Berlin Heidelberg New York

Springer is a part of Springer Science+Business Media

springer.com

© Springer-Verlag Berlin Heidelberg 2008

Typesetting: Camera-ready by author, data conversion by Scientific Publishing Services, Chennai, India
Printed on acid-free paper   SPIN: 12445684   06/3180   5 4 3 2 1 0

Foreword

Information technology has now pervaded all sectors of legal activities, and the very modern concepts of e-law and e-justice show that automation processes or more generally the use of computers to facilitate the legal practitioner, the judge, public administration and above all the citizen are ubiquitous. In spite of some reluctances shown in the past, the law field is experiencing nowadays a new *computer turn*.

Legal professions might have been facilitated in this evolution by the correlative revival of theoretical legal fields such as the study of legal philosophy, logics and reasoning, legal linguistics and legistics, which provide the structural basis to the development of artificial intelligence and law. But one of the most significant trends today is the wish to transpose the use of technology in each field of professional and private life to the legal field as well.

Current cross-border developments of human, economic and social activities add, moreover, the necessity to deal with foreign systems, mostly in foreign languages. The current extensions of European judicial cooperation make practitioners look forward to finding solutions in technology. During the last 18 months, Ministers of Justice of the European Union have showed an unbroken will to push forward the development of a European Justice Portal, to be opened within the next two years.

Needs and expectations today appear to be huge and it seems that every day there is a new legal field where solutions are partly expected from technology. European policies on transparency and information society, for instance, require the use of technology and its steady improvement.

European funding has already stimulated research in the field of computers and law and should continue to do so. The European Union as a system of law—functioning in the national systems of its current 27 member states as well as in its own legal system in 23 languages—needs reliable technological advances to implement its policies aiming at economic and social growth in a liveable environment.

On the level of the institutions of the European Union, organizing access to legal information and documents in 23 languages reveals itself as a new challenge. Nevertheless, strong documentary structures of the databases as well as the use of XML formats today offer a real potential for further development in retrieval and reuse of information.

The research gathered in this volume aims at building knowledge above content, at passing from the simple data retrieval to knowledge retrieval and at showing how artificial intelligence technologies are growing mature in the field of law. All projects share a common point in being supported by the European Union.

Computable Models of the Law presents under the subtitle "Languages, Dialogues, Games, Ontologies" not only research projects which seem to produce consequences in a distant future, but projects which can already find implementation areas and meet the needs of the community of lawyers and citizens.

April 2008   Pascale Berteloot

Preface

The origins of this book go back to a workshop held at the European University Institute of Florence on December 1 and 2, 2006. The theoretical purpose of that workshop was to start a fruitful discussion on the different ways of understanding and explaining contemporary law, for the purpose of building computable models of it (namely, models enabling the development of computer applications for the legal domain).

We realized that we cannot take for granted a single or unique way of modeling legal knowledge, namely, that there are multiple ways of identifying and circumscribing the "law" to be modeled, and multiple ways of representing legal contents into automatically processable information structures. The idea, then, was to get a better understanding of the theoretical assumptions of the different approaches underlying current EU projects on artificial intelligence and law, in order to explore future links and cooperation.

The practical purpose of the meeting was twofold. First, the meeting was meant to share some results obtained through different EU projects focused on computation, argumentation, law and normative systems. Secondly, by doing so, we thought that we could draw a general picture—the European state of the art in the field—and foster future synergies among the universities, institutes, companies, lawyers and computer scientists who were developing EU projects on artificial intelligence and law.

Actually, that workshop was just a starting point. During the next year several new contributions were received, discussed and reviewed. As a result, the volume contains 20 papers on the hot topics under research in the EU projects: legislative XML, legal ontologies, Semantic Web, search and meta-search engines, Web services, system's architecture, dialectic systems, dialogue games, multi-agent systems (MAS), legal argumentation, legal reasoning, e-justice and online dispute resolution.

Contributions have been provided by several ongoing (or recently finished) European projects on computation and law: ALIS, ArguGrid, ASPIC, DALOS, ESTRELLA, OpenKnowledge, SEAL, and SEKT. Some important national projects have provided their results as well: the Dutch BEST and DURP; and the Spanish Metabuscador, and OCJ-Iuriservice.

The final volume is divided into five sections: (i) Knowledge Representation, Ontologies and XML Legislative Drafting; (ii) Knowledge Representation, Legal Ontologies and Information Retrieval; (iii) Argumentation and Legal Reasoning; (iv) Normative and Multi-agent Systems; (v) Online Dispute Resolution.

We would like to thank all the contributors for their work and the patience they have shown with the editors, and to acknowledge the various publicly funded

R&D projects (One-Lex, E-sentencias FIT-350101-2006-26, Iuriservice SEJ2006-10695, and Metabuscador FIT-350100-2007-161) that have made publishing this book possible.

April 2008   Pompeu Casanovas
                                                                Giovanni Sartor
                                                                 Núria Casellas
                                                               Rossella Rubino

Table of Contents

Computable Models of the Law

I   Knowledge Representation, Ontologies and XML Legislative Drafting

II   Knowledge Representation, Legal Ontologies and Information Retrieval

III   Argumentation and Legal Reasoning

IV   Normative and Multi-agent Systems

V   Online Dispute Resolution

Computable Models of the Law and ICT: State of the Art and Trends in European Research

Giovanni Sartor[1,2], Pompeu Casanovas[3], Núria Casellas[3], and Rossella Rubino[2]

[1] European University Institute, Badia Fiesolana, Via dei Roccettini 9,
50016 San Domenico di Fiesole, Florence, Italy
giovanni.sartor@eui.eu
[2] CIRSFID Via Galliera, 3, 40121 Bologna - Italy
rossella.rubino@unibo.it
[3] Institute of Law and Technology (UAB)
Law Faculty, Campus Universitat Autnoma de Barcelona (Ed. B)
08193 Bellaterra, Barcelona, Spain
{pompeu.casanovas,nuria.casellas}@uab.es

1 Introduction

This volume is devoted to the presentation of several research contributions from some significant European research projects in the domain of *legal technologies*. In this domain European research has been particularly active in the last years, often achieving global leadership. This is due to the commitment of individual researchers, research centers and universities, but also to the support of the European Union that in various research programs has devoted a significant attention (and some relevant financing) to legal technologies, considering them a decisive aspect of E-government and a crucial resource for the development of the information society.

The projects here considered have emerged on the basis of the intersection of two converging trends: on the one hand the diffusion of information technologies in legal activities and on the other hand the development of computational models of legal knowledge and cognition. Exactly this intersection makes such projects feasible and significant, providing the background for the development of effective and innovative legal technologies. Therefore, before introducing the projects and contributions included in this issue we shall shortly present these trends.

2 The Diffusion of ICT in the Legal World

In recent years, technology has pervaded all domains of legal practice, playing an essential role in the daily routines of legal professionals. From the early text processors to the state-of-art tools of web 2.0 environments, legal professionals have adopted a number of technological solutions to work faster and more efficiently. Innovation continues to be incessant and the ICT market for the legal domain keeps growing at a nice clip. Today, court chambers and law firms tend to be more and more paper-free, and nearly all judges, lawyers, and legal assistants

P. Casanovas et al. (Eds.): Computable Models of the Law, LNAI 4884, pp. 1–20, 2008.
© Springer-Verlag Berlin Heidelberg 2008

perform their information searches online. Citizens may also obtain legal advice and file some cases through the Internet, where a growing number of dispute resolution services (also known as ODR) are also available. As in many other professional areas, a new paradigm is emerging where notions such as flexibility, synchronicity, or collaboration reign supreme.

Legal practice has always been an intensive knowledge task. Nevertheless, the technological innovations of the past few years have changed the way lawyers, judges, prosecutors, and legal clerks prepare their cases or draft documents and decisions. In some areas—notably, document management, case management, time recording systems, and legal information search—technology has provided software systems and databases that have already become standard in the legal profession.[1] With the coming to age of the Internet era, all these tools have progressively adapted to the needs of the online environment, offering more and more utilities and services. Therefore, it is not surprising that the IT expenditure of law firms, courts, or justice departments shows an upward trend. To quote some recent examples, the 2006 American Bar Association Tech Report reports that the average law firm spends 6% to 7% of gross revenue on technology-related expenses [3]. The Annual Technology Survey by the magazine Law Firm Inc. found that the 200 American largest firms spend on average almost $33K per lawyer, with an increase in IT spending in 2007 of about 15 percent–from $9.7 million in 2006 to $11.2 million in 2007 [4]. The 2007 ILTA Purchasing Technology Survey (surveying 467 ILTA member firms) specifies that firms under 200 lawyers are those which register "higher implementation rates for case management, courtroom technology, docketing software, imaging/scanning/OCR, patch management, and records management software" [5,6]. It is important to note that the law firm's sizes influence the use or distribution of ICT technologies. Compared to larger firms, smaller firms generally have fewer staff per attorney and do not usually arrange a standardised replacement cycle for PCs or laptops and buy, instead of lease, new technology. Larger firms are more likely to use virtual server software and to have metadata removal software for cleaning documents and video conferencing equipment [7].

Annual technology surveys cover an ever wide range of ICT topics – i.e. computing technologies in law offices and courtrooms, online research, electronic data discovery, web-based and mobile communications, etc. Interestingly enough, tech reports themselves have nowadays become part of the legal marketplace since, in nearly all cases, access to full reports is fee-based or restricted to members of legal associations. Let us summarise some of the basic trends reported in those technology surveys.

2.1 Legal Information Search

The vast majority of the legal profession performs legal information searches online. Although there are a number of legal information providers, Thomson

---

[1] Bibliography on lawyering and technology, and legal internet sites for lawyers is increasingly growing as well. For a comprehensive view on bibliography see [1]. See for a comprehensive listing of Internet sites [2].

(Westlaw), Reed Elsevier (Lexis-Nexis) and Wolters Kluwer, known as the Big Three, together control about 85 percent of the legal information market [8]. Outside of this specific market, almost all respondents to the 2007 ABA Tech Survey (96 percent) state that they rely on the Internet for news, at least once a week [8]. More specifically, the top five online resources mentioned are third party web sites (72 percent), e-mail newsletters (58 percent), e-mail discussion lists (38 percent), e-mail case alert services (37 percent), and online advance sheet services (30 percent) [8].

2.2 Electronic Data Discovery

Electronic data discovery (EDD) or e-discovery "is the process of collecting, preserving, reviewing and producing electronically stored information in response to a regulatory or legal investigation" [9]. The 2007 ABA Legal Technology Survey reports that 16 percent of firms receive EDD requests three to eleven times a year, and another 13 percent two times a year or less (while 57 percent of attorneys have never received EDD requests on behalf of their clients, this average was 62 percent in 2006 and 73 percent in 2005) [10]. Litigation support software is currently developing new tools to match these demands, because e-discovery is "the new reality of litigation" and because "the attorney is ultimately responsible for mistakes to the same extent he or she would be for errors in conducting traditional discovery" [11].

2.3 Web-Based Communications

Web-based communications have also been adopted in the daily practice of law. According to data from the 2006 AM LAW annual survey, web conference software ranks first, with 67 percent of the firms surveyed using it for administrative meetings, client meetings, in-house training programs, and communication with colleagues [12]. In contrast, 46 percent of firms report that instant messaging is prohibited, due to internal policies to minimize the risk of inappropriate exchanges in terms of e-discovery.

Moreover, technology helps lawyers to become mobile. Twenty-five percent of solo practitioners use wireless modems [13], and, according to the 2004-05 ABA LTS, eighty-eight percent of lawyers have access the Internet while away from the office. Not surprisingly, security is a top concern. Firms are placing a greater emphasis on disaster planning, data protection and other safeguards.

2.4 Collaborative Tools

Firms also tend to increase their use of extranets to communicate with clients, but other collaborative technologies such as intranets and wikis are less common. In contrast, used less frequently are legal blogs (or blawgs),[2] According to surveys, lawyers still do not use this tool: on the one hand, only 5 percent of 2007

---

[2] See [14] for an opinion mining analysis of the legal blogosphere.

ABA respondents have a blog, which is generally maintained by a single lawyer or a group of lawyers of the firm; on the other hand, over half of respondents never read blogs for current awareness (22 percent less than once a month, 12 percent one to three times per month, and 12 percent once or more in a week). The use of RSS feeds is even less frequent: 83 percent never use them and only 5 percent one or more times a week. Finally, the use of podcasts is still minimal: 3 percent of respondents use podcasts for current awareness one or more times per week, versus 80 percent of lawyers who never use podcasts.

2.5   Metadata and XML Technologies

Metadata is increasingly becoming an issue. On one hand, in 2006, for example, the ABA Committee on Ethics and Professional Responsibility ruled that opposing counsel could look at the metadata to check for changes or comments inside the documents [15].[3]

On the other, the current adoption and development of standards for legal information, electronic court filing, court documents, transcripts, criminal justice intelligence systems, etc. has become the core activity of a number of initiatives and projects. To quote some examples, the non-profit OASIS Legal-XML (a subgroup within OASIS) was created in 1998 to develop "open, non-proprietary technical standards for structuring legal documents and information using XML and related technologies" (LegalXML 2007). Technical committees within LegalXML work on areas such as court filing, e-contracts, e-notary, international justice, lawful intercept, legislative documents, and online dispute resolution. In Europe, the LEXML community defines itself as a "European network searching for the automatic exchange of legal information" [17]. Developing European standards includes learning lessons from previous national projects such as Norme in Rete (Italy), Metalex (the Netherlands), LexDania (Denmark), CHeXML (Switzerland), or eLaw (Austria).[4]

2.6   Technologies in Courtrooms and Judicial Offices

Technology has been one of the main concerns of national and European Courts, but the degree of its application to court offices and sentencing shows a great variety of solutions.[5] Several types of systems have been identified. Richard Susskind,

---

[3] See [16] for a description of the new legal problems of metadata, watermarking documents, digital times stamping, and *clickwrap* agreements (end-user licenses that appear on the computer, which a user must accept prior to downloading, installing or using a software application or online service).

[4] See, for the state of the art, [18].

[5] "Each country seems to have developed its own system strarting from scratch, outsourcing many phases of the projects due to a lack of technical expertise in te ministries of justice or court service agencies" [19]. See the 2000 European Report (9 European countries) on civil procedures as well: http://ruessmann.jura.uni-sb.de/grotius/english/.

e.g., describes the following: (i) management information systems, to help monitor the throughput and performance of courts; (ii) case administration systems, to support and automate the administrative work of court staff; (iii) judicial case management, including case tracking, case planning, telephone and video conferencing, and document management, intended for direct use by judges; (iv) judicial case management support systems, being the systems used by court staff in support of judges who are involved with case management; (v) non-judicial case management, to help court staff progress those many cases which are not disposed of judicially [20].

In 2001, M. Fabri and F. Contini advanced three cycles of ICT application to justice: (i) exploratory cycle; (ii) establishment of governance structures; (iii) evaluation and e-justice. A great deal of hopes has been put in the path towards a virtual courtroom and e-justice. *E-justice* may be defined as "the exploration and the exploitation of the possibilities of integration between the whole of judicial procedures and the web" [21]. But, as many scholars have pointed out, promises have only half-been fulfilled due to the diversity of legal cultures and the complexity of the organisational and judicial tasks [22,23]. This perspective has recently been corroborated by more recent comparative studies.[6] Technology per se does not ensure better organisational or sentencing results in judicial settings, and it constitutes, still, a challenge for the governance of European legal systems.

However, we think that this is precisely the reason why advanced research in Artificial Intelligence and Law matters. Only through the emergence and full development of the Semantic Web, well developed ontologies, dialectical systems for legal reasoning, and a better understanding of the possibilities offered by the implementation of multi-agent systems (MAS) it will be possible to fill the gap between organisational constraints, complexity of legal decision making, and implementation of web services for the judiciary.

Moreover, as shown by several projects that will be described later on (Section 4), the synergy between courts, other ways of alternative dispute resolution (ADR-ODR), and a user-based orientation of the computational models may contribute to a better and easier access of citizens to justice.

3   The Development of Computational Models of the Law

As we have observed at the beginning, research developed in European projects has merged the development of effective technologies for the legal practice, with the definition of computable models for representing legal knowledge. Some of these models have already been transferred into widely available tools for the legal practice (as those presented above), others have led so far only to a few applications, not still available to the majority of the potential users, others, finally are still in the research phase. Legal knowledge can be coded in different

---

[6] Contini and Cordella point out that "ICTs are in fact involved in the continuous interplay between human beings, formal rules, and technologies", and M. Fabri stresses that "not much appears to have been done in terms of listening to the 'voice of the customers'" [24,25].

natural and artificial languages and contained in multiple formats (text, audio, video, graphics, etc.). It can be developed, shared and conveyed in different ways by individuals, professional groups, corporations, political organisations, companies, or citizens at large. For developing advanced ICT applications for the legal domain we need to understand legal knowledge in context and we need to represent part of this knowledge in such a way that it can be used by a computer system. This calls for the interdisciplinary cooperation between different kind of scientific and technical expertise (lawyers, judges, jurists, social scientists, organisation scientists, legal theorists, computer scientists, engineers). In the following pages we will just mention some of the aspects these models address.

3.1   Models of Legal Documents

The computable representation of legal documents has been addressed since the 60's, when the first electronic systems for legal documentation where developed (see [26]). However, in the last few years the advent of the Internet and in particular, the advent of the semantic web have brought about a real revolution: the old discussion on distinguishing fixed field and free text and indexing legal materials in various ways has been supplanted by the discussion on how to structure legal texts according to shared standards, and how to enrich textual information with metadata of different kinds, possibly organised according to ontologies (see [27]). These metadata should not only facilitate access to the text, but also enable advanced manipulations over them, such as providing access to point-in-time legislation. Moreover, metadata can be extended to include representations of norms and concepts and thus provide the links to further levels in the representation of legal knowledge. Thus a strong interaction exists between the scientific inquiry into the study of legal documents and the development of XML-based model that are adopted in the practice, in the development of legal information systems and in the management of legal workflows, for instance in the legislative and in the judiciary processes (DALOS, SEAL).

3.2   Models of Legal Norms

Also the computable representation of legal norms now has a considerable history, where the first attempts go back to the beginning of the 70's and a considerable breakthrough was provided by the use of logic programming in the 80's (see [28]). More recently a considerable amount of new research has been accomplished, including the development of working rule-based systems, the definition of logics for legal norms as defeasible rules, models of the dynamics of normative systems, models of the use of norms in legal argumentation (see [29]). These developments not only have supported new applications for the legal domain, but also have provided a much needed impulse to innovation in legal theory and legal logic [30]. Recently studies of modeling legal norms have merged with studies on the interaction of autonomous agents: the issue of the norm-based governance of autonomous agents provides a challenging domain of inquiry where different areas of law, legal theory, cognitive psychology and computing can merge their efforts [31]. The projects presented in this book have taken into account these developments, but also have provided significant contributions to them (ESTRELLA, ALIS, ARGUGRID). These projects, to which some companies active in legal knowledge-based systems participate) show how the computer-supported application of legal rules is making its way into the legal practice (especially in the public administration) though much effort is still needed to bridge academic inquiries and applications for the legal profession.

3.3   Models of Legal Concepts

The computable modeling of legal concepts was originated in the framework of legal databases, where large efforts were devoted especially in the 60's and the 70's to indexing techniques based on conceptual relationships (such as thesauri, structured keywords, and so on). This area of research, after some years of partial neglect, has been approached with a renewed interest in the 90's and after, due to the emergence of the Internet, and the need to provide conceptual schemata to enable access and integration of heterogeneous resources. In the last years, much effort has been devoted to developing legal ontologies, namely machine-processable representations of legal concepts, which hopefully can be used as shared frameworks for multiple users [32]. Different approaches have been used for this purpose, such as moving from (automated) linguistic analysis of legal texts, from the analysis of legal practice, or from the available conceptual constructions in legal doctrine and legal theory (see [33]). Different models have also been adopted with regard to the integration between specifically legal conceptual resources and general conceptual resources, concerning the integration between legal ontologies and top level general ontologies, and the connection between legal concepts and common-sense concepts or concepts of other specific disciplines (DALOS, ESTRELLA, METASEARCH, SEKT). Technologies for the management of legal concepts are already in use in a number of applications (such as document management, electronic tutoring, etc.).

3.4   Models of Legal Cases

A fundamental aspect of legal knowledge is represented by cases, and in particular by judicial precedents. The research on the computable representation of precedents was started in the US already in the 70's, in the framework of research in artificial intelligence. More recently, the issue of the representation of cases has taken new strands: on the one hand the representation of cases has been connected to the logical representation of norms and to legal argumentation (ESTRELLA), on the other hand it has been connected to the issue of enriching case record with machine processable metadata (CASELEX).

3.5 Models of Legal Interaction

Traditionally efforts toward the representation of law have aimed at representing normative information. Recently, however, a distinct effort has been produced toward modeling also human interaction based upon the law. In this regard we can distinguish different lines of research. One deals with modeling legal dialogues, namely the argument-based interactions where legal issues are addressed in different contexts (see [34], [35]). Here the focus is on dialectical protocols establishing what argumentative moves are allowed to each party, and when the dialogue is to terminate, with what outcomes (ESTRELLA). While theoretical models of legal dialogues are not easily matched to real interactions, and into systems effectively managing such interactions [36], they are already inspiring systems for OCR. A different, but complementary approach consists in focusing on the interests and motivations of the parties, and to consider what actions they will rationally take toward one another considering their interests but also their expectations concerning the actions of the other parties (BEST). This leads to the use of game theory in modeling legal interactions (ALIS). Finally information on legal interaction also includes broader sociological analyses of legal problem-solving in social settings (SEKT).

4 The European Projects on Technology and Law Included in This Book

The current volume does not cover all European or EU-supported projects on legal technologies and computable models of the law, but provides a significant sample of them, showing some achievements obtained so far and indicating some promising directions of research. In the following pages we shortly present each of the projects indicating its main objectives.[7]

4.1 ALIS - Automated Legal Intelligent System

ALIS is a STREP Project funded by the European Commission under the 6th Framework. The Consortium is coordinated by ORT France (Dr. Michel Rudianski) and the project partners are: Imperial College London, Sineura SPA, Atos Origin SA, CBKE (Research Centre for Legal and Economical Aspects of Electronic Communication), SIVECO Romania SA, Exalead SA, Technical University Darmstadt, Alma Consulting Group, CIRSFID (University of Bologna) and the Gesica Paris Friedland law firm.[8]

ALIS aims at reducing the distance between citizens or companies and the legal system (regulations), by easing their access and use. Attention is paid to develop a system that will be: inclusive (open to citizens, companies and

---

[7] Information regarding projects has been extracted directly from their web pages and available project reports. Please refer to the cited webpages for further and accurate information regarding the projects.

[8] 01-01-2006/30-06-2009 (027968), http://www.alisproject.eu/.

industries); pan-European (open and usable in the whole European Union); time- and cost-effective; ergonomic and intelligent (capable of developing reasoning adapted both to the user and to the case under consideration).

ALIS is directed to offer easy management of legal knowledge in order to facilitate compliance with existing laws and regulations of governmental action, prevent conflicts when possible, propose methods for alternative dispute resolution and facilitate development and evolution of consistent legal and regulatory systems. To do so, ALIS investigates the combination of the recent advances in Theory, Artificial Intelligence and Law & Regulation Corpus Structuring Semantics in order to build efficient modeling tools.

4.2 ARGUGRID - Argumentation as a Foundation for the Semantic Grid

The Coordinator of this project is the Department of Computing of the Imperial College London (Dr. Francesca Toni) and the other members of the Consortium are: The University of London (Department of Computer Science, Royal Holloway), the Pisa University (Dipartimento di Informatica), the Institute of Communication and Computer Systems (National Technical University of Athens), the School of Engineering and Technology (Asian Institute of Technology), InforSense LTD, GMV S.A., and cosmoONE Hellas Market-site S.A. ARGUGRID is a STREP Project funded by the EC under the 6th Framework.[9]

The project aims at making an impact upon the Grid research area via the new model, architecture and platform and to impact business and business practices, by empowering Grid-enabled business application where multiple service providers and requesters exist. Also, although it focuses on e-business application scenarios, its results are outreaching to all kinds of applications. ARGUGRID is directed to making two novel contributions to grid computing:

1. To define the overall architecture for interfacing service-oriented workflows with argumentative agent technology, through the definition of semantic descriptions of workflows and the development of tools that map the results of the agent negotiation and planning phase into executable workflows.
2. To develop a grid-based platform enabling the formation of virtual organisation for the communication and interaction of agents. Using Peer-to-Peer and Overlay Network techniques and standardised communication protocols.

4.3 ASPIC - Argumentation Service Platform with Integrated Components

ASPIC is a STREP Project funded by the European Commission under the 6th Framework. The scientific Coordination of the project belongs to Cancer Research UK (Prof. John Fox). The consortium is formed by: Singular Logic S.A. (Greece), Zeus Consulting S.A. (member of the LogicDIS Group, Greece), University of Ljubljana (AI Laboratory, Slovenia), Technical University of Catalonia

---

[9] 01-06-2006/31-05-2009 (035200), http://www.argugrid.org.

(Knowledge Engineering and Machine Learning Group, KEMLg, Spain), Institut de Recherche en Informatique de Toulouse (University Paul Sabatier, France), University of Liverpool (Department of Computer Science, UK), University of Utrecht (Intelligent Systems Group, Institute of Information and Computing Sciences, The Netherlands) and City University of New York (USA).[10]

ASPIC is focused on knowledge-based services for the Information Society, based on semantically rich logic formalisms called argumentation systems. Initially, ASPIC will develop a common framework to underpin the services that are emerging as core functions of the argumentation paradigm. These include reasoning, decision-making, learning and communication. The end goal of the project is to offer a suite of software components based on this framework and to develop a platform for integrating these components with knowledge (e.g. semantic web) resources and legacy systems. The ASPIC consortium includes partners experienced in using argumentation systems in eHealth, eCommerce and eGovernment applications and these domains will provide practical domains for testing and validating technology components.

4.4   BEST - BATNA Establishment Using Semantic Web Technology

The BEST project is conducted by the Computer/Law Institute (Dr. Arno Lodder) and the AI Department of the Vrije Universiteit von Amsterdam and funded by the Netherlands Organisation for Scientific Research. The project is part of the ToKeN research programme, an interdisciplinary programme in which cognitive and computer science focus on fundamental problems of interaction between a human user and a knowledge and information system.[11]

The objective of the BEST project is to provide laymen, who want an insight into the legal aspects of their disputes over damages, with information regarding their position for negotiation. Knowledge regarding the expected outcome of a court proceeding, arguments for out-of-court settlements and the BATNA (the Best Alternative to a Negotiated Agreement) could offer citizens the opportunity to determine how much room for negotiation (if any) is available when settling the damage and to decide for other forms of dispute resolution other than litigation. This information is to be provided through intelligent disclosure of existing case-law using semantic web technology: listing relevant case-law to the situation at hand through ontology-based search and navigation.

4.5   DALOS - Drafting Legislation with Ontology-Based Support

DALOS is an e-Participation project funded by the EC under the e-Participation Preparatory Action launched on 1st January 2007. Since the coherence and the alignment of legislative language could contribute to improve the quality and clarity of the legislative texts produced in the European Union, DALOS aims

---

[10] 01-01-2004/01-01-2007 (IST-002307), http://www.argumentation.org/.

[11] 01-02-2005/01-07-2010 (634.000.436B), http://www.best-project.nl.

at providing legal drafters with enhanced linguistic and knowledge management tools. Therefore, legislative drafters and decision makers would have control over the multilingual language of the European legislation, and over linguistic and conceptual issues involved in the transposition of that legislation, which would contribute to the harmonisation and coherence of legislative texts. The project uses the results obtained and the ontological-terminological resources developed within the LOIS project, so legislative drafters may query linguistic and ontological resources in order to locate the appropriate and standardised term which corresponds to a specific legal concept. The resources will be made available through a standard interface and upgraded also by using T2K (Text-2-Knowledge, an ontology learning tool) and GATE for advance language analysis, data visualisation and information sharing in different languages. The multilanguage linguistic-ontological resources will be integrated and made accessible within xmLegesEditor, a legislative drafting environment able to implement legislative XML standards. The Coordinator of this e-Participation project is the Institute of Legal Information Theory and Techniques, ITTIG (Dr. Daniela Tiscornia) and the other partners are: the Institute of Computational Linguistics (ILC-CNR), the Minister of Reforms and Innovations in Public Administration/National Center for Information Technology in Public Administration (MRIPA-CNIPA), The Department of Computer Science of the University of Sheffield (USFD), the Institute of Law and Technology (IDT-UAB), Leiden University (UNI-Leiden), the European University Institute (EUI), CELI, the Leibniz Center for Law (UvA) and the Camera dei Deputati (Italy).[12]

DALOS will cooperate with other eParticipation projects, to avoid duplication of testing activities and to make the lexical resources accessible within other pilot editors developed (SEAL), or to share dissemination and exploitation processes (SEAL, LEXIS).

4.6 ESTRELLA - European Project for Standardized Transparent Representations in Order to Extend LegaL Accessibility

ESTRELLA is a STREP Project funded by the EC under the 6th Framework. The main technical objectives of the ESTRELLA project are to develop a Legal Knowledge Interchange Format (LKIF), building upon emerging XML-based standards of the Semantic Web, including RDF and OWL, and Application Programmer Interfaces (APIs) for interacting with legal knowledge-based systems. The Coordinator of the Consortium is Universiteit van Amsterdam (Dr. Tom van Engers) and the other partners are: the University of Liverpool, Università di Bologna (CIRSFID), Fraunhofer FOKUS, RuleWise B.V., Rule-Burst (EUROPE) Limited, KnowledgeTools International Gmbh, Interaction Design Ltd, SOGEI (Società Generale d'Informatica S.P.A.), CNIPA (Centro Nazionale per l'Informatica nella Pubblica Amministrazione), Hungarian Tax and Financial Control Administration, Budapesti Corvinus Egyetem, Ministero dell'Economia e delle Finanze (Italy), Consorzio Pisa Ricerche SCARL[13].

---

[12] 01-01-2007/30-04-2008, http://www.dalosproject.eu.

[13] 01-01-2006/30-06-2008 (027655), http://www.estrellaproject.org

ESTRELLA will support legal document management and legal knowledge-based systems, in an integrated way, in order to provide a complete solution for improving both the quality and the efficiency of public administration processes which require the application of complex regulations.

The outcome of this project will facilitate a market of interoperable components for legal knowledge-based systems, allowing public administrations and other users to choose among competing development environments, inference engines, and other tools, freely. Translators between the LKIF format and the existing proprietary formats of LKBS vendors participating in the project will be developed and, thus, vendor neutrality and independence will be achieved and demonstrated. European and national (from two European countries) tax related legislation will be modeled and used in the pilot applications for the demonstration and validation of the ESTRELLA platform.

4.7   OPENKNOWLEDGE

OpenKnowledge is a three-year project co-funded by the European Commission within the 6th Framework and it aims at providing a new form of peer-to-peer knowledge sharing in open environments through: 1) interaction model routing, 2) context maintenance, 3) dynamic ontology matching, and 4) visualisation. The project is based on the ideas that the open WWW has been successful on a global scale because of the participation costs (costs at a basic level are low and the individual benefit of participation is immediate, increasing rapidly as more participants join in), although the same cannot be said about systems based on semantics. Therefore, OpenKnowledge focuses on semantics related to interaction as they could be acquired at low cost (participation) and could be used instead of a priori semantic agreements.[14] The Coordinator of this project is the University of Edinburgh (Informatics, Dr. David Robertson) with the following partners: the Knowledge Media Institute (Open University), AI Vrije Universiteit Amsterdam, The Artificial Intelligence Research Institute (IIIA-CSIC), Electronics and Computer Science (University of Southampton) and the Department of Information and Communication Technology (University of Trento).

The areas of bioinformatics and emergency response are the testbeds for this research. So far, several results have been achieved: the definition of an interaction modeling language, a working prototype, the establishment of scenarios from the bioinformatics and emergency response areas and the production of initial specifications for dynamic ontology mapping, good enough answer and trust analysis and visualisation.

4.8   METASEARCH - Semantic Legal Metasearch Project

The Metasearch project ("R&D of a legal semantic metasearch engine, result clusterer and automatic classifier of legal sources") is a Spanish PROFIT project

---

[14] 01-01-2006/31-12-2008 (027253), http://www.openk.org/.

funded by the Ministry of Industry, Tourism and Trade. The consortium includes Wolters Kluwer Spain (coordinator), the Institute of Law and Technology (IDT-UAB) and Intelligent Software Components, S.A. (iSOCO).[15] This project complements and follows another PROFIT project -E-Sentencias [E-Rulings]-coordinated by iSOCO, with the same members of the former consortium (plus the UAB School of Engineering, ETSE).[16]

The project aims at developing a complex and complete system for search, indexing and automatic mark-up of legal documents in order to enable user semantic search and retrieval. Legal ontologies, specific relevance algorithms and a system for automatic markup and indexing will be used to allow the semantic search.

4.9 SEAL - Smart Environment for Assisting Legislative Drafting

The coordinator of the SEAL project is the Leibniz Center for Law (Dr. Tom M. van Engers). The University of Bologna (CIRSFID), the Institute of Legal Information Theory and Techniques (ITTIG), Be Informed, O&I Management Partners and the Italian and Austrian Parliaments are the other partners of this project.[17] SEAL is an e-Participation project funded by the EC under the e-Participation Preparatory Action launched on 1st January 2007. The SEAL project aims at providing stakeholders of the legislative process (i.e. legal drafters) with a supporting environment that enables the construction of legal drafts through the use of drafting patterns and the creation of different connections from and to relevant existing legal sources. A repository with existing laws, draft versions and amendments will be made available, together with easy to use access methods. Collaborative support will be offered by groupware facilities.

The project, which includes the collaboration of three parliaments, aims at developing an integrated working environment for legislative drafters both within the parliament and the ministries. The MetaLex regulation-drafting environment (MetaVex, developed at the Leibniz Center for Law) is one of the three environments being evaluated in SEAL, an open source WYSIWYG editor for legislation based on Vex and the Eclipse IDE. The other two environments are the xmLegesEditor (owned/provided and maintained by CNR-ITTIG) and the Norma Editor (owned under licence and maintained by CIRSFID University of Bologna).

4.10 SEKT- Semantically Enabled Knowledge Technologies

The SEKT integrated project, co-funded by the EU 6th Framework programme, had a duration of 36 months and involved a large consortium of academic institutions and business companies, bringing together some of Europe's leading contributors to the development of knowledge technologies, data-mining systems

---

[15] 1-1-2007/1-12-2008 (FIT-150500-2002-135 and FIT-350100-2007-161).

[16] 01-01-2007/31-03-2008 (FIT-350101-2006-26), http://esentencias.isoco.net.

[17] 01-01-2007/30-04-2008, http://www.eu-participation.eu/seal.

and natural language processing technologies. The SEKT consortium was formed by: British Telecommunications Plc. (Dr. John Davies, Project Coordinator), Institute AIFB (University of Karlsruhe, Prof. Dr. Rudi Studer, Technical Coordinator), Empolis GmbH, the Jozef Stefan Institute, the University of Sheffield, the University of Innsbruck, Intelligent Software Components, S.A. (iSOCO), Ontoprise GmbH (Intelligente Lsungen fr das Wissensmanagement), Sirma AI Ltd., the Vrije Universiteit Amsterdam, the Institute of Law and Technology (IDT-UAB) and Kea-pro GmbH.[18]

The SEKT vision was to develop and exploit the knowledge technologies which underlined Next Generation Knowledge Management, integrating fundamental research, development of components and input from real world case studies in the public and private sectors: Siemens case study (Improving Individual Productivity), British Telecom case study (Reducing Overheads of Knowledge Creation and Maintenance), and the Legal case study (Decision Support for Legal Professionals). The aim of SEKT was to develop and exploit semantically-based knowledge technologies to support document, content and knowledge management, towards the design of appropriate utilities to users in three main areas: digital libraries, the engineering industry, and the legal domain, providing users with quick access to the relevant pieces of information. Ontologies, as key technology for the Semantic Web, were developed in the different areas.

The Legal Case Study (IDT-UAB and iSOCO) was focused on the improvement of Iuriservice, a web-based intelligent FAQ support system for judges. Judges from the Judicial School may input questions to the system in natural language and obtain access to a database of experience-based answers (organised as question-answer pairs) to practical day-to-day questions. The search system is enhanced using ontologies and semantic distance calculation. The members of the consortium provided: 1) methodological support for ontology construction (Institute AIFB), 2) ontology learning tools (Text2Onto from the Institute AIFB and OntoGen from the Jozef Stefan Institute), 3) upper-ontological references (Sirma AI Ltd.), 4) consistency checking and multi-version reasoning (Vrije Universiteit Amsterdam), 5) ontology alignment support (University of Innsbruck), 6) user needs, benchmarking, usability and business benefits (Kea-pro GmbH) and 7) the use of the General Architecture for Text Mining, GATE (University of Sheffield).

5   Content of This Book

The content of the volume is divided into five different parts. The main topics are roughly the following: (i) XML legislative drafting tools and methods, (ii) legal ontologies and system functionalities (e.g. information retrieval), (iii) argumentation and legal reasoning, (iv) norms and electronic (or virtual) institutions, (v) online dispute resolution and justice. This division is not absolute. Strictly speaking, e.g., topics such as knowledge representation or legal ontologies are present in all sections. The intended distribution of papers is only a proposal to facilitate the reading, stemming from legal XML. We will briefly summarise them.

---

[18] 01-01-2004/31-12-2006 (IST-506826), http://www.sekt-project.com.

5.1 Knowledge Representation, Ontologies and XML Legislative Drafting

The first two contributions give an overview of the MetaLex XML and LKIF formats and the MetaVex tool. A. Boer, R. Winkels and F. Vitali describe two XML standard proposals: MetaLex XML, directed to impose a standardised view on data regarding the publication process of legal information (for the purposes of software development) and LKIF, a legal representation language that consists on a reusable and extensible core ontology which allows reasoning together with an interchange format for legal knowledge representation languages. Then, S. van de Ven, R. Hoekstra, R. Winkels, E. de Maat and A. Kollar describe MetaVex, the MetaLex regulation-drafting environment (VEX stands for "Visual Editor for XML"), as an independent open source editor for legislative drafters and parliamentarians to facilitate the legislative process and support the creation of documents complying with the MetaLex standard (but adaptable to different XML schemas).

From the DALOS project, E. Francesconi and D. Tiscornia outline the design of the ontological-linguistic resource being developed, as well as the methodologies for its construction. The DALOS resource is based on the LOIS database (one of the wider lexical resources currently available in the legal field with 35.000 concepts in five European languages). The main purpose is directed to foster the quality of legislative drafting.

The last contribution included in this section, by J.A. de Oliveira Lima, M. Palmirani and F. Vitali present, within the context of the ESTRELLA project, an application of the FRBRoo document model for defining an information ontology of legal resources that takes into account the dimension of time. FRBRER is an entity-relationship model for the organisation of bibliographical records, and FRBRoo is a new version that uses the object oriented approach. This model can be applied to legal resources and the inclusion of the time dimension allows a more precise modeling within the legislative process workflow.

5.2 Knowledge Representation, Legal Ontologies and Information Retrieval

The second part of this volume includes contributions related to legal knowledge representation and the use of legal ontologies. T. van Engers, E. Hupkes, R. Winkels and A. Boer describe Legal Atlas, a software component intending to improve the access to spatial regulations. The tool shows to which geospatial objects a concept in the legal source applies and which (parts of) legal sources apply to selected geospatial objects. The concept of space is central and three different standards have been used to link, through a common RDF-based format, spatial regulations and maps (MetaLex, GML and IMRO2006). The Legal Atlas Ontology (in OWL format) is also described.

The following two contributions relate to the legal case study of the SEKT Project and refer to two different analyses. J. Voelker, S. Fernandez Langa and Y. Sure describe the adaptation of Text2Onto, a framework for ontology learning and data-driven ontology evolution that can be used to automatically generate ontologies from textual resources. Machine learning and natural language processing techniques are used to extract ontology entities and relationships from open-domain unstructured text. This adaptation offers support towards ontology learning from Spanish legal rulings, and facilitates the construction process of the ontology developed within the SEKT Legal Case Study. Z. Huang, S. Schlobach, F. van Harmelen, N. Casellas and P. Casanovas describe the use of MORE towards an experimental analysis of the properties of the version space of OPJK, by checking for stability, novelty and monotonicity. MORE is a multi-version ontology reasoning system, based on a temporal logic approach, and the different versions of an ontology are considered as a sequence of ontologies connected to each other via change operations. Ontology modelers might gain insight into their modeling process with the analysis of the results.

Finally, within the METASEARCH project, A. Sancho Ferrer, J. Manuel Mateo Rivero and A. Mesas Garca close this section with the presentation of a research towards the development of a semantic search engine that optimizes the search experience of the customers of the Wolters Kluwer-La Ley legal publishing databases. Taking into account studies regarding the search behaviour of users, the contribution then offers some experiments and results regarding the improvement of both precision and recall. This Metasearch engine is being implemented effectively for WK-La Ley customers.

5.3   Argumentation and Legal Reasoning

Contributions regarding argumentation and legal reasoning are gathered in the central part of the volume. Within the ESTRELLA project, A.Z. Wyner, T.J.M. Bench-Capon, and K. Atkinson distinguish in their contribution between three senses of "argument" -arguments, cases and debates- and the relations between them. T.F. Gordon, from the same project, presents the syntax and argumentation-theoretic semantics of LKIF language for modeling legal rules. According to the author, LKIF may enable four kinds of legal knowledge to be encoded in XML: arguments, rules, ontologies and cases. This paper illustrates an example based on German family law showing how LKIF rules can be used with the Carneades argumentation system to construct, evaluate and visualize arguments about a legal case.

F. Toni, within the ARGUGRID project, claims that assumption-based argumentation can serve as an effective computational tool for argumentation-based epistemic and practical reasoning and presents formal mappings from frameworks for epistemic and practical reasoning onto assumption-based argumentation frameworks. Also within the framework of the ARGUGRID project, M. Morge presents an Argumentation Framework implemented in Prolog for practical reasoning in legal disputes (arguments are defined as tree-like structures), which suggests different alternative courses of action and provides explanations for the choices.

Under the ALIS project, three more contributions focusing on intellectual property rights (IPR) as use case are introduced in this section. M. Rudnianski and H. Bestougeff propose bridging the theoretical framework of argumentation and a specific type of qualitative games called games of deterrence. In this qualitative games players can only distinguish between acceptable and unacceptable outcomes. Focusing on statements, the authors analyse argumentation between two parties as a game of deterrence in which established properties can be used to solve the argumentation issue. They show how the issue of defeasibility can be approached by such an analysis. R. Riveret and A. Rotolo aim to provide a Temporal Deontic Defeasible Logic adopting an analytical approach and argumentation semantics. Within the ongoing academic discussion on time and norms, authors provide a representation of legal reasoning and address temporal non-monotonic reasoning and handling of legal temporal status. Finally, G. Contissa reflects a feasibility test carried out to model a representation of legal knowledge in the area of IPR using the RuleBurst rule-based system technology. The work is focused on Italian Copyright law, with the aim to develop a method that could be extended and applied, in a subsequent stage, to other IP legislations in Europe.

5.4   Normative and Multi-agent Systems (MAS)

The fourth part of the book is dedicated to the analysis of the concept of norms and the use of norms in multi-agent systems. The contribution by R. Rubino and G. Sartor presents the concept of source-norm (norms establishing what other norms validly belong or do not belong to a normative system). Authors provide a taxonomy of source-norms, by distinguishing between enactment-recognizing source-norms and practice-recognizing norms, and between fundamental and dependent source-norms. Authors also show how source norms can support self-regulated institutions, namely institutions composed by agents that not only obey rules, but also determine what rules are part of the institution's normative system and that create new rules. Source norms have been represented by using the logic of the PRATOR system for defeasible argumentation and their application has been tested through the ASPIC Argumentation Engine and the ESTRELLA inference engine.

A. Perreau de Pinninck, C. Sierra and M. Schorlemmer, from the OPEN-KNOWLEDGE project, present a new distributed mechanism that ostracises norm violating agents in an open MAS to attain norm compliance. In MAS, sets of norms may be added to restrict some of the available actions with the objective of improving agent coordination. However, autonomous agents have the choice whether or not to support certain norms and to abide by them, thus it may be worthwhile for an agent not to abide by a norm and profit at the expense of the other agents that follow it. This contribution offers results from several simulations based on interactions that consist of the prisoner's dilemma game.

5.5   Online Dispute Resolution

Finally, the analysis and description of alternative dispute resolution methods and applications close the book. First, pointing at the citizens' use of technology,

E.M. Uijttenbroek, A.R. Lodder, M.C.A. Klein, G.R.Wildeboer, W. van Steenbergen, P.E.M. Huygen, R.L.L Sie and F. van Harmelen present some results
from the BEST project. Different retrieval experiments and results using a
thesaurus-based statistical indexing technique are shown, directed to the development of a system that supports laymen by retrieving relevant case law on
liability issues. This information could then be used by laymen to make informed
choices based on the knowledge of the best alternative to a negotiated agreement
(BATNA) in order to proceed towards litigation or other private dispute resolution processes (negotiation, mediation, arbitration).

C. Cevenini and G. Fioriglio, within the ALIS project, examine the current
state-of-the-art of the use of Information and Communication Technologies in
judicial and alternative dispute resolution procedures in Italy. Some of the procedures and applications revised are: the Italian On-LIne Civil Trial, Squaretrade,
Risolvionline and Cybersettle.

The contribution by P. Casanovas and M. Poblet ends up the volume exploring the broad conceptual background of *relational justice*. Relational Justice is
defined as "the justice produced through cooperative behavior, agreement, negotiation, or dialogue among actors in a post-conflict situation". This work has
been developed for the EU COST Action A21 Restorative Justice Developments
in Europe (2002-2006).[19] Relational concepts of justice may be used to build
up ontologies for the new emerging field of ODR in transnational and global
law. This paper constitutes a first attempt to identify and describe the main
concepts that are being employed in the field (empathy, apology, forgiveness).
It is stated that Artificial Intelligence techniques may handle these concepts as
legal concepts as well.

Acknowledgements

We would like to acknowledge various publicly funded R&D projects
(One-Lex, E-sentencias FIT-350101-2006-26), Iuriservice SEJ2006-10695, and
Metabuscador FIT-350100-2007-161).

References

1. AAVV: Thirty-ninth selected bibliography on computers, technology, and the law.
   Rutgers Computer and Technology Law Journal 33 (2007)
2. Coggins, T.L.: Legal, factual and other internet sites for attorneys and others. 12
   Richmond Journal of Law and Technology 17, 1–25 (2005)
3. Ikens, L.: The 2006 ABA Tech Report: trends in courtroom technology: has the
   picture changed?. Technical report, LJN's Legal Tech Newsletter January, 2007
   (2007) [accessed 10 september], http://www.ljnonline.com/alm?lt
4. Law Firm Inc.: American Law Technology Survey 2007 (2007) [accessed December
   10, 2007], http://lawfirminc.law.com

---

[19] http://www.cost.esf.org.

5. ILTA: Technology purchasing survey (2007a) (2007) [accessed December 10. 2007], http://www.iltanet.org/pdf/2007PurchasingSurvey.pdf
6. ILTA: Telecommunications survey (2007b) (2007) [accessed october 14, 2007], http://www.iltanet.org/communications/
pub_detail.aspx?nvID$=$000000011205\&h4ID=000000872605
7. IOMA: Increasing margins: Technology trends that partners can use to improve ROI in 2007. PRL (2006)
8. Svengalis, K.: Legal information: Globalization, conglomerates and competition monopoly or free market (2007) [accessed october 14, 2007], http://www.rilawpress.com/AALL2007.ppt
9. Murphy, B.: Roundtable discussion: E-discovery (2007) [accessed september 10, 2007], http://www.kmworld.com/Articles/ReadArticle.aspx?ArticleID=37331
10. American Bar Association (ABA): Legal technology survey report results (2007) [accessed 10 september 2007],
http://www.abanet.org/tech/ltrc/survstat.html
11. Morris, F.J.: E-Discovery: Best practices for employment lawyers. what support do you need? how do you work with E-Discovery experts? In: Current Developments in Employment Law, ALI-ABA, Santa Fe, NM (2005)
12. Violino, B.: Digital dialogue: firms connect with online collaboration and wireless tools (2007)[accessed october 14, 2007],
http://lawfirminc.law.com/display.php/file%3D/texts/0907/amlawtech
13. Krause, J.: Solos lead the wireless way. ABA Journal E-Report 4, 3 (2005)
14. Conrad, J., Schilder, F.: Opinion mining in legal blogs. In: Eleventh International Conference on Artificial Intelligence and Law, ICAIL 2007, Stanford, CA, June 4—8, pp. 231–236. ACM Press, New York (2007)
15. Krause, J.: Acrobat 8 does flips for attorneys. ABA Journal E-Report 6, 3 (2007)
16. Taylor, E.E., Mallie, M.J.: New technologies in IP transactions. In: Institute, P.L. (ed.) Handling Intellectual Property Issues in Business Transactions 2008. PLI/Pat, vol. 928, Practising Law Institute (2008)
17. Vicente Blanco, D., Martínez González, M.: Spain on Going Legislative XML Projects. In: Biagioli, C., Francesconi, E., Sartor, G. (eds.) Proceedings of the V Legislative XML Workshop, pp. 23–38. European Press Academic Publishing, Firenze (2007)
18. Biagioli, C., Francesconi, E., Sartor, G.: Proceedings of the V Legislative XML Workshop. European Press Academic Publishing, Firenze (2007)
19. Fabri, M.: Introduction: state of the art, critical issues, and trends of ICT in European judicial systems. In: Fabri, M., Contini, F. (eds.) Justice and Technology in Europe: How ICT is changing the judicial business, pp. 1–18. Kluwer Law International, The Hague (2001)
20. Susskind, R.: The challenge of the information society: Application of advanced technologies in civil litigation and other procedures. Report on England and Wales. Technical report (2000), http://ruessmann.jura.uni-sb.de/grotius/english/
21. Contini, F.: Conclusion: Dynamics of ICT diffusion in European judicial systems. In: Fabri, M., Contini, F. (eds.) Justice and Technology in Europe: How ICT is changing the judicial business, pp. 317–331. Kluwer Law International, The Hague (2001)
22. Fabri, M., Contini, F. (eds.): Judicial Electronic Data Interchange in Europe: Applications, Policies, and Trends. IRSIG-CNR Lo Scarabeo, Bologna (2003)
23. Velicogna, M.: Justice systems and ICT. What can be learned from Europe? Utrecht Law Review 3, 129–147 (2007)

24. Contini, F., Cordella, A.: Italian justice system and ICT: matches and mismatches between technology and organisation. In: Cerrillo, A., Fabra, P. (eds.) E-Justice: Using Information Communication Technologies in the Court System, IGI Global, Hershey (2008)

25. Fabri, M.: The Italian style of e-justice in a comparative perspective. In: Cerrillo, A., Fabra, P. (eds.) E-Justice: Using Information Communication Technologies in the Court System, IGI Global, Hershey (2008)

26. Bing, J.: Handbook of Legal Information Retrieval. North Holland, Amsterdam (1984)

27. Benjamins, V.R., Casanovas, P., Breuker, J., Gangemi, A. (eds.): Law and the Semantic Web: Legal Ontologies, Methodologies, Legal Information Retrieval, and Applications. Springer, Berlin (2005)

28. Sergot, M.J., Sadri, F., Kowalski, R.A., Kriwaczek, F., Hammond, P., Cory, H.: The british nationality act as a logic program. Communications of the ACM 29, 370–386 (1986)

29. Prakken, H., Sartor, G. (eds.): Logical Models of Legal Argumentation. Kluwer, Dordrecht (1997)

30. Sartor, G.: Legal Reasoning: A Cognitive Approach to the Law. Treatise on Legal Philosophy and General Jurisprudence, vol. 5. Springer, Berlin (2005)

31. Conte, R., Falcone, R., Sartor, G.: Agents and norms (special issue). Artificial Intelligence and Law 7, 1–113 (1999)

32. Breuker, J., Gangemi, A., Tiscornia, D., Winkels, R.E.: Ontologies for law. Artificial Intelligence and Law 19, 239–455 (2004) (Special issue)

33. Ajani, G., Peruginelli, G., Sartor, G., Tiscornia, D. (eds.): The Multilanguage Complexity of European Law. European Press Academic Publishing, Firenze (2007)

34. Gordon, T.F.: The Pleadings Game. An Artificial Intelligence Model of Procedural Justice. Kluwer, Dordrecht (1995)

35. Prakken, H.: Coherence and flexibility in dialogue games for argumentation. Journal of Logic and Computation 15, 1009–1040 (2005)

36. Sartor, G.: A teleological approach to legal dialogues. In: Law, Rights and Discourse. Themes from the Legal Philosophy of Robert Alexy, Hart, Oxford (2007)

MetaLex XML and the Legal Knowledge Interchange Format

Alexander Boer[1], Radboud Winkels[1], and Fabio Vitali[2]

[1] Leibniz Center for Law, University of Amsterdam, The Netherlands
[2] Dept. of Computer Science, University of Bologna, Italy

Abstract. Electronic government invariably involves XML and electronic law: legislation is as essential to public administration as the ball is to a ball game. This paper gives an overview of two XML standard proposals dealing with two complementary aspects of electronic legislation – the documents themselves as a carrier, and an institutional reality they represent – in a coherent way: MetaLex XML and the Legal Knowledge Interchange format (LKIF). MetaLex XML is well on its way to becoming formal and de facto standard for legislation in XML. LKIF is yet to be submitted as a proposed standard. LKIF includes some interesting innovations from an AI & Law perspective.

1 Introduction

Electronic government invariably involves XML and electronic law: legislation is as essential to public administration as the ball is to a ball game. Publication of legislation, and the development of tools for working with legislation is at the moment still a jurisdiction-specific enterprise, even if it is standardized at the jurisdiction level. What is required is a jurisdiction-independent XML standard that can be used for interchange, but also - maybe more importantly - as a platform for development of generic legal software.

For vendors of legal software this opens up new markets, and for the institutional consumers of legislation in XML it solves an acute problem: how to handle very different XML formats in the same IT infrastructure. Increasing legal convergence between governments in the European Union, and the growing importance of traffic of people, services, goods, and money over borders of jurisdictions has led to an increased need for managing legislation from different sources, even in public bodies and courts. EU tax administrations for instance need access to all VAT regimes of other member countries to correctly apply EU law, and EU civil courts may nowadays for instance be confronted with the need to understand foreign law on labour contracts to decide on cases involving employees with a foreign labour contract choosing domicile in the country where the court has jurisdiction.

Over the last decade, legislators have begun to adopt XML standards for the formal sources of law they manage, and there is even some activity to standardize on a supranational level. Since these legislator's standards however generally

P. Casanovas et al. (Eds.): Computable Models of the Law, LNAI 4884, pp. 21–41, 2008.
© Springer-Verlag Berlin Heidelberg 2008

speaking have an institutional status, coordination between countries requires cooperation between governments, and this process moves too slowly from a consumers point of view, and for reasons largely irrelevant to the consumer.

This paper gives an overview of two XML standard proposals dealing with two complementary aspects of electronic legislation – the documents themselves as a carrier, and an institutional reality they represent – in a coherent way: MetaLex XML and the Legal Knowledge Interchange format (LKIF). MetaLex XML is well on its way to becoming formal and de facto standard for legislation in XML. LKIF is yet to be submitted as a proposed standard. LKIF includes some interesting innovations from an AI & Law perspective.

MetaLex XML positions itself as an interchange format, a lowest common denominator for other standards, intended not to necessarily replace jurisdiction-specific standards in the publications process but to impose a standardized *view* on this data for the purposes of software development at the consumer side. The MetaLex schema is based on best practices from amongst others the previous versions of the MetaLex schema, the Akoma Ntoso schema, and the Norme in Rete schema. Other important sources of inspiration are i.a. LexDania, CHLexML, FORMEX, R4eGov, etc. In addition to these government or open standards there are many XML languages for publishing legislation in use by publishers. Standards like PRISM, in which major publishers are involved, are also a source of inspiration.

The MetaLex XML standard recently moved forward significantly, with the adoption of part of it as a CEN[1] prenorm, and its adoption by several industry projects. Many of the participants of the CEN workshop have also been involved in the Legislative XML workshops (see for instance the archive of the frontpage of the MetaLex website[2] for previous calls for participation and online proceedings and presentations). In the process of standardization MetaLex changed significantly compared to its previous incarnations (versions up to 1.3.1).

While MetaLex is an enabling technology for Legal Knowledge Based Systems (LKBS), amongst other uses, LKIF directly addresses the interchangeability of legal knowledge representation.

The interpretation of law is a lot harder to standardize than its manifestion in XML, but it has great potential in the market. Legal knowledge representation is – or should be – by its very nature a continuous affair for public administrations, because they simply must accomodate changes to legislation and changes in interpretation following from court verdicts, regardless of whether the change fits in organizational policy or not. In the absence of standards for knowledge representation, public administrations either have to accept vendor lock-in for any LKBS they deploy, or value the LKBS as a system with a potentially very short lifecycle. The absence of a standard in this sense limits the size of the market for LKBS.

---

[1] Comité Européenne de Normalisation; European Committee for Standardization; Europäisches Komitee für Normung.

[2] http://www.metalex.eu

Some vendors of LKBS – KnowledgeTools, RuleBurst, and RuleWise – are involved in the standardization effort, but a standard for legal knowledge representation is obviously also of great interest for the academic community. In this paper we also explain a major design decision of LKIF that generated a lot of discussion among those involved in the specification of LKIF.

2 MetaLex

MetaLex is the subject of earlier publications, e.g. [1,2]. MetaLex is a generic and extensible framework for the XML encoding of the structure of, and metadata about, documents that function as a source of law. It aims to be jurisdiction- and language-neutral, and is based on modern XML publishing concepts like a strict separation between text, markup, and metadata, building on top of structure instead of syntax, accommodation of transformation pipelines and standard APIs, as well as emerging Semantic Web standards like RDF and OWL.

MetaLex, whose first version dates from 2002 (cf. [1]), has been redesigned from scratch in the CEN standardization workshop, taking into account lessons learned from Norme in Rete[3] – the Italian standard for legislation – and Akoma Ntoso[4] - the Pan-African standard for parliamentary information, and has been submitted as a norm proposal to the CEN.

A partial CEN Workshop Agreement (CWA) now exists. It does not yet constitute a complete, workable XML standard. This partial agreement contains agreements about the abstract content models supported by the standard, the way metadata is added to a document, and a generic model for organizing metadata in RDF. Additional agreements are on the agenda.

The MetaLex workshop aims to use the distinctions made by the IFLA Functional Requirements for Bibliographic Records (FRBR), which aims to distinguishes content and form aspects of bibliographic entities, between bibliographic entities as 1) works, 2) expressions, 3) manifestations, and 4) items (see figure 1). This distinction in four levels, which is strictly implemented in MetaLex, distinguishes the different levels of abstraction at which one can think about documents roughly as follows:

- A **bibliographic object** is a bounded representation of a body of information, designed with the intent to communicate, preserved in a form independent of a sender or receiver. A bibliographic work, expression, manifestation, and item are bibliographic objects.
- A **bibliographic work** is a bibliographic object, realized by one or more expressions, and created by one or more persons in a single creative process. We recognize the work through individual expressions of the work, but the work itself exists only in the commonality of *content* between and among the various expressions of the work.

---

[3] http://www.normeinrete.it/
[4] http://www.akomantoso.org

- An **bibliographic expression** is a realization of one bibliographic work in the form of signs, words, sentences, paragraphs, etc. by the author of that work. Any change in *content* constitutes a gives rise to a new expression.
- A **bibliographic manifestation** embodies one expression of one bibliographic work. The boundaries between one manifestation and another are drawn on the basis of both content and physical form. When the production process involves changes in physical form the resulting product is considered a new manifestation. Thus, a specific XML representation, a PDF file (as generated by printing into PDF a specific Word file with a specific PDF distiller), a printed booklet, all represent different manifestations of the same expression of a work.
- A **bibliographic item** exemplifies one manifestation of one expression of one work: a specific copy of a book on a specific shelf in a library, a file stored on a computer in a specific location, etc.

Work, expression, and manifestation are intentional objects, i.e. they exist only as the object of one's thoughts and communication acts, and not as a physical object. An item is a physical object. Note however that items stored on a computer can be easily copied to another location, resulting in another item, but still an instance of the same manifestation. This makes adding metadata about the item *to* the item in principle impossible. On the Internet generally speaking only the *uniform resource locator* (URL) is an item-specific datum. The item level is therefore not very relevant to XML standards.

The proposed standard is primarily concerned with identification of legal bibliographic entities on the basis of literal content, i.e. on the expression level, and prescribes a single standard manifestation of an expression in XML. Different expressions can be versions or variants of the same work. In addition there is the aspect of role, that relates the bibliographic entity to specific contexts of use: this is consistently treated as metadata.

2.1   Scope of the Standard

The *CEN Workshop on an Open XML Interchange Format for Legal and Legislative Resources (MetaLex)*, declares, by way of its title, an interest in legal and legislative resources, but the scope statement of the first workshop agreement limits the applicability of the proposed XML standard to sources of law and references to sources of law.

As understood by the workshop, the source of law is a writing that can be, is, was, or presumably will be used to back an argument concerning the existence of a constitutive or institutional rule in a certain legal system, or, alternatively, a writing used by a competent legislator to communicate the existence of a constitutive or institutional rule to a certain group of addressees. Because the CEN Workshop is concerned only with an XML standard, it chooses not to appeal to other common ingredients of definitions of law that have no relevant counterpart in the information dimension.

Source of law is a familiar concept in law schools, and may be used to refer to both legislators (fonti delle leggi, sources des lois), legislation and case law (fonti

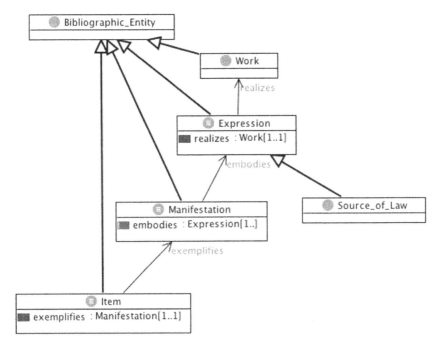

Fig. 1. A taxonomy of bibliographic entities in MetaLex

del diritto, sources du droit), custom, etc. It should be noted that many romance languages make a distinction between the legislator as source of law, by way of speaking or writing, and the law as source of right(s), which is presumably what the existence of the law brings about. In its broadest sense, the source of law is anything that can be conceived of as the originator of legal rules. In the context of MetaLex it strictly refers to communication in writing, and in a sense covers the *fonti del diritto* in Italian and *sources du droit* in French. There are two main categories of source of law in writing: legislation and case law.

The notion of a legislative resource includes legislation, and all writings produced by the legislator explaining and justifying legislation. The legislator is a legal person: it exists separately from any natural persons and organizations involved in the process of drafting and evaluating legislation. It is the formally correct completion of certain processes, usually dictated by law, that makes the legislator the formal author of a writing, and at the same time identifies the addressees to whom it applies. Obviously, the persons and organizations involved in the process of legislating may produce writings that are clearly precursors or legally required ingredients to the end product. These writings are also included in the notion of a legislative resource, but in this case it is not easy to give straightforward rules for deciding whether they are, or are not to be considered legislative resources. Different jurisdictions will have different theories on this subject.

2.2 MetaLex Content Models

A MetaLex XML element is characterized by a name, a content model, and zero or more attributes. According to the philosophy of descriptive markup (cf. the Text Encoding Initiative[5]), the name of an XML element is usually semantically-charged (i.e. it provides a hint as to the meaning of the text fragment, or its role within the whole of the document). Additional information about the content of the element goes into attributes. The *content model* is an algebraic expression of the elements that may (or must) be found in the content of the element. Generic elements, on the other hand, are named after the content model: they are merely a label identifying the kind of content model.

All XML vocabularies contain a mix of descriptive and generic elements, and, depending on the foreseen uses of the documents, emphasize one of the approaches. For instance, vocabularies with precise procedural semantics (e.g. XSLT, SVG) do not depend on generic elements, while vocabularies intended for diverse content (for instance XHTML) employ generic elements. Consider for instance that in XHTML 2.0 both a and img elements are being replaced or phased out in favour of generic substitutes using attributes.

Current validation languages (e.g. XML Schema) do not allow validation rules to be associated to attribute values, so element names are currently the only way to associate validation rules to documents. This is a cause of pollution of principles, forcing semantically-charged elements to assume a rigid content model, while generic elements take care of odd situations that where not foreseen when the content models where designed.

Legislative drafting technique has a long tradition, and often its own standards of what legislative documents should look like. This makes descriptive markup combined with strict content models very tempting. On the other hand, there are so many exceptions that can be found in concrete examples we sometimes just want to give up on precise description altogether and resort to generic elements, in particular because there should be not one *iota* of difference between the original expression of the legislator and the XML manifestation of that expression.

The approach of the workshop is to provide for a complete and automatic interchangeability of approaches, from generic to descriptive and vice versa. These are the fundamental content models of MetaLex:

container a container of a sequence of other elements;
hcontainer a hierarchical container of nested elements with titles and numbers;
block the largest structure where text and inline elements mix freely, e.g., paragraphs and other (usually vertically-organized) containers of both text and smaller structures;
inline an inline container of text and other inline elements (e.g., bold); and
milestone an empty element that can be found in the text (as opposed to meta).

---

[5] http://www.tei-c.org/P4X/SG.html

Specialized content models are for instance `root`, `mcontainer`, `meta`, `anchor`, and `date`. Sharing content models is achieved by using two special attributes, `name` and `type` that provide information about the meaning and the content model of the element. The following elements are for instance equivalent from the point of view of the standard:

```
<clause metalex:type="metalex:container" metalex:name="clause"/>
<clause metalex:type="metalex:container"/>
<metalex:container metalex:type="metalex:container" metalex:name="clause"/>
<metalex:container metalex:name="clause"/>
```

This approach is different from the language extensions (implemented using substitution groups) of legacy MetaLex (1.3.1 and before): no central registry of extensions is used. The MetaLex attributes can be thought of as processing instructions that can be embedded in existing XML standards.

2.3   Conformance of Elements

Conformance in the strict sense means 1) validation of XML documents against a schema that includes the MetaLex XML schema, 2) the theoretical possibility of obtaining an XML document that uses solely MetaLex generic elements and validates against the MetaLex XML schema by way of simple substitution, and 3) conformance to the MetaLex CWA written guidelines. Any XML encoding is *transformation conformant* if instances can be transformed automatically into conformant MetaLex XML instances.

The process of declaring a concrete element conforming to the MetaLex norm works as follows:

1. You must use one of the abstract content models for the element;
2. You may define a restriction of the corresponding concrete type;
3. You may not define an extension to the content model of a concrete type;
4. You may define an extension of a concrete type for the purpose of adding attributes;
5. You must define the elements as a substitution group of one of the abstract elements and you must identify a type which is either one of the provided concrete types, or the restriction of the content model or extension of attributes of a concrete type that you have defined.

To easily define an element conforming to the standard that can be used in XML manifestations of sources of law, define a non-abstract complex type, for instance a restriction `articleType` of `hcontainerType` (see figure 2), and create an element belonging to the substitution group of one of the abstract elements according to the subtype specified, for instance:

```
<xsd:element name="article" substitutionGroup="e:abs-hcontainer"
    type="articleType" />
```

```
<xsd:complexType name="articleType">
    <xsd:complexContent>
        <xsd:restriction base="e:hcontainerType">
            <xsd:sequence>
                <xsd:element ref="e:absHtitle" maxOccurs="unbounded"/>
                <xsd:sequence maxOccurs="unbounded">
                    <xsd:element ref="e:absContainer"/>
                </xsd:sequence>
            </xsd:sequence>
            <xsd:attributeGroup ref="e:globnumb"/>
        </xsd:restriction>
    </xsd:complexContent>
</xsd:complexType>
```

Fig. 2. `articleType` is a restriction of `hcontainer`

Existing vocabularies can usually be redefined in terms of MetaLex content types. It is not sensible to give an example of *a* MetaLex XML instance here because no such notion exists: MetaLex is intended as a metaschema for other schemas that define concrete XML vocabulary.

2.4   Metadata

MetaLex precribes what counts as a MetaLex metadata statement, how it is stored inside a MetaLex document, and what classes of entities and which predicates (properties) MetaLex distinguishes: its *ontology*. The RDF ontology is of course extensible. The ontology classifies:

bibliographic entities: the work, expression, manifestation, and item level, and content models;
reference: type of reference between bibliographic entities;
activities: actions and thematic links, and thematic roles of bibliographic entities in at least the actions creation, enactment, repeal;
agent and competence: the agents and institutional instruments (legislative power, etc.) used in legislative activity.

MetaLex `meta` elements are used to embed metadata that can alternatively be stored in the form of Resource Description Framework[6] (RDF) statements in RDF documents. Elements derived from the `meta` content model are carriers of RDFa [7] attributes, and are therefore RDFa statements. All entities are identified using URI.

As an example of MetaLex metadata we include here a mechanism which is currently still a proposal; The XML document declares what it is a manifestation of by way of metadata. Assuming `about=""` (i.e. empty string URI reference[8])

---

[6] http://www.w3.org/RDF/
[7] RDF Annotation; `http://www.w3.org/TR/xhtml-rdfa-primer/`
[8] Note that URI, which is absolute, and URI reference (cf. IETF 3986), which is absolute or relative, and can therefore be empty, are different. URI are globally unique, but URI references are not: only after resolution to a URI they are globally unique.

refers to the document itself, the following declares a standard manifestation, expression, and work level XML base (using a proposed naming convention):

```
<meta id="m1" about="" rel="metalex-owl:exemplifies"
    href="/tv/act/2004-02-13/2/tv">
<meta id="m2" about="/tv/act/2004-02-13/2/tv" rel="metalex-owl:embodies"
    href="/tv/act/2004-02-13/2">
<meta id="m3" about="/tv/act/2004-02-13/2" rel="metalex-owl:realizes"
    href="/tv/act/2004-02-13">
```

The RDF reading of **m1** is as follows: **m1** is a statement that states that the (referent of) `metalex:exemplifies` of (the referent of) (`empty string`) is (the referent of) `/tv/act/2004-02-13/2/tv`.

It is also possible to for instance directly state the type of a bibliographic object with the MetaLex OWL vocabulary, although this is presumably rarely useful:

```
<meta id="m4" about="/tv/act/2004-02-13/2/tv" rel="rdf:type"
    href="metalex:Manifestation">
```

Read for **meta** in the examples above any appropriate element that permits RDF/A metadata attributes. At the moment (i.e. in the existing agreement) this is any element conforming to the **meta** content model. The URI references in the examples are relative, conforming to the proposed naming convention: the base is set by the processing environment.

In the current CEN agreement we propose to use a simple categorization of thematic roles loosely based on Judith Dick's representation of legal arguments (cf. [3]) for actions and events affecting the lifecycle of sources of law. Each *occurrent* has one or more participants: Figure 3 shows the classification of participants. The *patient* is for instance immanent and product of the action, and undergoes some structural change as a result of the action: at the level of bibliographic entities this applies to the work, while the expression usually takes the role of result or instrument. The instrument is immanent and source of the action, and is not changed during the action: this is for instance the *modifying* expression in a modification of a work, which results in a new consolidation. One of the greater qualities of thematic classification of participants is that it is largely impervious to differences in legal theory.

2.5   Citation and Reference

Citation and reference are not yet covered by the CEN prenorm. A reference is something that refers to or designates something else, or acts as a standin for a relation between two things: the referrer and the referent. The current proposal is that all references conform to the (`referrer, predicate, referent`) RDF triple data model, and are represented as RDF or RDF/A.

The following are examples of inline reference and citation conformant to the current *proposal*:

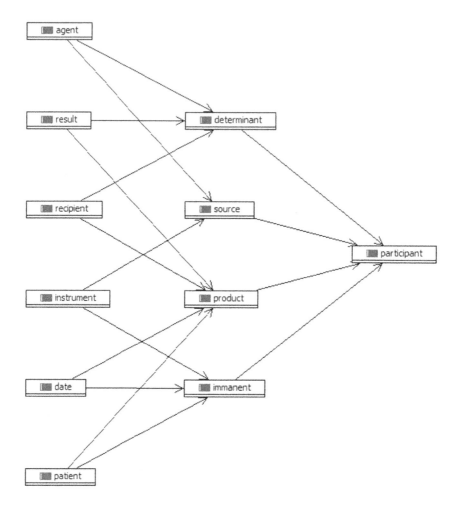

Fig. 3. Each MetaLex occurent has one or more participants. The figure shows a taxonomy of participants.

```
<meta about="#x" rel="metalex-owl:cites"
    href="http://gov.tv/tv/act/2004-02-13/2/tv#y">

<meta about="#x" rel="metalex-owl:refersTo"
    href="http://gov.tv/act/2004-02-13/concepts#theft">
```

Note that the reference is not to an item, but to an expression or work. The standard defines no specific content model for reference and citation: the user is free to define inline element structures of his own.

In the sense intended here a reference is an XML element (directly or indirectly) containing text, and the text refers to something else. A citation is an expression that refers to something intralinguistic, i.e. in practice to another

XML element (directly or indirectly) containing text. Other references refer to something extralinguistic, i.e. something other than text, recoverable from the context in which the document was produced. *Article 1*, *the first article* and *the previous article* are examples of citation, and *the Minister*, *the Republic*, *the accused*, and *We, Beatrix*, etc. are for instance examples of relevant references to other things.

It is important to distinguish between two different aspects of citation and reference:

1. The purely operational aspect, which holds that a reference or citation is an element that has the right attributes and property value and "points somewhere"; and
2. The semantic interpretation, which holds that reference/citation is the meaning of the content enclosed by the XML element: even if there is no metadata identifying the target of the reference/citation, it still remains a reference/citation because this is inherent in the meaning of the content.

While the second reading is ultimately the correct one, MetaLex is based on the design principle that the schema identifies structure and attributes, and not the meaning of text: giving meaningful names to elements is therefore left to the user. While defining inline elements for citation is undoubtedly useful, it should not be part of an abstract meta standard.

3   The Legal Knowledge Interchange Format

The LKIF proposal for standardising legal knowledge representation for legal knowledge-based systems (LKBS) is the main product of the Estrella project[9]. LKIF is intended to serve two main purposes: 1) as a reusable and extensible core ontology, application programmer interface, and inference engine specification for legal decision support systems, knowledge management systems, and argumentation support systems; and 2) as an interchange format for existing (proprietary) legal knowledge representation languages.

The requirements for LKIF are derived from several sources: 1) a survey of research on computational models of legal reasoning and argumentation, from the field of Artificial Intelligence and Law; 2) an analysis of the business requirements articulated by the participating vendors, and from the logical reconstructions of the logics used by these vendors; and 3) feedback and comments by the participating user organisations and members of the observatory board. The Estrella consortium includes a number of companies experienced in building legal decision support systems, academic partners from the Computer Science & Law field, and some public bodies. The objective is to produce a proposal to a standardisation body, probably the CEN, in 2008.

LKIF is a knowledge representation language for legal arguments, rules, terminological axioms, and cases. LKIF can be characterized as an ontology of law for

---

[9] A sixth framework IST project (IST-2004-027655), see http://www.estrellaproject.org

the Semantic Web, and as a knowledge representation language specifically suitable for legal reasoning in its own right. On the language level LKIF combines
existing Semantic Web technology - RDF and the Web Ontology Language[10]
(OWL) – and a new LKIF Rules language extending (the semantics of) RDF
and OWL for dealing with presumptive inferences. The LKIF Rules language
has an *argumentation-theoretic semantics*: its semantics is defined in terms of
argumentation schemes.

The distinction between ontology and rules mirrors a difference between the
knowledge representation paradigms used by the three participating vendors in
the consortium: RuleWise's UML models (RuleWise) are intended as ontology,
to be extended by a specific operationization in the form of rules for specific
LKBS, while RuleBurst's and KnowledgeTools's languages directly define an
LKBS knowledge base and appear to be more accurately captured by LKIF
Rules.

Parts of LKIF, its rule semantics and its ontology, were described earlier in
[4,5,6]. This publication gives an overview of project results sofar.

3.1   Interface between LKIF and MetaLex

LKIF naturally interfaces with MetaLex, although it can be used with any
document format that identifies sources of law and their relevant parts with
identifying URI *on the expression level*. Both LKIF and MetaLex metadata are
accessible as RDF data: MetaLex metadata is therefore directly accessible in an
LKIF processing environment and does not have to be duplicated.

If LKIF is used in combination with another document format that does require extraction and duplication in RDF of relevant metadata, then the MetaLex
ontology can be used to standardize the RDF which is in this case stored separately from the originating document in the LKIF-based knowledge base.

MetaLex, at least the XML element side of it, limits itself to describing content
models, i.e. the purely syntactical view of the text, while LKIF only models what
the text is about, i.e. purely semantics of the text. It is common to make a direct
mapping between knowledge representation structures and the structure of the
text for purposes of maintaining isomorphism (cf. [7]), or because the knowledge
representation is based on a linguistic analysis of the text (cf. for instance [8,9]),
as in the following simplistic and fictional example:

```
<rule>A <antecedent>motorcycle</antecedent> is
a <consequent>vehicle</consequent>.</rule>
```

The metadata mechanism of MetaLex, and the possibility of naming one's own
inline elements, make it possible to emulate the same mechanism in MetaLex
and LKIF. The `rule` element can refer to the LKIF axiom MOTORCYCLE ⊑
VEHICLE using RDFa attributes, as in the previous metadata examples, while
the `antecedent` element refers to the term MOTORCYCLE and the `consequent`
refers to the term VEHICLE.

---

[10] http://www.w3.org/2004/OWL/

Prescribing such a mechanism is however beyond the scope of MetaLex and LKIF, respectively.

3.2   The LKIF Ontology

The LKIF ontology is a standard OWL ontology, based on description logic (DL), or alternatively description logic programs (DLP), semantics, and can be used separately from the LKIF rules in Semantic Web applications if required. The main purpose of the ontology is to constrain use of terminology in LKIF applications, and ontology is intended to be the part of the knowledge representation most amenable to reuse outside the context of the original LKBS, although the applicability of this doctrine in law generates a lot of discussion: Matters of terminology do demonstrably become subject of legal argument. The ontology is explained in greater detail in [6,10].

Important in the LKIF ontology, and legal knowledge representation in general, are the concepts of obligation and permission, as one would expect, and social roles. The deontic notions are underconstrained to accommodate differences of opinion on their interpretation, and play a considerably less central role in LKIF than in legal knowledge representation in general. Figure 4 shows an entity-relationship diagram with some deontic notions and their relationships. They meet the following criteria, given that $O(\alpha \mid \beta)$ means that α is obligatory when β:

1. What is obligatory is permitted. The axiom $O(\alpha \mid \beta) \rightarrow P(\alpha \mid \beta)$ is true.
2. The impossible and the meaningless are not obligatory: $\neg O(\alpha \mid \alpha)$ and $\neg O(\neg \alpha \mid \alpha)$ are true. Taken from [11].
3. There are no conflicting obligations. The obligations $O(\alpha \mid \beta)$ and $O(\neg \alpha \mid \beta)$ are inconsistent: $\neg(O(\alpha \mid \beta) \wedge O(\neg \alpha \mid \beta))$ is true.
4. The obligation $O(\alpha \mid \beta)$ and permission $P(\neg \alpha \mid \beta)$ are inconsistent: $\neg(O(\alpha \mid \beta) \wedge P(\neg \alpha \mid \beta))$ is true.
5. The sentences $O(\alpha \mid \top)$, $O(\beta \mid \alpha)$, $O(\neg \beta \mid \neg \alpha)$ are only satisfied by the ordering identified by [12].

Social roles usually have complementary roles, and this constitutes the basis for a relational view on roles. The complementarity of roles is the consequence of (and results in) mutual expectations on behavior. This is the basis of normative control. In law we find relations between legal rights and legal duties, privileges and liabilities, etc. The legal system also makes assumptions about the working of the mind. As a default, agents are held responsible for their actions, as the execution of actions happen normally under conscious control (i.e. with intention). Moreover, the legal system also makes the assumption that not only the effects, but also certain side-effects are foreseeable, for which an agent can also be held responsible.

The core concepts of the ontology mostly describe *intentional* entities, entities that exist because they are intended to exist by agents. These form the necessary ingredients for representing the *institutional* entities, entities recognized and intended by collectives of agents, that are typical of the domain of law, which could

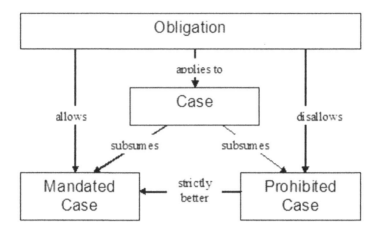

Fig. 4. An entity-relationship diagram describing the conceptual structure of obligations and prohibitions in the LKIF ontology

be characterized as formalized and institutionalized social order. The LKIF ontology also explcitly acknowledges a separated physical, mental, social, and abstract world, and the role of conventional metaphors based on the physical world in the construction of the mental, social, and abstract world (cf. generally [13]).

The strong bias of the LKIF ontology towards intentional entities is strikingly different from metaphysically inclined top ontologies such as SUMO or Sowa's upper ontology (cf. [14]), in which intention plays a less central role, but shows similarities with for instance the DOLCE ontology (cf. [15]) and Dennett's distinction in [16] between intentional, design and physical stance.

The intentional stance introduces a model of the mind to account for changes brought about by agents. The agent *acts*, i.e. he initiates processes – physical, social, or mental – that bring about changes that are intended. From here it is only a small step to recognizing that the agent can also perform physical "formal" acts (e.g. signing something) to effect institutional change by communicating to others one's intention to make that change. In DOLCE the distinction between agentive and non-agentive, which has a similar impact, is relatively prominent. The main criticism of DOLCE, voiced in [10,17], is that it is rather a representation of the terms used for describing knowledge, than a representation of knowledge itself.

One usage of LKIF is to use it as a basis for one's own ontology extending LKIF for some domain. Note that the ontology consists of terminological axioms: all claims can be considered defeasible in law, but that the proposition represented by the claim terminologically entails some other proposition is not. Ontology is not falsifiable: it is an agreement about use of terminology. Because the OWL axioms are in principle not falsifiable they should be used with due care.

3.3   LKIF Rules

Arguments instantiated by application of an LKIF rule are defeasible, i.e. presumed to be contingent. LKIF Rules semantics is based on the notion of defeasible rules as a type of *argumentation scheme*, particularly one argumentation scheme for each conclusion of the rule. Applying an LKIF rule is instantiating an LKIF argument from one of these schemes.

Figure 5 shows a simple argument structure generated by the application of LKIF rules.

Argumentation schemes generalize the concept of an inference rule to cover plausible but non-deductive forms of argument. The semantics of LKIF rules is

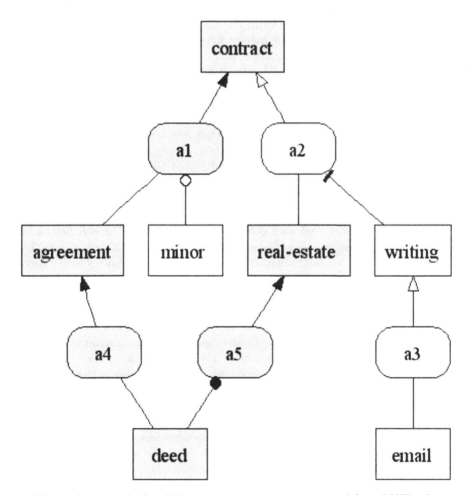

Fig. 5. An example GraphViz argument structure generated from LKIF rules

based on concepts also found in Carneades (cf. [18]). While the issue of plausible – for instance inductive, defeasible, or autoepistemic (basing one's conclusions on the presence or absence of knowledge about something) – inference is obviously relevant to any domain of knowledge, it is especially salient in law because of the importance of explicit dialectical considerations in legal procedure. To ignore it is to ignore obvious hints like the use of keywords like *unless* in the original source of law.

Argumentation schemes can also be used to bridge the gap between a prima facie interpretation of the law, ignoring dialectical and procedural considerations, and the ways it is operationalised in specific administrative, and adversarial settings. In this case the same text would map to different interpretations in the form of LKIF rules, possibly in addition to a terminological interpretation. The argumentation schemes can be presumed to be less reusable out of the original context than the ontology, at least if used this way, but are on the other hand a much closer fit to the actual behaviour of LKBS.

The concept of burden of proof in an adversarial setting provides us with good examples: if you for instance make a false claim that may harm someone's reputation, you are guilty of defamation, unless you made it in good faith and the reasonable belief that it was true. Logically speaking the unless could be easily replaced with a (but) not, and turned into a terminological axiom, but by doing so you fail to take account of implied burden of proof: it is up to the plaintiff to argue that the claim is false and harmful, up to the defendant to argue it was made in good faith, and up to the plaintiff that it was not reasonable to believe that it was true. It is valid to *argue* that a claim is defamatory because it is false and harmful, even if the argument may be defeated by arguments made by the counterparty.

Administrative procedure can be interpreted in a similar way: while tax and social security legislation is for instance prima facie mostly terminological, public bodies in reality mostly deal with claims from taxpayers or the beneficiaries of social security and engage in a kind of structured dialog with their clients. For most taxpayers, the tax administration taxes wages in advance based on the *presumption* that it can correctly guess how much income the taxpayer will generate over the fiscal year. At the end of the year it invites the taxpayer to claim an income, and the tax administration may then proceed to ask the taxpayer to substantiate claims with evidence or collect evidence from third parties. The tax administration eventually makes a decision, against which the taxpayer may then appeal, etc. In each of these setting the parties work with different operationalizations of the same law at different stages in the process. A decision that may be based on presumption in an earlier phase, must be backed by hard evidence in a later one, etc.

As usual in OWL, any syntactic node can carry an identifier. The following is an example of an LKIF rule in OWL, with one statement, (`MadeInGoodFaith ?c`), expanded inline, and the others only referred to by way of their identifier (`Statement2`, `Statement3`):

```
<lkif:Rule rdf:ID="Defamation2">
  <lkif:antecedent>
    <lkif:PositiveStatement rdf:ID="Statement1">
      <lkif:subject rdf:resource="#c"/>
      <lkif:predicate
       rdf:resource="http://www.w3.org/1999/02/22-rdf-syntax-ns#type"/>
      <lkif:object rdf:resource="#MadeInGoodFaith"/>
    </lkif:PositiveStatement>
  </lkif:antecedent>
  <lkif:unless rdf:resource="#Statement2"/>
  <lkif:consequent rdf:resource="#Statement3"/>
</lkif:Rule>
```

3.4   Rules Versus Ontology

LKIF rules are considerably more expressive than OWL, in particular with respect to the use of variables. In some cases users will inevitably resort to a rule even where a terminological axiom was intended. This – unfortunately in our view – makes it hard to maintain that there is any clear conceptual distinction, as there are independent technical reasons to make something a rule instead of a terminological statement. There is a risk that users will perceive LKIF rules simply as a general purpose more expressive extension to OWL to be used whenever needed. The difference in expressiveness is unavoidable: LKIF needs more, and OWL cannot be much more expressive without losing its usefulness as a Semantic Web integrative technology. For OWL real time consistency checking is an important design issue. For the LKIF Rules tractability is much less of an issue, considering its semantics definition.

In practical terms the knowledge engineer is free to draw the boundary between ontology and rules wherever he likes. It is possible to exclusively use rules, but the price to be paid is in its potential for reuse: alignment of ontologies is generally considered to be the first step to integration of knowledge bases, and by using rules one – firstly – signals that the information is considered contingent by its author, and secondly one loses subsumption as an organizing principle. On the other hand, putting obviously defeasible reasoning policies in the ontology is obviously very bad practice, will eventually cause inconsistencies, and will – considering OWL's role as a Semantic Web integrative technology – generally result in the ontology, and any rules dependent on it, not being used.

The distinction in the end reflects a different degree of entrenchment. Even if one takes the position that in the end everything is defeasible, one will usually want impose some entrenchment ordering so that in the case of conflict some things give way more easily than others. This position is taken by Boer in [19], who notes the importance to Legal Knowledge Engineering of a *rhetorical hierarchy* in legal argumentation to the effect that one should prioritize technical arguments over normative arguments, normative arguments of epistemological arguments, and epistemological arguments over ontological arguments. Also outside the legal field, raising ontological issues ("Oh, that depends on what you mean with X") in an intellectual discussion that started out with a technical disagreement is generally interpreted as a sign of weakness of one's position.

The rules versus ontology distinction can be used to model a number of phenomena knowledge engineers encounter where they may want to impose such an entrenchment ordering prioritizing stronger forms of argument to weaker ones: "classical not" versus a negation as failure, not knowing whether something is the case versus knowing something is not the case, constitutive rules versus essential rules of an institution, etc. Relations between these distinctions have been explored, and some have been shown to be special cases or alternative formulations of each other: e.g. defeasible (cf. [18]) and constitutive rules can be operationalized as argumentation schemes, negation as failure, unless, and assuming as autoepistemic expressions (cf. [20]), essential rules as terminological axioms, etc.

3.5   Integrated LKIF Semantics

It is not likely that we would be able to come up with a model-theoretic semantics for LKIF rules that is compatible with OWL semantics and pleases most potential users. The argumentation scheme-based account of LKIF rules semantics sidesteps the issue by defining the semantics of LKIF rules only in relation to its valid use in argumentation structures. This does not preclude the usage of LKIF rules to exchange rules compatible in form that in the originating LKBS are used with a compatible model-theoretical interpretation, for instance prolog clauses.

Conversely, we do not at present account for OWL in terms of argumentation schemes, although this is certainly feasible. It is not unlikely that an integrated semantics definition for LKIF will be developed later that explains description logic in terms of argumentation schemes instead of extending the model-theoretic interpretation of OWL. There are two problems in our view: 1) not all possible ontological inferences make sense in an argumentation structure, and 2) ontological commitments are by definition nondefeasible. The current working agreement is to allow treatment of a terminological entailment between two propositions as an argument (by implication opening it to counter-argument). Enforcing the deeper entrenchment of terminology can be achieved on the level of evaluation of the argument structure if desired, by discounting challenges on terminological arguments.

4   Future Developments

MetaLex is under active development by the CEN workshop. New proposals are scheduled for 2008. Proposals are circulated by way of the publicly accessible workshop website[11]. The partial agreement was adopted by the workshop in the understanding that it will be augmented with additional agreements on ontological formalization, citation and reference, time and versioning, and components and component inclusion.

The workshop requested more rigorous formalization of the four ontological levels (work, expression, manifestation, item) at which a bibliographic entity

---

[11] http://www.metalex.eu/

exists, and what properties belong to which level. The TC will make a definitive list of properties of expressions that distinguishes version, variant, consolidation, original, translation, and localization: for instance language, author, version-period (infinity is default value), date of creation, authoritiveness. Variants are for instance expressions that partially or totally overlap in version-interval. A version is usually temporal-diachronic, and a variant synchronic.

Central to versioning in law is the problem of identifying the expression of which a manifestation is a representation at a certain manifest date (version date of the xml file). When (interval) was it law? Secondary problem is the distinction between an XML expression existing, being in force, and efficacy and applicability intervals. Note that at any time before an expression ends its existence in the body of law, metadata is not final. Very important is the date of creation of the XML manifestation and item: metadata about an expression is either prediction or hindsight. Hindsight is more accurate, but prediction more relevant.

As is implied by the FRBR definitions of bibliographic entities, the parts of a bibliographic entity are also bibliographic entities. Any part, except the top level container, of a standard MetaLex XML manifestation can be implemented as an inclusion pointer to an external object on the expression, manifestation, and item level. On the manifestation level one for instance makes choices about object names and media formats (a tiff, jpeg, pdf image etc.). In some cases a text that is (or could be) embodied by a MetaLex manifestation (e.g. a chinese appendix of a treaty) is embodied alternatively by a media object.

LKIF has to reach a stable form in 2008. The issue of integrated LKIF semantics is the most pressing open problem, although it has sofar not hindered the progress of the Estrella project. In addition to a syntax and semantics specification, the Estrella project also develops application programmer's interfaces for manipulation of LKIF knowledge bases that are also intended to become part of LKIF. In addition we are running a number of small pilot projects to establish the usability of LKIF for modeling law, including the game of Nomic and EC council directive 90/434/EEG, and the ontology – particularly the vocabulary for describing legal cases – is still under development.

The MetaLex workshop is open to participation, but in the understanding that the voting members of a CEN workshop are *organizations* and not individuals. Individuals wanting to contribute to discussions are of course welcome on the workshop wiki mentioned earlier. LKIF has not reached this stage yet, but we welcome feedback on the LKIF ontology[12].

Acknowledgements

This work was performed as part of Estrella, supported by the European Commission (IST-2004-027655).

---

[12] http://www.estrellaproject.org/lkif-core/

References

1. Boer, A., Hoekstra, R., Winkels, R., van Engers, T., Willaert, F.: ^{META}lex: Legislation in XML. In: Bench-Capon, T., Daskalopulu, A., Winkels, R. (eds.) Legal Knowledge and Information Systems (Jurix 2002), pp. 1–10. IOS Press, Amsterdam (2002)
2. Boer, A., Hoekstra, R., Winkels, R., van Engers, T.: ^{META}lex: Jurisdiction and Language. In: Palmirani, M., van Engers, T., Wimmer, M.A. (eds.) Proceedings of the E-Government Workshop in conjunction with JURIX 2003, December 2003, pp. 54–66. Universitätsverlag Rudolf Trauner (2003)
3. Dick, J.P.: Representation of legal text for conceptual retrieval. In: ICAIL 1991: Proceedings of the 3rd international conference on Artificial intelligence and law, pp. 244–253. ACM Press, New York (1991)
4. Gordon, T.F.: Constructing arguments with a computational model of an argumentation scheme for legal rules. In: Proceedings of the Eleventh International Conference on Artificial Intelligence and Law, pp. 117–121 (2007)
5. Rubino, R., Rotolo, A., Sartor, G.: An owl ontology of fundamental legal concepts. In: van Engers, T.M. (ed.) Legal Knowledge and Information Systems. Jurix 2006: The Nineteenth Annual Conference, December 2006. Frontiers in Artificial Intelligence and Applications, vol. 152, pp. 101–110. IOS Press, Amsterdam (2006)
6. Breuker, J., Boer, A., Hoekstra, R., van den Berg, K.: Developing content for LKIF: Ontologies and frameworks for legal reasoning. In: van Engers, T.M. (ed.) Legal Knowledge and Information Systems. Jurix 2006: The Nineteenth Annual Conference, December 2006. Frontiers in Artificial Intelligence and Applications, vol. 152, pp. 169–174. IOS Press, Amsterdam (2006)
7. Bench-Capon, T., Coenen, F.: Exploiting isomorphism: development of a kbs to support british coal insurance claims. In: Sergot, M. (ed.) Proceedings of the third International Conference on AI and Law, New York, pp. 62–69. ACM, New York (1991)
8. Biagioli, C., Francesconi, E., Passerini, A., Montemagni, S., Soria, C.: Automatic semantics extraction in law documents. In: ICAIL, pp. 133–140 (2005)
9. de Maat, E., Winkels, R., van Engers, T.: Making Sense of Legal Texts. In: Grewendorf, G., Rathert, M. (eds.) Formal Linguistics and Law, Mouton, De Gruyter, Berlin. Trends in Linguistics - Studies and Monographs (TiLSM) (in press, 2008)
10. Hoekstra, R., Breuker, J., Di Bello, M., Boer, A.: The LKIF Core ontology of basic legal concepts. In: Casanovas, P., Biasiotti, M.A., Francesconi, E., Sagri, M.T. (eds.) Proceedings of the Workshop on Legal Ontologies and Artificial Intelligence Techniques (LOAIT 2007) (June 2007)
11. Makinson, D.: On a fundamental problem of deontic logic. In: McNamara, P., Prakken, H. (eds.) Norms, Logics and Information Systems. New Studies in Deontic Logic and Computer Science. Frontiers in Artificial Intelligence and Applications, vol. 49, pp. 29–53. IOS Press, Amsterdam (1999)
12. Chisholm, R.M.: Contrary-to-duty imperatives and deontic logic. Analysis 24, 33–36 (1963)
13. Lakoff, G., Johnson, M.: Metaphors We Live By. University of Chicago Press, Chicago (1980)
14. Sowa, J.F.: Knowledge representation: logical, philosophical and computational foundations. Brooks/Cole Publishing Co., Pacific Grove (2000)

15. Gangemi, A., Guarino, N., Masolo, C., Oltramari, A., Schneider, L.: Sweetening ontologies with dolce. In: Gangemi, A., Guarino, N., Masolo, C., Oltramari, A., Schneider, L. (eds.) EKAW 2002. LNCS (LNAI), vol. 2473, pp. 223–233. Springer, Heidelberg (2002)
16. Dennett, D.: The Intentional Stance. MIT Press, Cambridge (1987)
17. Hoekstra, R., Breuker, J., Di Bello, M., Boer, A.: LKIF Core: Principled ontology development for the legal domain. In: Breuker, J., Casanovas, P., Klein, M., Francesconi, E. (eds.) Law, Ontologies, and the Semantic Web. IOS Press, Amsterdam (2008) (submitted for review)
18. Gordon, T.F., Prakken, H., Walton, D.: The Carneades model of argument and burden of proof. Artificial Intelligence 171(10-11), 875–896 (2007)
19. Boer, A.: The Consultancy Game. In: Breuker, J.A., Leenes, R., Winkels, R.G.F. (eds.) Legal Knowledge and Information Systems. Jurix 2000: The Thirteenth Annual Conference. Frontiers in Artificial Intelligence and Applications, pp. 99–112. IOS Press, Amsterdam (2000)
20. Motik, B., Horrocks, I., Rosati, R., Sattler, U.: Can OWL and logic programming live together happily ever after? In: Cruz, I., Decker, S., Allemang, D., Preist, C., Schwabe, D., Mika, P., Uschold, M., Aroyo, L.M. (eds.) ISWC 2006. LNCS, vol. 4273, pp. 501–514. Springer, Heidelberg (2006)

MetaVex: Regulation Drafting Meets the Semantic Web

Saskia van de Ven, Rinke Hoekstra, Radboud Winkels, Emile de Maat,
and Ádám Kollár

Leibniz Center for Law, Faculty of Law, University of Amsterdam
PO Box 1030, 1000 BA, Amsterdam
{s.vandeven,hoekstra,winkels,e.demaat,a.i.kollar}@uva.nl

Abstract. Currently almost all legislative bodies throughout Europe use general purpose word-processing software for the drafting of legal documents. These regular word processors do not provide specific support for legislative drafters and parliamentarians to facilitate the legislative process. Furthermore, they do not natively support metadata on regulations. This paper describes how the MetaLex regulation-drafting environment (MetaVex) aims to meet such requirements.

Keywords: XML, RDF(S), OWL, metadata, regulation, drafting, semantic web.

1 Introduction

Legislative drafting and designing amendments to existing or new legislation are important parts of the work done by national parliaments, regional assemblies, city councils and ministries in Europe. Currently almost all of these legislative bodies use general purpose word-processing software to create legal documents.

However, these regular word processors generally do not provide users with targeted support to facilitate the legislative process. They are often badly integrated with legacy systems that support storage, search and publishing facilities, and provide no streamlined environment for drafting and discussing legislation and other kinds of regulations. Such an environment would integrate workflow, search facilities, tools to support group dynamics (including versioning and distribution) and features that facilitate publication. It should provide access to other legal sources through intra- or Internet for direct referencing (see below), but also for background information. A legal drafter changing a particular law might for instance be interested in certain cases or commentaries that point out weaknesses in the current version. These cases and commentaries will be published and maintained by other organisations than the one that employs the legal drafter. In other words, the drafting environment should be able to cope with distributed sources.

Legislative drafting is a complex process that takes place in a political and dynamic environment, which involves many stake-holders. Since a new or adapted regulation is often connected to existing laws, the drafters and other stake-holders should be aware of relationships between the law under construction and those existing legal sources. Legal drafting practice has learned that legal quality can benefit from the use of specific legal drafting patterns.

P. Casanovas et al. (Eds.): Computable Models of the Law, LNAI 4884, pp. 42–55, 2008.

The SEAL project (Smart Environment for Assisting the drafting and debating of Legislation)[1] develops a supportive environment that enables easy construction of legal drafts using drafting patterns and creation of connections from and to existing legal sources. The infrastructure will provide access to a repository with existing laws, draft versions and amendments and will offer easy to use access methods. Collaboration between stake-holders will be supported by groupware facilities such as automated signalling functions and routing of drafts and amendments.

This environment will be developed for three European parliaments. An initial working environment is foreseen in the end of 2007. This will be tested, refined and implemented in co-operation with the parliaments and legislation drafters during the project.

The MetaLex regulation-drafting environment (MetaVex) is developed at the Leibniz Center for Law and is one of the three environments being evaluated in SEAL. The other two environments are the xmLegesEditor: owned/provided and maintained by CNR-ITTIG [1] and The Norma Editor: owned under licence and maintained by CIRSFID University of Bologna [8]. In the following sections we identify the requirements, introduce the XML document standard underlying the system, and describe its current status.

2 Requirements

MetaVex aims to streamline the legislative process by addressing the problems discussed in the introduction. In this section we describe the requirements against which the environment is evaluated. These criteria can be summarized as follows:

Look and Feel. The editing environment should provide a look and feel similar to normal word processors. Document editing should be done in a WYSIWYG[2] interface; this way legal drafters can create document structure and content without knowledge of specific commands or technical notations.

Drafting Patterns. Legislative drafters should be supported by the editor in complying to prescribed legal drafting patterns. Offering users suggestions and predefined phrases in the form of templates improves and speeds up the process of generating document structure and content.

Referencing. The use of references to other legal sources is an important way in which drafters add structure and meaning to a document. The editor should facilitate the frequent use of these references and offer ways to validate the legal sources they cite. References should be *detailed*, i.e. point to the smallest relevant element of a regulation.

Metadata. A way to store extra information about a document e.g. author, version, modification, should be provided. Possibilities to add information concerning document structure as well as content is regarded as an advantage.

Version Management. The environment should offer support to manage document versions, starting from the first draft until and beyond the time at which the document is published. This allows users to always be able to identify the latest version of a document.

---

[1] SEAL is a project in the e-Participation initiative of the European Commission.
[2] What You See Is What You Get

Groupware. By using groupware facilities, drafters can collaborate on the same project. These facilities will not only consist of sharing comments or amending existing legislation, but will allow for elaborate authorisation and accountability management.

Workflow. Workflow support should be an integral part of the environment to be able to divide tasks into sub-tasks, assigning them to people and keeping track of progress.

Storage. Users should be able to store documents in a *local data repository*, providing them with advanced search mechanisms. It should also be possible to connect to a server with various types of clients over the *internet*, e.g. by using a browser.

Publishing. The environment should allow straightforward publishing of texts in legacy formats. This allows publishing of legal drafts in an early stage, which makes it possible to interact with the public (businesses, citizens and interest groups) during the drafting process.

3 Syntax and Semantics: MetaLex

MetaVex is a regulation-drafting environment for MetaLex documents: texts are saved as XML documents that comply with the MetaLex format for legal sources. This standard provides a generic and easily extensible framework for the XML encoding of the structure and content of legal documents. It addresses many of the requirements introduced in the previous section, as is described in e.g. [3]. In this section the advantages of MetaLex will be explained. Section 4 will address how users of MetaVex can benefit from these advantages, unless stated otherwise.

MetaLex is currently undergoing a CEN standardisation process. It is input to the CEN workshop on an Open XML interchange format for legal and legislative resources[3]. The MetaLex/CEN schema is based on best practices from amongst others the previous versions of the MetaLex schema, the Akoma Ntoso schema [11], and the Norme in Rete[4] DTD. A first version of this schema was adopted as part of a CEN workshop agreement on 6 December 2006[5].

The use of a standard interchange format enables *public administrations* to link legal information from various levels of authority and different countries and languages. Moreover, the standard will enable *companies* that are active in the field of legal knowledge systems to connect to and use legal content in their applications, which allows them to support a much larger market. An open interchange format will also protect customers of such companies from vendor lock in. Finally, the standard will help to improve transparency and accessibility of legal content for both citizens and businesses.

MetaLex provides extensive mechanisms to add metadata both to specific parts of a document and to the document as a whole. Every element of a legal text can be uniquely identified through a URI, and annotated with information regarding e.g. its version, publication date, validity interval, efficacy, language, jurisdiction, and authority. Furthermore, the standard introduces the possibility for marking references,

---

[3] http://www.cenorm.be/cenorm/businessdomains/businessdomains/isss/activity/ws/_metalex.asp
[4] http://www.nir.it
[5] http://www.metalex.eu/wiki/

both to elements of (other) regulations and to individual entities not part of a regulation, such as institutions or concepts defined by the regulation.

As the standard is primarily intended as an interchange format, annotating legal texts with metadata allows a single MetaLex document to contain several versions of a text. MetaLex not only includes an event-based model for managing multiple versions of legal documents *through time* [4], but multiple *language* versions of the same text can be included in just one MetaLex document as well.

All metadata statements in MetaLex conform to the triple model of RDF[6]. This means that any MetaLex metadata can be used to generate an RDF triple: statements about entities are interpreted as subject, predicate, object triples. And conversely, because every MetaLex element has a unique identifier, it is possible to make external statements in RDF referring to any element of a legal text.

MetaLex provides a strong connection to other semantic web standards as well, such as RDF Schema and OWL[7] as both have RDF/XML syntax. A MetaLex XML document can be translated into OWL by means of XSL transformations (XSLT's). An XSLT provides a mapping between an XML source document and a desired destination format. For instance, we can translate any MetaLex XML document into HTML, XSL:FO and RDF/OWL using such stylesheets. Consider the following piece of MetaLex XML which denotes an article:

```
<Article id="a1">
   <IndexDesignation>
       <Category>
          <TextVersion xml:lang="en">Article</TextVersion>
       </Category>
       <Index>
              <TextVersion xml:lang="en">1</TextVersion>
       </Index>
   </IndexDesignation>
</Article>
```

Using the standard metalex2owl.xsl transformation, we can produce the following RDF/XML code:

```
<metalexrdf:Article rdf:about="http://www.metalex.nl/ec/2002/58#a1">
   <metalexrdf:properStructuralSuccessor
   rdf:resource="http://www.metalex.nl/ec/2002/58#a2"/>
       <metalexrdf:properStructuralMember>
              <metalexrdf:IndexDesignation>
                  ...
              </metalexrdf:IndexDesignation>
       </metalexrdf:properStructuralMember>
</metalexrdf:Article>
```

As RDF is order-independent this kind of transformation would contain the risk of messing up the original document order. Document order is important when legislative documents are concerned, therefore the XSLT should address the order of all document elements explicitly. To achieve this goal the successor for each element in the original document is identified and passed through to the destination document in RDF. As a result the MetaLex RDF encoding will contain explicit *sequences* to

---

[6] The Resource Description Framework. See http://www.w3.org/RDF/
[7] The Web Ontology Language. See http://www.w3.org/2004/OWL

represent the sequences of articles, parts, sentences etc., preserving the order of the document elements present in the original XML file. The container membership property "structuralMember" and the sequence property "structuralSuccessor" are used to represent these kind of sequences. More information about transforming MetaLex XML into RDF/OWL and issues regarding this subject can be found on the MetaLex website[8].

By integrating MetaLex with the semantic web standards RDF, RDFS and OWL, metadata both on the elements of legal texts themselves, as on the *contents* of those texts can be described. OWL and RDFS can be used to describe the contents of legal texts: the *concepts* that occur in them, but also their *normative* content. These formal representations of the semantics of legal texts can be used to perform elaborate legal reasoning, such as consistency checking, legal assessment etc. and for building knowledge-based applications which can be used by citizens to gain advice on complex legal issues. OWL provides additional expressive power, which can be used to describe not only the *content* or *domain* of a regulation, but also the *authority* through which a regulation is enforced, and the history and background of *modifications* of the regulation, as is described in [12]. The MetaLex CEN schema defines a general framework for describing events and actions in OWL. More information about this framework can be found at the MetaLex CEN Wiki[9]. At this moment MetaVex does not support the use of the mentioned semantic web standards yet, as is discussed in section 5.

The MetaLex CEN workshop furthermore adopted the RDFa[10] standard for embedded metadata. RDFa does not have its own namespace: the significance of XML elements and attributes to RDFa processors is determined entirely by names. An RDFa element is defined as any XML element that contains one or more RDFa attributes: about, property, rel, href, instanceof or content. The following example show an article in MetaLex XML:

```
<Article id="a1">
   <IndexDesignation>
        <Category>
                <TextVersion xml:lang="en">Article</TextVersion>
        </Category>
        <Index>
                <TextVersion xml:lang="en">1</TextVersion>
        </Index>
   </IndexDesignation>
</Article>
```

This article could be augmented with RDFa in the following way:

```
<metalex:Article id="a1" about="http://www.metalex.nl/ec/2002/58#a1">
instanceof="metalexrdf:Article"
property="metalexrdf:properStructuralMember">
        <IndexDesignation instanceof="metalexrdf:IndexDesignation"
        property="metalexrdf:properStructuralMember">
                <Category instanceof="metalexrdf:Category"
                property="metalexrdf:textversion">
                        <metalex:TextVersion
                        instanceof="metalexrdf:TextVersion"
                        xml:lang="en">Article
                        </metalex:TextVersion>
```

---

[8] http://www.metalex.eu/information/guidelines
[9] http://www.metalex.eu/wiki/
[10] http://www.w3.org/2006/07/SWD/RDFa/

```
        </metalex:Category>
        <metalex:Index instanceof="metalexrdf:Index"
        property="metalexrdf:textversion">
                <metalex:TextVersion
                instanceof="metalexrdf:TextVersion"
                xml:lang="en">1
                </metalex:TextVersion>
        </metalex:Index>
    </metalex:IndexDesignation>
</metalex:Article>
```

An RDFa processor can be used to generate RDF triples from the RDFa elements. The real power of RDFa is that it enables you to add semantic values to XHTML documents. This starts with adding some simple statements, but can develop to using full RDF power inside of an XHTML document. At this moment the use of RDFa is not yet beneficial when used within MetaVex, as is mentioned in section 5.

More importantly, MetaLex allows formal representations of legislation to refer to and be grounded in the documents containing the official texts. An example is LKIF, the Legal Knowledge Interchange Format [2], currently being developed in the ESTRELLA project[11], a vendor neutral representation format for legal knowledge. Existing Semantic Web initiatives are aimed at modelling concepts (OWL "ontology") and rules (RuleML, RIF). The LKIF builds on but goes beyond this generic work to allow further kinds of legal knowledge to be modelled, including: meta-level rules for reasoning about rule priorities and exceptions, legal arguments, cases and case factors, values and principles, and legal procedures. It is based on a layered approach, providing a method of using OWL and RIF and contains two sublanguages, for defeasible rules and for subjunctive betterness. Furthermore, the LKIF is grounded in a core ontology of basic legal concepts: LKIF Core [6]. The ontology covers a base level of components required for explaining epistemological, situational, and mereological patterns as they occur in legal reasoning.[12]

An example of the flexibility of the MetaLex schema is the combination of regulations, maps and spatial planning adopted in the Legal Atlas tool [13][13]. The MetaLex region attribute can be used to refer to the geographical region to which rules in a regulation can be applied: i.e. it specifies geographical jurisdiction. In Legal Atlas, this is used in combination with RDF and GML[14] to show spatial planning regulations both as maps and as texts.

4   MetaVex

MetaVex is a platform independent open source editor, and shares a large part of its codebase with the Visual Editor for XML (Vex)[15]. It is developed within the Java Eclipse[16] development platform, which allows future development of plug-ins and add-on functionality.

---

[11] ESTRELLA: European project for Standardized Transparent Representations in order to Extend Legal Accessibility, IST-2004-027655, http://www.estrellaproject.org

[12] http://www.estrellaproject.org/lkif-core

[13] http://www.leibnizcenter.org/projects/current/legal-atlas

[14] Geography Markup Language.

[15] Vex is currently no longer under active development. See http://vex.sourceforge.net/

[16] http://www.eclipse.org

MetaVex is specifically intended to support the creation of documents complying with the MetaLex standard for legal sources, but flexible enough to be easily adapted to different XML schemas.

Target users of MetaVex will be drafters and members of parliament. These users cannot be expected to be familiar with editing an XML-structure directly. For this reason, the editor offers a WYSIWYG interface, which does not require any knowledge of or experience in creating XML-code. In fact, the editor shows close resemblance to a conventional word processor, and at the same time allows a user to alter or create content while keeping the integrity of the underlying structure intact.

Since XML documents themselves do not carry information about how to display the document MetaVex uses CSS[17] to determine formatting. A user will be able to choose from different types of predefined formatting, but cannot change the formatting in line.

The use of XML as an underlying document structure makes it easy to validate the structure of documents created with MetaVex against the rules defined in the MetaLex schema file. This schema file restricts element and attribute names and allowed combinations. MetaVex uses this schema file to check which elements can be inserted at a certain position in a document, while sustaining a well-formed and valid XML structure.

To enforce this structure during the composing or editing of a document, the editor provides a context sensitive menu of the elements valid at a particular position within the document. Users can only insert elements available in this menu, or elements to which the content model of the current element is agnostic. This procedure ensures schema compliance at every stage of document creation using the editor.

This functionality is one of the major differences between MetaVex and normal word processors. When using a normal word processor, a user can just start typing and does not have to bother about adding specific text elements. The use of templates strongly reduces this difference by offering users a way to add new elements or whole blocks of elements at once: creating e.g. an article is similar to form-filling. MetaVex contains a pane offering the user specific templates to choose from. Furthermore the new document wizard offers a user the possibility to start a new document, based on a predefined template. This way the user does not have to start from scratch.

In the Netherlands, legislative documents are required to be composed according to what is prescribed in the Dutch Guidelines for Legal Drafting [5]. These guidelines do not only apply to technical aspects of writing legislation, but emphasise content too. MetaVex provides a set of templates that are structured according to these guidelines. These templates can provide not only structure, but standard content as well. Currently, the templates included in MetaVex follow Dutch guidelines and cannot be used to support drafting in other countries. However, other templates can be easily imported and used in MetaVex as well. This extensive use of templates not only offers guidance, but can also save users a lot of work.

The MetaVex user interface (see Figure 1) offers an "Insert Templates" panel that shows a list of the mentioned templates. Users can choose and click on one of the templates to insert a prebuilt collection of elements into the document. These elements together form for example a whole chapter or article. Not all templates are

---

[17] Cascading Style Sheets, see http://www.w3.org/Style/CSS

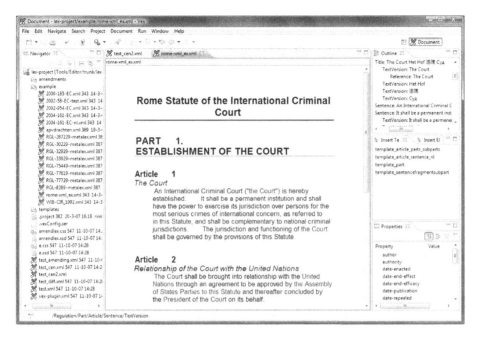

Fig. 1. Editing the statute of Rome in MetaVex

shown in the list, only the ones that can be inserted at the current cursor position, maintaining a valid and well-formed XML-structure underneath.

There is also an "Insert Elements" panel that shows a list of all single elements that can be validly inserted at the current cursor position. The same list is accessible through a context sensitive right-mouse menu. The left part of the screenshot in Figure 1 shows an "Outline" pane that displays the overall document structure as a hierarchy. This structure can be collapsed or expanded and allows a user to easily navigate through parts of the text. Conversely, the cursor position within the XML structure is reflected both in the selected element in this outline pane and in an XPath expression in the status bar. Furthermore, as mentioned in the previous section, MetaLex supports an extensive set of meta-data attributes which allow users to link many different kind s of extra information to a document. The user interface of MetaVex allows the user to edit the values of these attributes through the "Properties" panel. This panel displays a table of all attributes and their values available on the currently selected XML element. Finally, the "Navigator" panel shows a list of the files available in the current project. Each of these panels can be moved, closed or enlarged to suit a users' preference.

Most prominent in MetaVex is the editing panel. Multiple versions of the same text can be simultaneously edited in a single MetaLex document. The screenshot in Figure 2 illustrates this, showing the Statute of Rome (which introduces the International Criminal Court) in Dutch, English, Chinese and Russian language versions. As this can be confusing to users, MetaVex can hide irrelevant information: users can select a desired time interval or language version and hide other versions available in the

Fig. 2. Editing multiple language versions of the Rome Statute in MetaVex

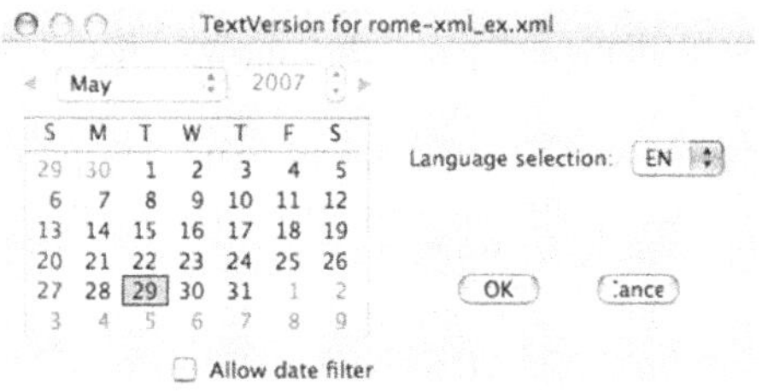

Fig. 3. Selecting a text version

document (see Figure 3). After the selection of for example the English text version, only the English version of the statute of Rome will be shown, as was shown in Figure 1. The selection of text versions is also used by the export functions of MetaVex. Any valid document can be exported to various common formats such as PDF and HTML using export wizards, which apply XSLT[18] transformations to the XML source to produce the desired output format.

Recently functionality is added to MetaVex to support the creation of amendments and the generation of a consolidated version from the original document and the accepted amendment. After a piece of legislation is drafted, there is the possibility to bring in proposals or amendments. Normally such an amendment is created separately

---

[18] eXtensible Stylesheet Language Transformations, http://www.w3.org/Style/XSL/

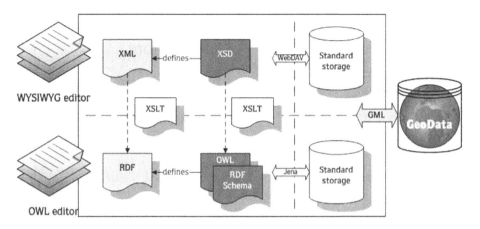

Fig. 4. MetaVex Architecture

from the original document that is being amended. The actual amendment consists of a list of proposed changes that will be applied to the draft bill once the amendment is accepted. Using the editor, amendments can be created just by adjusting the original text and the amendment will be generated automatically. At this moment this amendment support is not fully operational yet, but will be in the future. The next step will be to provide a function to automatically generate consolidated versions of legislation including the modifications resulting from the amending provisions.

Although MetaVex is an editor to alter or produce legislative documents, this does not mean that MetaVex is limited to cope with documents being addressed as "legal" only. In general, for all documents containing rules for a certain domain, a structured way of creating and filing those documents can certainly be useful. MetaVex can be used to suit the needs of many kinds of domains by providing structuring or formatting. For example, companies can develop their own general representation of data in the form of an XML schema file, according to which the intended documents will be composed and validated.

Figure 4 shows the MetaVex architecture and the way in which the various components will be integrated. In short, an XML document is constructed against an XML schema, and can be translated to RDF/OWL using XSLT transformations. MetaLex XML documents are stored in an XML storage facility (to be developed by one of the partners in the SEAL consortium) through a WebDAV interface. The RDF/OWL representation of the MetaLex document uses the vocabulary specified in the MetaLex OWL schema. It is stored using a Jena database backend. Currently, not all of the components shown in the picture have been developed. Future development issues are described in the discussion section, but first this section will end with providing some examples of the useful extras RDF offers in the context of MetaVex.

As mentioned before, links can be made from every element in a MetaLex document to a concept or identity in RDF. This is especially useful because of the frequent use of two kinds of references in legislative documents. First there are citations pointing to other structural elements of a document or another document as a whole, e.g.

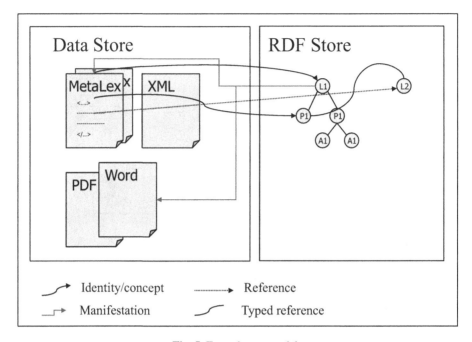

Fig. 5. From data to models

"Chapter 1, Article 2, second sentence" or "The Dutch traffic law". Second there are references to a concept, like 'civil servant' or 'boat'. When trying to validate the latter kind of reference an RDF model proves to be very useful. RDF can for example provide the intermediary concept 'boat' to temporarily resolve such a reference.

All documents containing information about the concept 'boat', can be linked to the concept 'boat' as well. Although resolving references might seem the most difficult, citations can be difficult to resolve as well. When a document contains a citation to a specific law and a new version of this law becomes enacted, the citation might need to be adjusted as well. Furthermore if a cited piece of legislation is not enacted yet, there is no actual document to link to. The RDF model can in this case contain a general concept of the cited law, containing all the versions that exist. These versions can be represented as subconcepts of the general one. This way it is possible to cite a document and all existing or even future versions. Not only different versions can be linked, different manifestations can be added as well, as is shown by the arrows in Figure 5 pointing from the RDF store to the Data store. Because RDF allows for typed references between concepts, relations between the different concepts can be expressed as well.

The advantage of storing such links is obvious: it makes it easier to find the correct version of a document. This is an important advantage for organisations that often have to deal with multiple versions of legislation, such as the Dutch Tax and Customs Administration, as is described in [12]. The new MetaLex/Cen standard will use a same sort of mechanism for storing various versions of a single document.

The use of RDF can also be advantageous in the context of search mechanisms and cataloguing. Ordinary search engines and catalogue systems make use of textual occurrences of a keyword. But when RDF comes in place, it is possible to search for all relevant pieces of legislation that are applicable on a certain subject, just following all of the provided links from the RDF concept to the XML sources and potential other sources. Linking to current handbooks, guides and regulations on legislative drafting can be considered as well.

5 Discussion and Future Work

As mentioned in the introduction, MetaVex is one of three environments being evaluated in SEAL. The other two environments are the xmLegesEditor and The Norma editor. The Norma editor and the xmLegesEditor are developed in Italy, in the context of the Norme-in-rete project. The Norma editor is designed as an add-on to Microsoft Word. This gives it the advantage that users can create and edit documents in a familiar environment. A disadvantage, however, is that it means that the Norma editor is not yet fully open source. This will change in the future, as the makers of the Norma editor aim to switch from MS Word to Open Office. Another disadvantage is that being dependable on another product in general has some risks. Whenever a new version of the other product is released, the editor has to adapt to this new version, which can be a time consuming process. The Norma editor already offers advanced support to automatic consolidation of documents and to the creation and validation of references. Users do not edit the XML structure directly, but the conversion to XML takes place after the editing has been done. A validation tool is used to parse the document structure and generate a valid and well-formed XML document. The user is notified when the validation tool detects inconsistencies.

In contrast, the xmLegesEditor is a native XML editor. It is rule-driven: it only allows operations that are valid with respect to the underlying document structure. The xmLegesEditor already contains functionality to extract and validate normative references and a parser to read and convert document into XML. It directly produces XML documents, compliant to the NIR DTD, but in the future it will also support XML schema.

Similar to the xmLegesEditor, the MetaVex editor is open source, rule-driven, and a native XML editor. Besides the functionality it already offers, there are quite some items on the requirement list that are currently not supported and will be discussed in the following part of this section.

MetaVex should enable intuitive construction and maintenance of references in legal documents, by functions for adding, editing and validating. [7] and [9] describe ways to automatically detect references in legislation with high accuracy. Embedding such functionality into MetaVex certainly would be a useful extension.

The use of MetaLex means a strong focus on semantic web technologies such as RDF(S) and OWL. For now it suffices to support these formats and edit OWL documents in a separate, already existing OWL-editor, such as Protégé[19] or TopBraid[20]. In the future, MetaVex will offer means to view and edit OWL-documents using an OWL-editor embedded inside MetaVex. The idea is to develop a separate view showing the RDF triples corresponding to a selected MetaLex element. The view should

---

[19] http://protege.stanford.edu
[20] http://www.topbraidcomposer.com

also enable users to edit the RDF triples directly. Admittedly, concurrent editing of the RDF and XML version requires relatively complex synchronisation.

As mentioned in Section 3, the MetaLex CEN workshop has adopted the RDFa standard for embedding metadata in MetaLex XML. Once MetaLex CEN is finalised, the MetaVex editor should offer functionality that supports the annotation of XML documents with the RDFa attribute set. Of course, this is closely related to a possibility for browsing an RDF graph. A recent development is GRDDL[21], a W3C recommendation, which specifies an even more flexible method for embedding RDF in arbitrary XML. GRDDL can be used to specify 1) per document, an XSL stylesheet for automatic extraction of RDF triples, or 2) a namespace-related stylesheet at XML Schema level. The advantage of GRDDL is that is more general, i.e. it is suited for any XML, and not just XHTML, and allows more flexible embedding of metadata, not restricted to the RDFa attribute set.

The MetaVex architecture (Figure 4) shows a connection to geodata. At this moment, MetaVex does not provide this connection yet, but it will provide one in the future. The connection will be similar to the approach of Legal Atlas.

So far nothing has been mentioned yet about MetaVex' storage mechanism and how certain features like versioning, security, groupware facilities etc. will be integrated. Although an implicit way of maintaining version information using MetaLex' attributes already exists, the ability to cope with several versions of a document in an explicit matter should also be taken into account. A content management system satisfying the requirements of section 2, will be developed within SEAL. However, MetaVex will commit to standards-based interfaces to open source RDF repositories such as Sesame[22] and Jena/Joseki[23].

MetaVex is still under construction and there is a lot of work that needs to be done. Nevertheless a solid, easy extendable and highly adaptable solution for editing XML-structured documents in a user-friendly environment already exists. As soon as all proposed features are fully present, MetaVex will lift the editing of legal documents to a whole new level by its unique combination of syntax and semantics.

References

[1] Agnoloni, T., Francesconi, E., Spinosa, P.: Xmlegeseditor, an OpenSource visual XML editor for supporting Legal National Standards. In: Biagioli, C., Francesconi, E., Sartor, G. (eds.) Proc. of V Legislative XML Workshop, pp. 239–252. European Press Academic Publishing (2007)

[2] Boer, A., Gordon, T., van den Berg, K., Di Bello, M., Förhécz, A., Vas, R.: Specification of the Legal Knowledge Interchange Format (LKIF). Deliverable 1.1, Estrella (2007)

[3] Boer, A., Hoekstra, R., Winkels, R., van Engers, T., Willaert, F.: METALex: Legislation in XML. In: Bench-Capon, T., Daskalopulu, A., Winkels, R. (eds.) Legal Knowledge and Information Systems, Jurix 2002, pp. 1–10. IOS Press, Amsterdam (2002)

---

[21] Gleaning Resource Descriptions from Dialects of Languages, see http://www.w3.org/TR/grddl/

[22] http://www.openrdf.org

[23] http://jena.sourceforge.net

[4] Boer, A., Winkels, R., van Engers, T., de Maat, E.: A Content Management System based on an Event-based Model of Version Management Information in Legislation. In: Gordon, T. (ed.) Legal Knowledge and Information Systems, Jurix 2004, pp. 19–28. IOS Press, Amsterdam (2004)

[5] Aanwijzingen voor de regelgeving (Directives for regulations), regulations for legislative drafting issued by the Prime-Minister, November 26, Stcrt (1992)

[6] Breuker, J., Hoekstra, R., Boer, A., van den Berg, K., Rubino, R., Sartor, G., Palmirani, M., Wyner, A., Bench-Capon, T.: OWL ontology of basic legal concepts (LKIF-Core). Deliverable 1.4, Estrella (2007)

[7] de Maat, E., Winkels, R., van Engers, T.: Automated detection of reference structures in law. In: van Engers, T. (ed.) Legal Knowledge and Information Systems. Jurix 2006: The Nineteenth Annual Conference. Frontiers in Artificial Intelligence and Applications, vol. 152, pp. 41–50. IOS Press, Amsterdam (2006)

[8] Palmirani, M., Brighi, R.: An XML Editor for Legal Information Management. In: Traunmüller, R. (ed.) EGOV 2003. LNCS, vol. 2739, pp. 421–429. Springer, Heidelberg (2003)

[9] Palmirani, M., Brighi, R., Massini, M.: Automated Extraction of Normative References in Legal Texts. In: Sartor, G. (ed.) Proceedings of the 9th International Conference on Artificial Intelligence and Law (ICAIL 2003), pp. 105–106. ACM Press, New York (2003)

[10] Shadbolt, N., Berners-Lee, T., Hall, W.: The Semantic Web Revisited. IEEE Intelligent Systems 21(3), 96–101 (2006)

[11] Vitali, F., Zeni, F.: Towards a country-independent data format: the Akoma Ntoso experience. In: Biagioli, C., Francesconi, E., Sartor, G. (eds.) Proc. of V Legislative XML Workshop, pp. 239–252. European Press Academic Publishing (2007)

[12] Winkels, R., Boer, A., de Maat, E., van Engers, T., Breebaart, M., Melger, H.: Constructing a semantic network for legal content. In: Gardner, A. (ed.) Proceedings of the Tenth International Conference on Artificial Intelligence and Law (ICAIL), June 2005, pp. 125–140. IAAIL, ACM Press, New York (2005)

[13] Winkels, R., Boer, A., Hupkes, E.: Legal Atlas: Access to Legal Sources through Maps. In: Winkels, R. (ed.) Proceedings of the 11th International Conference on Artificial Intelligence and Law (ICAIL 2007), pp. 27–36. ACM Press, New York (2007)

Building Semantic Resources for Legislative Drafting: The DALOS Project

Enrico Francesconi and Daniela Tiscornia

ITTIG-CNR, Florence Italy

Abstract. The DALOS Project aims at building an ontological-linguistic resource to be used in the multilingual EU legislative drafting process, as well as in a linguistically reliable national transpositions of EU directives. This paper outlines the design of the ontological-linguistic resource, as well as the main phases carried on for its implementation.

1 Introduction

This article describes the theoretical back-ground, the methodological steps and the mid-term outcomes of the Dalos Project (e-Participation, 2006/024). The DALOS project was recently launched within the "eParticipation" framework, the European Commission initiative aimed at promoting the development and use of Information and Communication Technologies in legislative decision-making processes. The aim of such an initiative is to foster the quality of the legislative drafting, to enhance accessibility and alignment of legislation at European level, as well as to promote the awareness and democratic participation of citizens in the legislative process. In particular, DALOS aims at ensuring that legal drafters and decision-makers have control over legal language at national and European level, by providing law-makers with the linguistic and knowledge management tools to be used in the legislative processes, in particular within the phase of legislative drafting.

The article is structured as follows: the institutional environment in which the e-participation program is located is briefly outlined in Section 2, along with the description of the specific tasks addressed by the program in the legislative field and of the barriers posed by language which impede citizens in really understanding the law. In Section 3, the approach to linguistic-ontological resource development in the DALOS project is described. In particular, in Section 3.1, the complexity of the multilingual legal scenario is addressed and reference to previous experiences is provided. In Section 4, the characteristics of DALOS knowledge and the specification of its Knowledge Organization System (KOS) are presented. In Section 5, the phases to implement the linguistic ontological resource are shown and, finally, some preliminary conclusions are made.

2 The Legislative Process as a Place for 'e-participation'

The shift of power from single Member States to the EU and the lack of a full consensus of the people generated the so called "democratic deficit" of European

P. Casanovas et al. (Eds.): Computable Models of the Law, LNAI 4884, pp. 56–70, 2008.

Union institutions that is, a lack of responsibilities, transparency and public policies[1] – which has become one of the crucial points for the legitimation of the Union[2].

What is missing is the so-called "legitimation of the people", through which legitimation of the Union is perceived as something above the nations, therefore, originating directly from the degree of approval shown by European public opinion. In this case, the Community bodies are based upon the fundamental requirement of a strongly rooted European "common identity". Likewise, the question of the political representation of the EU surfaces. This term defines the process of natural policies fundamental for all representative democracies, through which the building of governmental policies is linked to the needs and requirements of the people.

In this regard, the EU has had some serious difficulties in identifying the European nations. Recent empirical studies have shown that European citizens feel they belong to their towns, regions and nations, and, then, to Europe: thus, the "European link" seems to be the weakest[3].

With the commitment to improve the citizen's view of the EU's activities, the Commission launched the debate on European governance in 2001. In a White Paper[4] the Commission analysed all the rules, procedures and practices affecting how powers are exercised within the European Union. The aim was to adopt new forms of governance that bring the Union closer to European citizens, make it more effective, reinforce democracy in Europe and consolidate the legitimacy of its institutions. One of the main driving principle towards this goal is the implementation of better and more consistent policies associating civil society organisations with the European institutions.

This is in line with the European trend that is formalised in the principle of proximity. In fact, the principle of proximity means that the decision shall be taken as openly and as nearly as possible to the citizen.

The EU should permit and encourage citizens to take an active role in the policy-making and lawmaking process and, moreover, the EU should promote a type of iterative decision-making process that also permits wider visibility and consideration of instances involving the public scope.

How should such targets be reached? Citizens cannot hope to influence European decision-making unless they are first fully informed about what the EU institutions are doing and which are the relevant questions on the European agenda aimed at engaging citizens' interest and involvement.

True participation can only occur through full information and knowledge of the European social, institutional and regulatory context.

---

[1] Dehousse, R. *European Institutional Architecture after Amsterdam: Parliamentary System or Regulatory Structure?* in EUI Working Papers, RSC No.11, 1998.

[2] Blondel, J., Sinnott, R. and Svensson, P., *People and Parliament in the European Union: Participation,* Democracy and Legitimacy, Oxford, Clarendon Press, 1998.

[3] Beetham, D. and Lord, C., *Legitimacy and the European Union.* Harlow. Longman, 1998.

[4] See European Governance. A White Paper. COM(2001) 428.

Such knowledge can only be offered to citizens by helping them understand the European regulatory system and the implications that it may or actually does have on society.

2.1 Knowing the Law

In Europe, there is a *de facto* obligation for all Member States to provide citizens with a true opportunity of knowing the law. Thus, there is an actual right of citizens to legal information (formal knowledge) and, as well, there is a commitment of Member States towards them to adopt suitable means in order to allow them to understand the law in terms of rights and duties (substantial right).

Every institutional act is a document involving words, processed by a "producer" (the organizational body) and addressing several users. Thus, communication problems emerge, due to the ambiguity and imprecision of natural language, to the difficulties in grasping the meaning of the message and, as a consequence, the addressees do not comply with European Directives due to a lack of understanding. In the European legislator's perspective, concerning legislative participation, it is necessary to allow citizens to access "understandable" legal and legislative information in order to enhance the process of comprehending the law, on the one hand, and it is necessary to improve the quality and the readability of legislative texts, thereby also contributing to the "certainty of law", on the other. This explains why the drafting phase is included within the legislative process objective of the EC program, as a pre-condition where ICT technologies are expected to contribute to 'better legislation'.

Legislative procedures are indeed very complex. The number of institutions, procedural steps, readings and revisions simply make EU legislative procedures more complex than EU citizens are generally accustomed to, in relation to their national legislative procedures. The complexity of the legislative procedures makes it more obvious that measures are needed to clarify them as well as the *results* of those procedures. *Language*, being the key to legal domains in general and, therefore, also to Community legislation, is at the same time the most important obstacle to its proper understanding. The sheer volume of the *acquis* itself, let alone the number of relevant documents that play a role in drafting legislation form an almost insurmountable obstacle for the non-specialized public in understanding the European Union's impact.

In the European context, the problem relating to "legal language" is seriously increasing. In fact, although multilingualism must undoubtedly be considered a treasure-house of European culture, it is, at the same time, a source of innumerable problems when it comes to drafting, translating and interpreting the acts produced by the Community institutions in the various official languages.

2.2 Law and Language

The legal terminology used in the various legal systems, both European and non-European, expresses not only the legal concepts which operate there, but further reflects the profound differences existing between the various systems and the differing legal outlook of the lawyers in each system. Law and language are, in fact, connected in many ways. First of all, they have a similar structure: each has, at its heart, rules which are constitutive of a system and which ensure its consistency. A second aspect is the dependency of law on language, since regulatory knowledge must be communicated, and the written and oral transmission of social or legal rules passes through verbal expression. Therefore, legal conceptual knowledge is closely related to the use of language within the legal domain. This means that linguistic information plays an important role in its definition, which may lead to the assumption that there is, as in other terminological domains, a relatively high level of dependence between legal concepts and their linguistic realisation in the various forms of legal language [1].

Given the structural domain specificity of legal language and the concepts involved, we cannot talk about "translating the law" to ascertain correspondences between legal terminology in various languages, since the translational correspondence of two terms satisfies neither the semantic correspondence of the concepts they denote, nor the requirements of the different legal systems. Overall, there is a lack of a clear language level where the equivalence has been set up. In "translating law", we have to negotiate the distance between the statute and the law or, more generally, between the law and its verbalisation [2] [3].

The goal of DALOS is, indeed, to support this process: the basic idea is to provide law-makers with a semantic framework, where the use of words and the underlining meaning assumptions are made explicit. Such a tool will allow for a clear overview of the consolidated lexicon in a regulatory domain and of the semantic relations among concepts; it will facilitate the harmonization of legal knowledge and lexicons between the EU and Member States. It will also support the dynamic integration of the lexicon by the legislator as well as the monitoring of the diachronic meaning evolution of legal terminology. As a result, it will improve:

- the internal quality of European law, avoiding inconsistent definitions of EU legal terms as well as the contradictory use of legal terms within different legislative sectors (see the EC Directives on consumer law, like timeshare, distance contracts, unfair terms [4]).
- the external quality of European law, avoiding that a same legal concept can be expressed differently in a Directive and in the transposition law.

3 The Ontology-Based Approach

Nowadays the key approach for dealing with lexical complexity is the ontological one, by which we mean a characterisation (understood both by people and processed by machines) of the conceptual meaning of the lexical units and of their connection with other terms.

As discussed in Section 2, in legal language every term collection belonging to a language system, and any vocabulary originated by a law system, is an autonomous vocabulary resource and should be mapped through relationships of equivalence with the others. Based on the assumption that in a legal domain one cannot transfer the conceptual structure from one legal system to another, it is obvious that the best approach consists in developing *parallel alignment* with the same methodology and referring to a *shared conceptual model*. Therefore the methodological steps aim at defining:

– the structure (linguistic, semantic, formal) of the lexicon (Sect. 3.1);
– the definition of an overall layered framework (Section 4);
– the resources generation (Section 5), starting from the terms extraction procedures and the ontology building;
– the way to navigate(update, modify) the resource (still in progress).

As for the first point, different methods may be applied to build lexical repositories for law, depending on the characteristic of the domain, the data structure and on the result to achieve.

Among structured data different degrees of formalization can be distinguished:

– controlled vocabularies (such as thesauri, classification trees, directories, keywords lists),
– semantic lexicons as well as foundational, core, and domain ontologies.

Semantic lexicons are means for content management which can provide a rich semantic repository. Compared to *formal ontologies*, semantic lexicons, also called *computational lexicon* or *lightweight ontologies*, are generic and based on a weak abstraction model, with limited formal modelling, since the elements (classes, properties, and individuals) of the ontology depend primarily on the acceptance of existing lexical entries. In lexical ontologies constraints over relations and consistency are ruled by the grammatical distinctions of language (noun, verbs, adjectives, adverbs) and many of the taxonomic links might be not logically consistent as they expresses a generalization relation more than a sub-class link.

The integration of lexical resources (heterogeneous because belonging to different legal systems, or expressed in different languages, or pertaining to different domains) leads to different final results depending on the desired results:

– generate a single resources covering both (merging);
– compare and define correspondences and differences (mapping);
– combine different levels of knowledge representation, basically interfacing lexical resources and ontologies.

The methodological approach chosen in the DALOS project is the third one: it requires the definition of mapping procedures between semantic lexicons, driven by the reference to an ontological level where the basic entities which populate the legal domain are described. Such an approach has been followed to obtain a correspondence between terms of different languages as well to align corresponding terms towards a common conceptualization at a higher knowledge level.

3.1 A Semantic Lexicon for Law: The LOIS Database

The DALOS resource is based on one of the wider lexical resource currently available in the legal field: the LOIS database [5] composed by about 35.000 concepts in five European languages (English, German, Portuguese, Czech, and Italian, linked by English). The LOIS methodology is based on an existing de facto standard, the WordNet and EuroWord Net resources, WordNet [6] is a lexical database which has been under constant development at Princeton University. EuroWordNet (EWN) [7] is a multilingual lexical database with WordNets for eight European languages, which are structured along the same lines as the Princeton WordNet[5].

In LOIS a concept is expressed by a *synset*, the atomic unit of the semantic net. A synset is a set of one or more uninflected word forms (lemmas) with the same part-of-speech (noun, verb, adjective, and adverb) that can be interchanged in a certain context. For example {*action, trial, proceedings, law suit*} form a *noun-synset* because they can be used to refer to the same concept. More precisely each synset is a set of *word-senses*, since polysemous terms are distinct in different *word-senses*, e.g.: diritto_1(*right*) and diritto_2(*law*).

A synset is often further described by a gloss, explaining the meaning of the concept. English glosses drive cross-lingual linking.

In monolingual lexicons terms are linked by lexical relations: *synonymy* (included in the notion of synset), *near-synonym, antonym, derivation*. Synsets are linked by semantic relations of which the most important are *hypernymy/hyponymy* (between specific and more general concepts), *meronymy* (between parts or wholes), *thematic roles, instance-of*.

Cross-lingual linking is based on *equivalence* relations of each synsets with an English synset: these relations indicate complete *equivalence, near-equivalence,* or *equivalence-as-a-hyponym* or *hyperonym*. The network of equivalence relations, the Inter-Lingual-Index (ILI), determines the interconnectivity of the indigenous WordNets.

Language-specific synsets from different languages linked to the same ILI-record by means of a synonym relation are considered conceptually equivalent. The LOIS approach is not completely language-independent, since the equivalence setting passes throughout the English WordNet and the English translation of glossas support the localization process.

The lesson learned from the LOIS experience is that a limited language independence could be enough for cross-lingual retrieval tasks, but it could be a weak point when considering re-using, extending, updating the semantic connections, and mainly, as in the case of law-making, when it is necessary to abstract the intended conceptual content from its lexical representation. What is needed is "the distinction between conceptual modelling at a language-independent level and a language and culture specific analysis and description of discourse related units of understanding" [8].

These considerations led us to make clear distinction, when designing the overall model of DALOS and the system architecture, among types of knowledge,

---

[5] (see http://www.globalwordnet.org).

layers of knowledge representation as well as semantic relationships between knowledge elements.

4   DALOS Knowledge Organization and Features

In the DALOS model design it is particularly important to identify the type of knowledge to be described, so to avoid the common attitude to indiscriminately mixing *domain knowledge* and knowledge on the process for which it is used (drafting, reasoning, searching, etc.). Such a mixing prevents knowledge representations from being automatically reusable outside the specific context for which the knowledge representation was originally developed [9].

In particular, for the DALOS knowledge resource we want to avoid that the knowledge to be used as support for legislative drafting on a specific matter (domain knowledge) is mixed with the knowledge on the general process of drafting which, obviously, is matter independent (see also [10]). What is needed therefore is a knowledge and linguistic support giving a description of concepts, as well as their lexical manifestations in different languages, in specific domains independently on the way they are regulated.

For the aim of developing a project pilot, the "consumer protection" domain has been chosen. After the normative corpus selection (16 EU Directives, 33 Court of Justice Judgements and 9 Court of First Instance Judgements) on which the *bottom-up* resources implementation is based, currently the activities for domain knowledge specification are oriented to:

- the standards to be used for knowledge representation;
- the Knowledge Organization System (KOS).

As regards standards, the RDF/OWL standard for WordNet representation as approved by the W3C standards has been used for the linguistic resource, thus guaranteeing interoperability as well as scalability of the solution.

As regards KOS, on the basis of the arguments expressed above, the DALOS resource is expected to be organized in two layers of abstraction (see Fig. 1):

- the ontological layer containing the conceptual modelling at a language-independent level;
- the lexical layer containing lexical manifestations in different languages of the concepts at the ontological layer.

Basically the ontological layer acts as a knowledge layer where to align concepts at European level independently from the language and the legal systems, where possible. It should be also noted that the legal sources addressed by DALOS are composed by 'parallel corpora' (the different linguistic version of the the EU sources), thus avoiding the difficulties in multilingual crossing faced in LOIS.

Moreover the ontological layer allows to reduce the computational complexity of the problem of multilingual term mapping (N-to-N mapping). Concepts at the ontological layer act as a "pivot" meta-language in a N-language environment, allowing the reduction of the number of bilingual mapping relationships from a factor N^2 to a factor $2N$.

Concepts at the ontological layer are linked by *subsumption* (subclass-of) as well as by domain-dependent *object property* relationships. On the contrary the lexical layer aims at describing language-dependent lexical manifestations of the concepts of the ontological layer. At this level terms are linked by linguistic relationships as those ones used for the LOIS database (*hyperonymy, hyponymy, meronymy*, etc.). In particular, to implement the lexical layer, the subset of the LOIS database pertaining to the "consumer protection" lexicon will be used. Moreover this database will be upgraded by using further texts where to extract pertaining terms from.

The connection between these two layers is aimed at representing the relationship between concepts and their lexical manifestations:

- within a single-language context (different lexical variations (lemmas) of the same meaning (concept));
- in a cross-language context (multilingual variations of the same concept).

In the DALOS KOS such link is represented by the *hasLexicalization* (and its inverse *hasConceptualization)* relationship. Fig. 1 shows the Knowledge Organization model designed for DALOS.

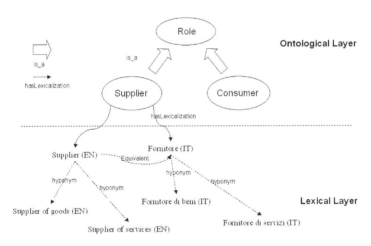

Fig. 1. DALOS knoledge organization

5   Implementation of the DALOS Resource

The DALOS ontological-linguistic resource is implemented through three main activities:

- Term extraction on the domain of "consumer protection" law from a set of selected texts by using NLP tools; this activity is aimed at upgrading the LOIS database (Lexical layer);

- Construction of a Domain Ontology on the "consumer protection" domain (Ontological layer);
- Semi-automatic connection between the Lexical layer (the LOIS database) and the Ontological layer by the *hasLexicalization* property implementation and its inverse *hasConceptualization* [Lexical layer ↔ Ontological layer].

This activity will be supported by semi-automatic tools and validated by humans.

The first activity (implementation of the Lexical layer) is being carried out using different NLP tools specifically addressed to process English and other EU language texts as GATE as well as Italian texts as T2K. GATE is a tool to support advanced language analysis, data visualisation, and information sharing in many languages, owned/provided and maintained by the Department of Computer Science of the University of Sheffield. T2K is a terminology extractor and ontology learning tool jointly developed by CNR-ILC and University of Pisa.

The second activity (construction of a domain knowledge at the Ontological layer) is an intellectual one which aims at describing the scenario to be regulated. In this context the use of an ontology is of primary importance. Laws in fact usually contain provisions [11] which deal with entities (arguments) but they do not provide any general information on them. Therefore a formalized description in terms of an ontology of the domain to be regulated will allow to obtain such additional general information on the entities a new act will deal with. Moreover, the use of an ontology, and particularly of the associated lexicon, allows to obtain a normalized form of the terms with which entities are expressed, enhancing quality and accessibility of legislative texts.

The third activity will deal with the connection between the two levels of abstraction (the Ontological layer and the Lexical layer). This activity is expected to be time consuming, since it will implement the legal concept alignment on the basis of their lexical manifestations in a multilingual environment. A tool to support such semi-automatic mapping is expected to be implemented within the project.

The current activities in DALOS are focussed in the terms extraction, related to the bottom-up construction of the Lexical layer and in the preliminary domain ontology building (Ontological layer implementation). Hereinafter such activities are described.

5.1   The Lexical Layer Implementation: The Italian Case

As discussed in the previous sections, the Lexical layer of the DALOS knowledge is mainly based on the lexical database of the LOIS project. However, hand–crafted lexical and ontological resources need to be continuously extended and refined in order to incorporate up–to–date knowledge: "ontology-learning" [12] from texts can be of some help in this direction[6]. To our knowledge, relatively few attempts have been made so far to automatically induce legal domain lexicons and ontologies from texts: this is the case, for instance, of [14] [15], [16]. In the DALOS

---

[6] Sections 5.1 is largerly based on the paper [13].

project, we decided to semi-automatically extend and customize pre-existing lexical and ontological resources on the basis of the terminological and ontological knowledge automatically acquired from texts belonging to the "consumer protection" domain with an ontology learning system. To this aim and for the Italian case, we used T2K (Text-to-Knowledge), a hybrid ontology learning system combining linguistic technologies and statistical techniques jointly developed by CNR-ILC and the Linguistics Department of the Pisa University [17].

T2K Architecture. T2K is a hybrid ontology learning system combining linguistic technologies and statistical techniques. T2K does its job into two basic steps:

1. extraction of domain terminology, both single and multi–word terms, from a document base;
2. organization and structuring of the set of acquired terms into proto–conceptual structures, namely
   - fragments of taxonomical chains, and
   - clusters of semantically related terms.

The two basic steps take the central pillar of the portrayed architecture, showing the interleaving of Natural Language Processing (NLP) and statistical tools. The approach to ontology learning adopted by T2K differentially exploits different levels of linguistic annotation of texts in an incremental fashion. Term extraction operates on texts annotated with basic syntactic structures: we use "chunking" technology to attain this level of basic syntactic structuring [18]. NLP requirements become more demanding when identified terms need be organised into conceptual structures. For this purpose syntactic information must include identification of dependencies among lexical heads (e.g. subject, object, modifier, etc.). The Italian parsing system underlying T2K is AnIta [19], a suite of linguistic tools in charge of the tokenisation of the input text, its morphological analysis (including lemmatisation), and syntactic parsing, which is in turn articulated in two different steps: "chunking", carried out simultaneously with morpho-syntactic disambiguation, and dependency analysis.

Term Extraction. Term extraction is the first and most–established step in ontology learning from texts. For our present purposes, a term can be a common noun as well as a complex nominal structure with modifiers (typically, adjectival and prepositional modifiers). As pointed out above, term extraction requires some level of linguistic pre–processing of texts.

T2K looks for terms in shallow parsed texts, i.e. texts segmented into an unstructured (non-recursive) sequence of syntactically organized text units called called "chunks" (e.g. nominal, verbal, prepositional chunks). Candidate terms may be one word terms ("single terms") or multi–word terms ("complex terms").

Secondly, the list of acquired potential complex terms is ranked according to their log–likelihood ratio [20], an association measure that quantifies how likely the constituents of a complex term are to occur together in a corpus if they were (in)dependently distributed, where the (in)dependence hypothesis is estimated with the binomial distribution of their joint and disjoint frequencies. Log-likelihood ratio in NLP is usually adopted for discovering collocations; we assume here that complex domain terms represent an instance of the more general class of collocations. It should be noted that in T2K the log-likelihood ratio is applied in a somewhat atypical way: instead of measuring the association strength between adjacent words, T2K measures it between the lexico–semantic heads of adjacent chunks.

The iterative process of term acquisition yields a list of candidate single terms ranked by decreasing frequencies, and a list of candidate complex terms ranked by decreasing scores of association strength. The selection of a final set of terms to be included in the TermBank requires some threshold tuning, depending on the size of the document collection and the typology and reliability of expected results. Thresholds define a) the minimum frequency for a candidate term to enter the lexicon, and b) the overall percentage of terms that are promoted from the ranked lists.

Different acquisition experiments have been carried out, by changing the minimum frequency threshold. Given the relatively restricted size of the acquisition corpus, better results were achieved setting the minimum frequency threshold to be equal to 3 for both single and multi–word terms. In the second extraction step, proto–conceptual structures involving acquired terms are identified. The basic source of information is no longer a chunked text, but rather a dependency–annotated text, including information about multi–word terms acquired at the previous extraction stage. We envisage two levels of conceptual organization. Terms in the TermBank are first organized into fragments of head–sharing taxonomical chains, With minimum frequency threshold set to 3, the numer of extracted *hyponymic* relations from the DALOS corpus is 911 referring to 172 *hyperonym* terms. The second structuring step performed by T2K consists in the identification of clusters of semantically related terms which is carried out on the basis of distributionally – based similarity measures. This is done by using CLASS, a distributionally – based algorithm for building classes of semantically related terms [21]. According to CLASS, two terms are semantically related if they can be used interchangeably in a statistically significant number of syntactic contexts.

For each target term, the set of the first 5 most similar terms is returned, ranked for decreasing values of semantic similarity. With the minimum frequency threshold set to 3, the number of identified related terms is 1,071 referring to 238 terminological headwords.

The proto–conceptual structures, i.e. the fragments of taxonomical chains of terms together with the clusters of semantically related terms, acquired during the term structuring step will provide useful input for both the construction of the DALOS domain ontology and the definition of the mapping between the lexical and the ontological layers (for a detailed description, see [13]). The outcome of the term extraction step will be exploited to extend the lexical coverage of the LOIS database (Lexical layer).

5.2   Ontological Layer Implementation

The domain ontology is populated by the conceptual entities which characterize the consumer protection domain. The first assumption is that all concepts *defined* within consumer law are representative of the domain and, as a consequence, that several concepts *used* in the definitional contexts pertain to the ontology as well, representing the basic properties, or in other words, the 'intensional meaning' of the relevant concepts.

In the consumer law domain the basic notion is that of 'commercial transaction' and of the 'legal roles' involved. An example is given in Fig. 2.

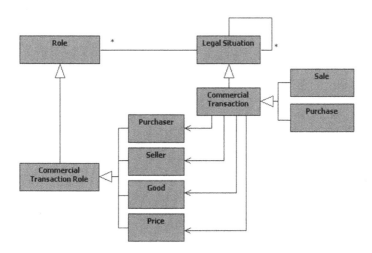

Fig. 2. Excerpt from the DALOS Consumer Law ontology

The DALOS domain ontology imports some basic notions, such as that of 'legal role' and 'legal situation', which are described in the so called *Core Legal Ontologies*. More precisely, in DALOS we rely the CLO[7] which specializes the DOLCE foundational ontology. The role of a *core legal ontology* is to separates entities/concepts which belong to the general theory of law from concepts proper of national legal systems or of a specific legal domain. It intends also to bridge the gap between domain-specific concepts and the abstract categories of formal upper level or foundational ontologies such as DOLCE.

The main entities in DOLCE (and consequently in CLO) are axiomatized, disjoint classes, characterized by meta properties, such as Identity, Unity and Rigidity. As for CLO, the most relevant distinction is between Roles (anti-rigid) and Types, which are rigid. For example, every instance of a role (e.g. seller, buyer, good) can possibly be a non seller, not good, etc. without loosing its

---

[7] The Core Legal Ontology (CLO) [22] is developed on top of *DOLCE* [23] [DOLCE+ library, available in OWL from: http://dolce.semanticweb.org/ Directly loadable from: http://www.loa-cnr.it/ontologies/DLP.owl] and Descriptions and Situations [24].

identity. Every instance of a type (e.g. a person) must be a person. A type can play more roles at the same time. For instance, a legal subject (either a natural or artificial person) can be a seller and a buyer. Domain-specific requirements are expressed by restrictions over ontological classes, for instance by defining 'consumer' as a role that can be *played by* 'natural person' only.

A further necessary assumption in the ontology construction is the representation of relationships between 'contexts' and 'concepts'; Fig. 3 shows the intended model: normative contexts are sub-class of 'legal text' and *part-of* 'article', the identified legal source; a specific context is an instance of normative context which regulate a class of legal situations, for instance 'payment'; as not all facts in the real world are relevant for law, the class of *legal situation* is the meta-class of all the possible instances of cases regulated by the law.

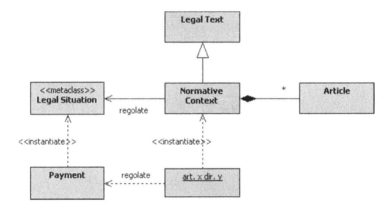

Fig. 3. Contexts and regulated situations

The relations between the classes of legal situations and their regulative counterparts is based on the reification principle defined in CLO. It enables an ontology engineer to quantify either on legal rules or relations (type reification) [25], or on legal facts (token reification). CLO extends the *Descriptions and Situations* vocabulary for reification. For example, *intensional specifications* like norms, contracts, subjects, and normative texts can be represented in the same domain as their *extensional realizations* like cases, contract executions, agents, physical documents.

6   Conclusions

The main purpose of the DALOS project is to provide law-makers with linguistic and knowledge management tools to be used in the legislative processes, in particular within the phase of legislative drafting. The aim is to keep control over the legal language, especially in a multilingual environment, as the EU legislation one, enhancing the quality of the legislative production, the accessibility and alignment of legislation at European level, as well as to promote awareness and democratic participation of citizens. In this paper the characteristics of the ontological-linguistic resource as well as the methodologies for its construction have been presented.

References

1. Rossi, P.: The language of law between the european union and the member states. In: Ajani, M.E.G. (ed.) Uniform Terminology for European Contract Law, p. 23 (2005)
2. Sacco, R.: Droit et langue. In: Rapports italiens au XV Congrès international de droit comparé, Milano (1998)
3. Pozzo, B.: Harmonisation of european contract law and the need of creating a common terminology. European Review of Private Law 6, 754–767 (2003)
4. Ajani, G., Lesmo, L., Boella, G., Mazzei, A., Rossi, P.: Multilingual conceptual dictionaries based on ontologies. In: Proceedings of the V Legislative XML Workshop, pp. 161–172. European Press Academic Publishing (2007)
5. Peters, W., Sagri, M., Tiscornia, D.: The structuring of legal knowledge in lois. Artificial Intelligence and Law 15, 117–135 (2007)
6. Fellbaum, C. (ed.): WordNet: An Electronic Lexical Database. MIT Press, Cambridge (1998)
7. Vossen, P., Peters, W., Díez-Orzas, P.: The multilingual design of the eurowordnet database. In: Mahesh, K. (ed.) Proceedings of IJCAI 1997 Workshop: Ontologies and multilingual NLP, Nagoya, Japan, August 23-29 (1997)
8. Kerremans, K., Temmerman, R.: Towards multilingual, termontological support in ontology engineering. In: Proceeding of Termino 2004, Workshop on Terminology (2004)
9. Breuker, J., Hoekstra, R.: Epistemology and ontology in core ontologies: Folaw and lricore, two core ontologies for law. In: Proceedings of EKAW Workshop on Core ontologies. CEUR 2004 (2004)
10. Biagioli, C., Francesconi, E.: A visual framework for planning a new bill. In: Quaderni CNIPA (Proceedings of the 3rd Workshop on Legislative XML), vol. 18, pp. 83–95 (2005)
11. Biagioli, C.: Towards a legal rules functional micro-ontology. In: Proceedings of workshop LEGONT 1997 (1997)
12. Buitelaar, P., Cimiano, P., Magnini, B.: Ontology learning from text: an overview. In: B., et al. (eds.) Ontology Learning from Text: Methods, Evaluation and Applications. Frontiers in Artificial Intelligence and Applications, vol. 123, pp. 3–12 (2005)
13. Lenci, A., Montemagni, S., Pirrelli, V., Venturi, G.: Nlp-based ontology learning from legal texts. a case study. In: Casanovas, P., Biasiotti, M.A., Francesconi, E., Sagri, M.T. (eds.) Proceedings of LOAIT 2007 - II Workshop on Legal Ontologies and Artificial Intelligence Techniques, pp. 113–129 (2007)
14. Lame, G.: Using nlp techniques to identify legal ontology components: concepts and relations. In: Benjamins, V.R., Casanovas, P., Breuker, J., Gangemi, A. (eds.) Law and the Semantic Web. LNCS (LNAI), vol. 3369, pp. 169–184. Springer, Heidelberg (2005)

15. Sais, J., Quaresma, P.: A methodology to create legal ontologies in a logic programming based web information retrieval system. In: Benjamins, V.R., Casanovas, P., Breuker, J., Gangemi, A. (eds.) Law and the Semantic Web. LNCS (LNAI), vol. 3369, pp. 185–200. Springer, Heidelberg (2005)
16. Walter, S., Pinkal, M.: Automatic extraction of definitions from german court decisions. In: Proceedings of the COLING 2006 Workshop on Information Extraction Beyond The Document, Sidney, pp. 20–28 (2006)
17. Dell'Orletta, F., Lenci, A., Marchi, S., Montemagni, S., Pirrelli, V.: Text-2-knowledge: una piattaforma linguistico-computazionale per l'estrazione di conoscenza da testi. In: Proceedings of the SLI 2006 Conference, Vercelli, pp. 20–28 (2006)
18. Federici, S., Montemagni, S., Pirrelli, V.: Shallow parsing and text chunking: a view on underspecification in syntax. In: Proceedings of the Workshop On Robust Parsing,
19. Bartolini, R., Lenci, A., Montemagni, S., Pirrelli, V.: Hybrid constrains for robust parsing: First experiments and evaluation. In: Proceedings of LREC 2004, Lisbon (2004)
20. Dunning, T.: Accurate methods for the statistics of surprise and coincidence. Computational Linguistics 19(1) (1993)
21. Allegrini, M.S., P., Pirrelli, V.: Example-based automatic induction of semantic classes through entropic scores. Linguistica Computazionale (2003)
22. Gangemi, A., Sagri, M., Tiscornia, D.: A constructive framework for legal ontologies. In: Benjamins, Casanovas, Breuker, Gangemi (eds.) Law and the Semantic Web. Springer, Heidelberg (2005)
23. Masolo, C., Gangemi, A., Guarino, N., Oltramari, A., Schneider, L.: Wonderweb deliverable d18: The wonderweb library of foundational ontologies. tech. rep (2004)
24. Masolo, C., Vieu, L., Bottazzi, E., Catenacci, C., Ferrario, R., Gangemi, A., Guarino, N.: Social roles and their descriptions. In: Welty, C. (ed.) Proceedings of the Ninth International Conference on the Principles of Knowledge Representation and Reasoning, Whistler (2004)
25. Gangemi, A.: Ontology design patterns for semantic web content. In: Gil, Y., Motta, E., Benjamins, V.R., Musen, M.A. (eds.) ISWC 2005. LNCS, vol. 3729. Springer, Heidelberg (2005)

Moving in the Time: An Ontology for Identifying Legal Resources

João Alberto de Oliveira Lima[1], Monica Palmirani[2], and Fabio Vitali[3]

[1] Senate of Brazil
70165-900 Via N2, Anexo C, Brasília, Brazil
joaolima@senado.gov.br
[2] CIRSFID - University of Bologna
via Galliera 3, 40100 Bologna, Italy
monica.palmirani@unibo.it
[3] Department of Computer Science - University of Bologna
Mura Anteo Zamboni 7, 40100 Bologna, Italy
fabio.vitali@unibo.it

Abstract. The paper presents an application of the $FRBR_{OO}$ document model for defining an information ontology of legal resources that takes into account the dimension of time. FRBR-based paradigms are used within several existing projects in computer support of activities in the legal domain, but they are mostly oriented to bibliographic organization of documents without a real modeling of the peculiar characteristics of the legal domain. Also, all of them refer to the current version of the FRBR model, called $FRBR_{ER}$. Yet, in these years the FRBR model is undergoing a major revision and a new version using an object-oriented approach is being developed. Thus we first have updated the model of legal resources to rely on the new object-oriented model, called $FRBR_{OO}$. More importantly, consistency problems were corrected and the time dimension was introduced, which came very useful when considering the legal domain and dealing with legal resources. Therefore, while it is not in the scope of this paper to define an abstract model of the norms (e.g. obligations, permissions, etc.) or the representation of the norms (e.g. rules model), we rather focus our attention on the abstract description of the normative acts lifecycle and how it is possible to fill the gap between the rule modeling and the structure of the legal resources.

Keywords: URI, FRBR, Ontology, temporal model.

1 Introduction

The application of the $FRBR_{ER}$ model in the organization of the various levels of abstraction of legal resources [11] has influenced many initiatives which deal with the organization of legal and legislative information such as the *Akoma Ntoso [24]*, *LexML Brasil*, *CEN Metalex* [17] and *Norme in Rete* [1] projects. This paper revisits this theme and discusses the application of the new $FRBR_{OO}$ model to legal resources.

P. Casanovas et al. (Eds.): Computable Models of the Law, LNAI 4884, pp. 71–85, 2008.
© Springer-Verlag Berlin Heidelberg 2008

This discussion has as its main focus the new definitions of the FRBR Group 1 entities and the analysis of the inclusion of time dimension in the new FRBR vision.

Several legal ontologies exist concerning the legal resources but mostly aimed at modeling the normative content [2] or the issues connected to the legal language [23]. On the other hand many different communities produced in the last ten years strong and robust international standards for describing the entities of the information resources (FRBR), sometimes oriented to manage the IPR issues (PREMIS) or the culture heritage resources (CIDOC CRM). Yet, we are currently missing an ontology to represent the legal resources identification, especially considering the dimension of time. There is in the state of the art a gap between the rule representation (e.g. knowledge rule base), the ontology on legal concepts and the identification of the legal resources. ESTRELLA project uses MetaLex/CEN standard for describing the XML structure of the Legal Resources, LKIF-core for defining the legal concepts in OWL, LKIF-rules for modeling the legal knowledge. Nevertheless there is not yet defined a common ontology, based on the information structure, for linking all these layers in considering also the dynamicity over the time. This paper starts to the URI naming convention of ESTRELLA project [17], and extends it for providing a preliminary step to a full modeling of legal knowledge framework.

Section 2 of this paper, after a brief introduction to the $FRBR_{ER}$ model, presents the definitions of Group 1 entities and lists some problems of the old model. Section 3 shows how the entities of the $FRBR_{ER}$ model were mapped to new classes in the new $FRBR_{OO}$ model and how the consistency problems were solved. Section 4, which is the main contribution of this paper, applies these new definitions to the legal domain and analyses the advantages and disadvantages. In section 5 we eventually produce our conclusions.

2 $FRBR_{ER}$

Initially described as entity-relationship model to the organization of bibliographical records, [10], the FRBR[1] is being revised by three working groups. One of these groups is working on the harmonization of the FRBR concepts with the CIDOC CRM[2] ontology [9], generating a new version which uses the object oriented approach[3]. This integration process between the main reference models of library and museum communities initiated in 2003 and not yet finalized has been a "good opportunity to correct some semantic inconsistencies or inaccuracies in the formulation of

---

[1] According to Doerr & Le Boeuf [5] the acronym FRBR (Functional Requirements for Bibliographical Records) "has now turned to a noun in its own right, used without particular intention to refer to 'functionalities', nor to 'requirements' but rather to the semantics of bibliographic records".

[2] According to Doerr & Le Boeuf [5] the acronym CIDOC CRM (Comité international de documentation Conceptual Reference Model) "is not particularly meaningful (CIDOC is affiliated to ICOM, the International Council of Museums). Just like FRBR, the acronym, rather meaningless by itself, has now turned to a noun in its own right."

[3] Like Doerr & Le Boeuf [5], this article used the acronyms $FRBR_{ER}$ to refer to the original model which uses the technique entity relationship and the acronym $FRBR_{OO}$ to the new model which uses the object orientation.

FRBR" [5]. Besides, this harmonization introduces in the FRBR the time dimension which is essential to the museum community as well as to the legal domain.

The FRBR$_{ER}$ model represented a great advance in the identification of the various abstraction levels of a work. It is a entity relationship model developed by IFLA in the period between 1991 and 1997 and published in 1998. This model is based on concepts which came from the evolution of the cataloguing discipline of the Library Science which happened mainly in the second half of last century.

The entities of the FRBR$_{ER}$ model are organized in three groups as follows:

- Group 1 entities (Work, Expression, Manifestation, and Item) – represent products of intellectual or artistic endeavor from the most abstract level (Work) to the physical (Item).
- Group 2 entities (Person, Corporate Body) – are responsible to the Group 1 entities according to the creation processes (Work), realization (Expression), production (Manifestation) or acquisition (Item), that is, a Person or Corporate Body can assume the roles of author (Work), editor (Expression), producer (Manifestation) or proprietary (Item).
- Group 3 entities (Concept, Object, Place and Event) – along with entities of Groups 1 and 2 serve as subject descriptors to the Work entity.

In spite of the importance of the Groups 2 and 3 entities, as a way to delimit the scope, this paper addresses only the entities of Group 1 as described below.

The FRBR$_{ER}$ model defines the entities of Group 1 represented by the entity-relationship diagram (Figure 1) like: a Work "is a distinct intellectual or artistic creation" [10, p.16]; an Expression "is the specific intellectual or artistic form that a work takes each time it is 'realized'" [10, p.18]; a Manifestation "is the physical embodiment of an *expression* of a *work*" [10, p.20]; and an Item "is a single exemplar of a *manifestation*" [10, p.23].

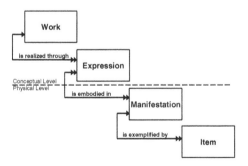

Fig. 1. FRBR$_{ER}$ – Group 1 Entities

While the entities Expression and Item have a relatively clear and consistent definition in the FRBR$_{ER}$, the definitions of Work and Manifestation allow divergent interpretations which contribute to the inconsistency of the model.

According to Doerr & Le Boeuf [5], the entity Work as defined in the FRBR$_{ER}$, "seemed to cover various realities with distinct properties"[4].

The entity Manifestation according to Doerr & Le Boeuf [5] "was defined in FRBR$_{ER}$ in such a way that it could be interpreted as something physical and conceptual at the same time". [5]

According to Doerr & Le Boeuf [5] the absence of the time dimension in the FRBR$_{ER}$ model has been reported in various papers such as Heaney [8], Fitch [6], Lagoze [12] and Doerr et al [4]. For example, Heaney [8] considers that: "Functional Requirements of Bibliographic Records largely ignores the time aspect, but I suggest there is much to be gained from an analysis, at each level (Work, Expression, Manifestation, Item) of: (a) how objects/entities of library interest exist over time; (b) how or whether they change over time (c) how or whether their existence is reflected in some sort of physical reality, tangible or not (d) whether the physical reality is continuous or intermittent"

3   CIDOC CRM and FRBR$_{OO}$

The CIDOC CRM model is a "formal ontology intended to facilitate the integration, mediation and interchange of heterogeneous cultural heritage information" (9). Developed since 1994 on an independent basis from the FRBR initiative this model was approved as international standard ISO 21127:2006[6]. A process started in 2003 with the aim to harmonize the CIDOC CRM with the FRBR: the final results produced a new version FRBR$_{OO.}$

In some cases, an entity gave origin to one class only like the example of the entity "Item" which was harmonized as "F5 Item" class, subclass of "E84 Information Carrier". In other cases, it was necessary to create various classes as a way to model each aspect which was under the umbrella of a single entity or that was not considered in

---

[4] The same authors clarify that:
"While the main interpretation intended by the originators of FRBR$_{ER}$ seems to have been that of a set of concepts regarded as commonly shared by a number of individual sets of signs (or 'Expressions'), other interpretations were possible as well: that of the set of concepts expressed in one particular set of signs, independently of the materialization of that set of signs; and that of the overall abstract content of a given publication" (2007, p. 11).

[5] They clarify this affirmation using the following terms:
"… it was defined at the same time as 'the physical embodiment of an expression of a work' and as an entity that 'represents all the physical objects that bear the same characteristics,' i.e., as both a physical artifact and a (mental) representation of physical artifacts (a set). The original Manifestation was likely to cover either a manuscript (in which case Manifestation overlaps with Item) or a publication (in which case Manifestation is both a Type and an Information Object)." (2007, p. 12).

[6] In 2003 a international working group was created with members from IFLA (International Federation of Library Associations and Institutions) and ICOM (International Council of Museums) with the objective to harmonize the FRBR$_{ER}$ with the CIDOC CRM ontology. In spite of this reviewing process is not concluded, the present version of the FRBR$_{OO}$ represents a considerable advance because it corrects inconsistencies of the old model besides adding to it the time dimension.

the old model. The following sections present how the entities Work, Expression and Manifestation of the old model $FRBR_{ER}$ were harmonized.

3.1 $FRBR_{OO}$ – Harmonization of Work Entity

According to Smiraglia [21], the concept of Work evolved since Panizzi A. (1841) leaving the role of a secondary entity in the first catalogues, which focused on the inventory function, having a more important role in modern catalogues after noticing that users of a information retrieval system is interested in the content and not in the support or a specific manifestation.

Even before the publication of the $FRBR_{ER}$ model some research argued about the need of the creation of an entity which grouped works derived from other work. For example, Yee [25], Carlyle [3] and Svenonius [22]) defended the creation of the entity "Superwork". Other researchers which defended the essence of the same idea nominated this entity "Bibliographic Family" [19], "Textual Identity Network" [13] and "Instantiation Network" [20].

The new $FRBR_{OO}$ crystallized the results of these researches creating the class "F21 Complex Work" which allows the grouping of Works according some criteria. The "F46 Individual Work" class has been defined to model the associated concepts with a specific group of symbols (Expression). In spite of "F1 Work" being the super-class of Work level, according to Doerr & Le Boeuf [5], it is the class "F21 Complex Work" which comes closer to the definition of Work of the $FRBR_{ER}$ model. The new

Table 1. Work related classes in $FRBR_{OO}$

Class	Scope note fragment (Doerr & Le Boeuf 2007)		
	Definition	Subclass of	Superclass of
F1 Work	This class comprises the sum of concepts which appear in the course of the coherent evolution of an original idea into one or more expressions that are dominated by the original idea. The substance of Work is concepts.	E28 Conceptual Object	F46 Individual Work F21 Complex Work
F46 Individual Work	This class comprises works that are realized by one and only one self-contained expression, i.e., works representing the concept as expressed by precisely this expression, and that do not have other works as parts.	F1 Work	F48 Aggregation Work
F21 Complex Work	This class comprises works that have more than one work as members.	F1 Work	F22 Serial Work
F43 Publication Work	This class comprises works that have been planned to result in a manifestation product type and that pertain to the rendering of expressions from other works.	F54 Container Work	F22 Serial Work
F22 Serial Work	This class comprises works that are, or have been, planned to result in sequences of manifestations with common features.	F21 Complex Work F43 Publication Work	

model defines six other classes related to the entity Work. Apparently complex, this new more detailed modeling allows us to represent each aspect which were under the umbrella of a single entity making the model easier to use. Table 1 presents a summary of these classes and for each class presents a fragment of the scope note and the corresponding superclasses and subclasses.

To obtain the complete definition of the new FRBR$_{OO}$, Doerr & Le Boeuf [5] should be consulted and, for the CIDOC CRM ontology, see ICOM [9].

3.2 FRBR$_{OO}$ – Harmonization of Expression Entity

While the substance of a Work is concepts, the substance of an Expression is signs. When a "Work" is done in its complete form by a group of symbols, there is an instance of class "F20 Self-Contained Expression". On the other case, when this group is incomplete, there is an instance of "F23 Expression Fragment" class. These classes are subclasses of "F2 Expression".

Table 2. Expression related classes in FRBR$_{OO}$

Class	Scope note fragment (Doerr & Le Boeuf 2007)		
	Definition	**Subclass of**	**Superclass of**
F2 Expression	This class comprises the intellectual or artistic realizations of works in the form of identifiable immaterial items, such as texts, poems, jokes, musical, or choreographic notations, movement pattern, sound pattern, images, multimedia objects, or any combination of such forms that have objectively recognizable structures. The substance of Expression is signs.	E73 Information Object	F20 Self-Contained Expression F23 Expression Fragment
F20 Self-Contained Expression	This class comprises the immaterial realizations of individual works at a particular time that are regarded as a complete whole. The quality of wholeness reflects the intention of its creator that this expression should convey the concept of the work. Such a "whole" can in turn be part of a larger "whole".	F2 Expression	F41 Publication Expression
F23 Expression Fragment	This class comprises parts of Expressions and these parts are not Self-contained Expressions themselves.	F2 Expression	
F41 Publication Expression	This class comprises the complete layout and content provided by a publisher (in the broadest sense of the term) in a given publication and not just what was added by the publisher to the authors' expressions.	F20 Self-Contained Expression	

Table 2 shows a summary of these classes and for each class presents a fragment of the scope note and the respective superclasses and subclasses. The class "F50 Performance Plan" has been excluded for not being on the scope of this paper.

3.3 FRBR$_{OO}$ – Harmonization of Manifestation Entity

The entity Manifestation which presented inconsistency problems in the old model is now represented by two distinct classes: one related to a conceptual vision (abstract) denominated "F3 Manifestation Product Type", subclass of "E55 Type" and the other related to a concrete vision (physical) "F4 Manifestation Singleton" which refers to author's original.

Table 3 presents a summary of these classes and for each class presents a fragment of the scope note and the respective superclasses and subclasses.

Table 3. Manifestation related classes in FRBR$_{OO}$

Class	Scope note fragment (Doerr & Le Boeuf 2007)		
	Definition	Subclass of	Superclass of
F3 Manifestation Product Type	This class comprises the definitions of any process products such as the publication.	E55 Type E72 Legal Object	
F4 Manifestation Singleton	This class comprises physical objects that each carry an instance of F2 Expression and that were produced as unique object, with no siblings intended in the course of its production.	E24 Physical Man-Made Thing	

4 FRBR$_{OO}$ and Legal Resources

Before beginning the analysis of how the new FRBR$_{OO}$ model can be applied to legal resources, it is important to establish operational definitions to the main legal concepts which will be analyzed. To do so, we are going to use the following definitions presented by Palmirani [15]):

Norm - A rule of conduct issued by a competent authority and prescribing or regulating behavior among individuals and within society. Its form of expression may be the written or the spoken word, but it may also be visual or be based on usage and custom (instance of the class F1 Work).

Normative provision - Any group of words or piece of writing expressing a norm or series of norms (instance of F23 Expression Fragment).

Normative document or act - An officially legislative written document through which a competent authority brings a norm into being (instance of F20 Self-Contained Expression).

Legal system - A set of norms belonging by some criterion to a single system and related to one another in different ways, as by hierarchy (one norm having a higher or lower standing than another), generality (more specific or more general), time (issued before or after another norm), and modification (one norm modifying the other norm or getting modified by it). (instance of F21 Complex Work). In addition "R12 has member (is member of)" property allows only "Individual Norm Works" as member of Legal System Class.

Normative system - The same legal system viewed from the outside is dynamic: it changes over time and can be represented in its evolution as a series of snapshots or film-stills in succession. The sequence in the time of legal systems, thus captured, we will call the normative system (instance of F21 Complex Work). In addition In addition "R12 has member (is member of)" property allows only "Legal System Works" as member of "Normative System" Class.

In this paper we intend to identify classes and instances concerning the external process of the legislative production. In other work we will present the classes and the instances concerning the parliament internal processes of the legal resources. The present session is organized as follows: initially we are going to identify classes and instances present in a specific edition of an official gazette and in an original signed normative document (4.1); furthermore, we will analyze the instances that take part in the process of modification of the norm in time (4.2); following that we will show how the concepts Legal System and Normative System are systematized in this new model (4.3); to conclude we will verify some special cases such as the treatment given to annexes (4.4) and to expressions of a norm using various languages (4.5).

4.1   From Signed Document to Official Publication

Figure 2 shows the "Work Conception" and "Expression Creation" events that creates an original document (Manifestation Singleton). The Item "Carrier Production Event" produces, from an original and publication work, various manifestations items.

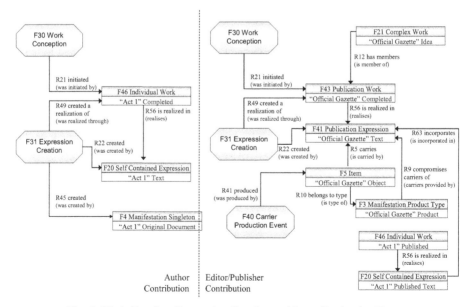

Fig. 2. Work Creation, Expression Creation and Items Production Events

Table 4. Original Signed Act

	Class	Instance
	F4 Manifestation Singleton	The original signed document is a physical object that carry an instance of F2 Expression.
	F20 Self Contained Expression (isA F2 Expression)	Normative text resulting from internal legislative process.
	F43 Individual Work	The concepts associated with the signed act document that is the abstract idea of the Act inside of the legal system.
	F21 Complex Work	In case of annexes

Table 5. Official Gazette Issue

	Class	Instance
	F5 Item	The Official Gazette issue
	F3 Manifestation Product Type	The publication product "Diário Oficial da União" issued in 6th August 2007 by Brazilian Official Press.
	F41 Publication Expression (isA F20 Self Contained Expression)	Complete layout and content provided by a publisher (including table of content, headings, expression from other works, etc).
	F43 Publication Work (isA F54 Container Work)	The concepts associated with the official publication issued in 6th August 2007 by Brazilian Official Press.
	F22 Serial Work (isA F21 Complex Work)	The periodical titled "Diário Oficial da União".
	F48 Aggregation Work	The Official Gezette issue is an aggregation of individual works or complex work concerning all the Acts included in.

Normally the "Act 1 Text" and "Act 1 Published Text" instance has the same content (set of signs), but, sometimes it is necessary to publish official communications with rectifications.

Table 4 relates classes and instances which can be identified in a signed official document. The original document has a work created by the legislative process whose authorship should be attributed to the actors involved in the legislative process since the moment of legislative initiative, the discussion period, deliberating, voting considering as well as the vetoes part which can also modify the normative text.

A specific edition of an official gazette contains various entities which coexist in the same support. On a first analysis it is possible to perceive the publication per se which is the result of an industrial process and generates a number of issues according to a determinate production plan. If we abstract the publication, it is possible to perceive the entities related to the shown legal resources.

Table 5 relates classes and instances which can be identified on a page of an official publication.

A physical issue of an official publication is an instance of class "F5 Item" because it carries a "F41 Publication Expression" and was produced by an industrial process. The textual expression of this manifestation is composed by the text expression of the official documents (Original Signed Act) together with the original content created by the Publisher such as table of contents, headings etc. In the case of legal resources, the text expression is originated from the signed text documents by the competent authority. Each signed official document (original) is represented by an instance of class "F4 Manifestation Singleton".

4.2   Legal Norm and Time

During the life cycle of a legal norm various actions can affect its content in relation to its form (Expression level) as to its subject matter (Work level). For example, the normative expression of a norm could be affected by actions of integration, modification or repealing. It is not the objective of this paper to discuss all action types which result from the modification of a legal norm. For a detailed view of this dynamics see Model for Legislative Consolidation [15] and [16]. This section shows how the new model FRBR$_{OO}$ permits to represent in a precise way the evolution of the norm in time.

The previous section showed that each legal norm, when we abstract the view of the official publication, is modeled as an instance of class "F46 Individual Work" with the corresponding instance of class "F20 Self Contained Expression". In the moment of an action of a norm modification, a derived work is created and it is represented by a new instance of "F43 Individual Work" class with the respective instance of "F20 Self Contained Expression" class. It is important to point out that the creation of a modified norm (derived) occurs in the date when the modifying norm acts and produce its effects in the destination document. In the case that this is the first event of modification of the norm, an instance of "F20 Complex Individual Work" class should be created with the objective to reference all instances of "F43 Individual Work" class of this norm. This dynamics is illustrated by Figure 3 which shows what happens when a norm ("Act 1") is altered by another norm ("Act 8"). In this example, the period called *vacatio legis* of the modifying norm is represented by the interval t_1-t_2 and the date of application of the modifying norm coincides with the date of entry into force.

4.3   From Norms to Legal and Normative Systems

The concepts of Normative and Legal Systems are implemented in our model using instances of class "F20 Complex Work". The set of legal resources which exist on a determinate date make a Legal System. The set of Legal Systems make a Normative System. This systematization is illustrated in Figure 4 which shows the elements of the previous example (Figure 3) grouped into Legal and Normative Systems.

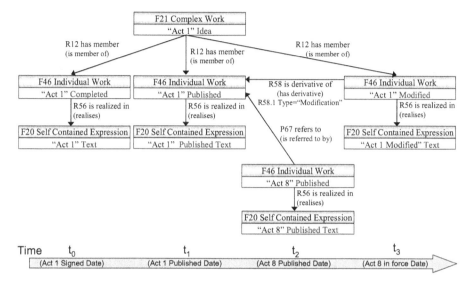

Fig. 3. Norm Dynamics

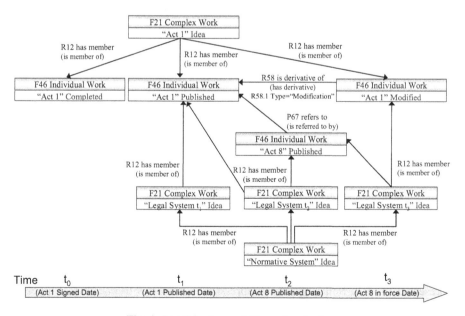

Fig. 4. Legal System and Normative System

4.4 Annexes

The publication of a normative text can be accompanied by complementary information in sections normally named "Annex". This kind of relationship can happen in a
recursive way, that is, an annex can have other annexes. Using the Ontology of Uni-
versals [7] terminology, an annex is considered a Role and not a Type. As roles have
a limited organizational relevance, to represent an annex it is necessary to analyze the
annex content entity. In the case that this entity has its own identity criteria such as
regulations or international treaties, the annex is classified as "independent annexes".
In the other case, when the entity presents complementary information and which are
dependent of the main part, such as tables, we classified it as "dependent annex". A
specific indicator of an "independent annex" is when the authoring process of the
annex is different from the authoring process of the main part. It is also possible a
hybrid situation in which the norm presents dependent and independent annexes.

When an independent annex happens, an instance of class "F21 Complex Work"
should be used as a way of grouping instances of "F43 Individual Work" which exist
to each component (main part and annexes). In the case of dependent annexes, the
property "R11 - is composed of (forms parts of) -" is used between "F20 Self Contained Expression" instances and there is no need, in this case, to define additional
instances at Work level. Figure 5 illustrates the two situations described above.

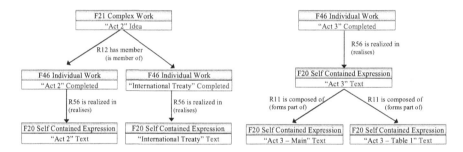

Fig. 5. Independent and Dependent Annexes

4.5 Legal Norm and Languages

When a norm is published in different languages, in cases such as legal translations or
simultaneous publications in multiple languages, even though each text has the same
value on a legal perspective, the legal resources are modeled with distinct instances of
class "F43 Individual Work" and the respective instances of "F20 Self Contained
Expression".

As class "F20 Self Contained Expression" does not have the property which asso-
ciates its group of signs to a specific language it is necessary to associate the entity
represented in "F20 Self Contained Expression" instance with one "E33 Linguistic
Object" instance. This is possible because FRBR$_{OO}$ is considered, using the terminology of Masolo *et al* [14], a multiplicative ontology, allowing co-localized entities.

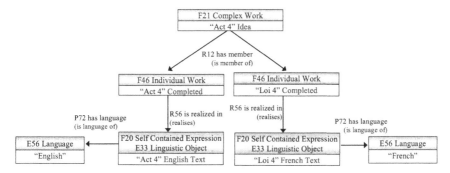

Fig. 6. Norm and Language

Figure 6 represents the modeling of norms published in different languages with the same legal validity.

5   Conclusion

The new $FRBR_{OO}$ model contributes in a significant way to the modeling of entities of the legal domain, which is a pre-requirement for any sophisticated legal knowledge modeling. The inclusion of the time dimension and the new definitions of the Work, Expression and Manifestation entities allow a more precise modeling of legal resources.

Our contribution clarifies not only the diachronic evolution of the legal resources in time, but it also puts the theoretical grounding for a future modeling of the relationships between the different entities participating to the legislative process workflow (e.g. bills, amendments). Our model is also applicable to all artefacts of the publishing process. Moreover the time dimension could support successful interconnections between different legal resources (e.g. between normative acts and case-law) that need precise point-in-time referencing.

This paper shows a model of the information objects of the legal domain and in particular focuses on the evolution of legal resources in time, defines a concept of normative systems and legal systems, and provides explicit support for modeling annexes and norms published in more than one official language.

Finally this paper aims to contribute inside of the Estrella project to fill the gap between the LKIF-rules representation and the MetaLex/CEN structure information using a model moving over the time.

Acknowledgments

This research was partially supported by ESTRELLA project CEC-FP6-IST Programme Contract Number: 027655 and by Brazilian Government under CAPES Grant BEX 4414/06-3. The paragraphs 2, 3, 3.1, 4.2, 4.5 are a contribution by Lima, the paragraphs 3.2, 3.3, 4.1, 5 are a contribution by Vitali, paragraphs 1, 4, 4.3, 4.4 are a contribution by Palmirani.

References

1. VV., A.: Studio di fattibilità per la realizzazione del progetto Accesso alle norme in rete. Informatica e Diritto, 1 (2000)
2. Breuker, J., Boer, A., Hoekstra, R., Sartor, G., Rubino, R., Palmirani, M., Gordon, T.F., Wyner, A., Bench-Capon, T.: Deliverable 1.4 of the European Project ESTRELLA - OWL Ontology of Basic Legal Concepts (LKIF-Core). Technical report, University of Amsterdam, Bologna, Liverpool and Fraunhofer FOKUS (2007)
3. Carlyle, A.: Ordering author and work records: An evaluation of collocation in online catalog displays. Journal of the American Society for Information Science 47, 538–554 (1996)
4. Doerr, M., Hunter, J., Lagoze, C.: Towards a core ontology for information integration. Journal of Digital Information 4(1) (2003-04-09),
   http://jodi.tamu.edu/Articles/v04/i01/Doerr/doerr-final.pdf
5. Doerr, M., Le Boeuf, P.: International Working Group on FRBR and CIDOC CRM Harmonization. FRBR object-oriented definition and mapping to FRBRER (version 0.8.1) (2007),
   http://cidoc.ics.forth.gr/docs/frbr_oo/frbr_docs/FRBR_oo_V0.8.1c.pdf
6. Fitch, K.: Aleg Data Model. Inventory. AustLit Gateway,
   http://www.austlit.edu.au:7777/DataModel/inventory.html
7. Guarino, N.: The role of Identity Conditions in Ontology Design. In: Workshop on Ontology and Problem Solving Methods. Springer, Heidelberg (1999)
8. Heaney, M.: Time is of the essence: some thoughts occasioned by the papers contributed to the International Conference on the Principles and Future Development of AACR. Oxford: Bodleian Library (1997),
   http://www.bodley.ox.ac.uk/users/mh/time978a.htm
9. ICOM/CIDOC Documentation Standards Group; CIDOC CRM Special Interest Group. Definition of the CIDOC Conceptual Reference Model: version 4.0, April 2004. Heraklion, Greece (2004), http://cidoc.ics.forth.gr/docs/cidoc_crm_version_4.0.pdf
10. IFLA Study Group on the functional requirements for bibliographic records. Functional requirements for bibliographic records: final report. K. G. Saur, Munich, Germany (1998),
    http://www.ifla.org/VII/s13/frbr/frbr.pdf
11. Lima, J.A.O.: An Adaptation of the FRBR Model to Legal Norms. In: Biagioli, C., Francesconi, E., Sartor, G. (eds.) Proceedings of the V Legislative XML Workshop, 2006. European Press Academic Publishing, Italia (2007)
12. Lagoze, C.: Business unusual: how "event-awareness" may breathe life into the catalog? In: Conference on bibliographic control in the new millennium. Washington: Library of Congress,
    http://lcweb.loc.gov/catdir/bibcontrol/lagoze_paper.html
13. Leazer, G., Furner, J.: Topological indices of textual identity networks. In: Woods, L. (ed.) Proceedings of the 62nd Annual Meeting of the American Society for Information Science, pp. 345–358. Information Today, Medford (1999)
14. Masolo, C., Borgo, S., Gangemi, A., Guarino, N., OltamariI, A.: Ontology library. IST-2001-33052 Deliverable 18, WonderWeb: Ontology Infrastructure for the Semantic Web (2003)
15. Palmirani, M.: Dynamics of Norms over Time: a Model for Legislative Consolidation. I Quaderni 18, 42–69 (2005)

16. Palmirani, M.: Modificatory Provisions for Legislative Consolidation, UNDESA Report (2006),
    `http://www.akomantoso.org/resources/modificatory-provisions-for-legislative`
17. Palmirani, M., Sartor, G., Rubino, R., Boer, A., De Maat, E., Vitali, F., Francesconi, E.: Guidelines for applying the new format. Deliverable 3.2, Estrella project, - Guidelines for applying the new format. Technical report, University of Bologna, Amsterdam, CNIPA (2007)
18. PREMIS, Data Dictionary for Preservation Metadata,
    `http://www.oclc.org/research/projects/pmwg/premis-final.pdf`
19. Smiraglia, R.P., Leazer, G.H.: Derivative bibliographic relationships: The work relationship in a global bibliographic database. Journal of the American Society and Information Science 50, 493–504 (1999)
20. Smiraglia, R.P.: Further reflections on the nature of 'A Work': An introduction. Cataloging & Classification Quarterly 33 (3/4), 1–11 (2002)
21. Smiraglia, R.P.: The History of "The Work" in the Modern Catalog. Cataloging & Classification Quarterly 35(3/4), 553–567 (2003)
22. Svenonius, E.: The intellectual foundation of information organization. Digital libraries and electronic publishing. The MIT Press, Cambridge, Mass (2000)
23. Peters, W., Sagri, M.T., Tiscornia, D.: The structuring of legal knowledge in LOIS. Artificial Intelligence and Law 15(2) (June 2007)
24. Vitali, F.: Akoma Ntoso Release Notes (2007), `http://www.akomantoso.org`
25. Yee, M.: What is a work? Part 3: The Anglo-American cataloging codes. Cataloging & Classification Quarterly 20(1), 25–46 (1995)

An Ontology for Spatial Regulations

Tom van Engers, Erik Hupkes, Radboud Winkels, and Alexander Boer

Leibniz Center for Law
University of Amsterdam
Amsterdam
The Netherlands
engers@uva.nl, ehupkes@uva.nl
winkels@uva.nl, boer@uva.nl

Abstract. The last decade improving access of legal sources using ICT and especially the Internet has lead to various internet portals for accessing textual sources, standardisation of those sources using W3C standards such as XML for describing the structure of such documents and meta standards such as MetaLex[1] [1,2]. In order to improve access to spatial regulations we should establish a successful marriage between geographical information systems based technology and a machine readable regulative framework, allowing connecting regulations as described in legal sources to an object oriented representation of the real world such as a zoning plan. This paper describes the architecture of an application developed for improving access to spatial regulations, integrating different sources such as GIS (Geographical Information Systems) information, maps and textual legal sources. This application called Legal Atlas uses a relatively compact ontology in OWL for combining spatial planning information in GML (Geographical Mark-up Language) with legal sources described in MetaLex XML. We will explain this ontology and the way it is used to support users in accessing spatial regulations, starting either from querying a map based interface of a text based one, i.e. starting from the spatial perspective or from the normative perspective.

1 Introduction

In crowded nations or regions such as the Netherlands spatial regulations are important domains of law. At municipal level this legal domain perhaps has the most influence on both the (local) government as well as the citizens. Spatial regulations determine if building permits may be granted, businesses can expand etc. Spatial regulations can be characterised as normal regulations, but although those regulations contain spatial elements - laws are for example bound to the jurisdiction that relates to location - in spatial regulations spatial elements obviously play a much more prominent role. In the legal practice this has lead to a dominant role for annotated maps as source of law. The last decade improving access of legal sources using ICT and especially the Internet has lead to various

---

[1] http://www.metalex.eu

P. Casanovas et al. (Eds.): Computable Models of the Law, LNAI 4884, pp. 86–104, 2008.

internet portals for accessing textual sources, standardisation of those sources using W3C standards such as XML for describing the structure of such documents and meta standards such as MetaLex[2] [1,2]. In order to improve access to spatial regulations we should establish a successful marriage between geographical information systems based technology and a machine readable regulative framework, allowing connecting regulations as described in legal sources to an object oriented representation of the real world such as a zoning plan. A major part of geospatial data used in the GIS world is collected by governments and actually represents normative statements, positions, and titles relating to space rather than representation of existing 'real' geographic features. The Leibniz Center for Law is involved in the DURP project (DURP is the acronym for Digitale Uitwisseling Ruimtelijke Plannen, in English: Digital exchange of spatial plans) that is aimed at developing a digital exchange format for spatial regulations. The DURP project is coordinated by the Dutch Ministry of Housing, Spatial Planning, and the Environment (VROM), and it involves diverse parties such as the Association of Cooperating Municipalities, the provinces and the Union of Water Control Boards. For the Leibniz Center for Law, being one of the founding fathers of the current CEN/MetaLex standard, having designed the MetaLex/NL XML schema [1] for 'regular' legal sources, involvement in the DURP project offered possibilities to apply models of law in combination with models of spatial functionality represented in geographic interfaces (maps). The MetaLex/NL XML interchange format has been extended in order to support exchange of spatial regulations, including the associated geospatial information.

We developed support software that is intended to improve access to spatial regulations. One component of that support software is called Legal Atlas [3]. This component builds upon past experiences in projects such as the ADDwijzer [4] project, in which it was showed that potentially valuable services can be delivered to citizens if only the legal sources of the spatial regulations would be available in the right form, and the already mentioned DURP project.

In order to establish the relationships between the regulations and the geographical objects we developed a relatively compact ontology in OWL for combining spatial planning information in GML with legal sources described in MetaLex XML. In this paper we will explain the architecture of the Legal Atlas application and we will especially focus on the role of the ontology in that application.

2   Linking Texts and Maps

In the previously mentioned ADDwijzer application the maps and text were linked by a hard coded mechanism with links in the texts and the map and using hyper links. In that application all relevant regulations were manually connected to regions on the map. A painstaking and tedious process, not only during design time. Such technical solution will make maintenance very labour intensive and therefore expensive both in terms of time and money. Unfortunately many

---

[2] http://www.metalex.eu

applications in the area of spatial planning use this type of solution and this is also the current solution in the area of spatial planning in the Netherlands. If a digital spatial plan is produced at all, the leading source is the map and one of the attributes of a region on that map is a link to an article about that region (e.g. "art23.htm"). There is no link from the text to the map. This simple method causes a number of problems. When either the text or map changes, one has to check by hand the consequences of this change. Is the connection to the region still valid? Should the region change? Should it be connected to other region(s) (as well), etc.

Retrieval of relevant information related to a particular item is hindered since it would require searching all other items and checking whether they link to that particular item. In the current DURP practice it is not even possible to have more than one piece of text linked to a region. Every region has only one 'destination' or 'purpose' (except for the so-called 'double destination' which designates an area overlapping other areas that has a different purpose, e.g. the area under high-voltage cables crossing agrarian and other regions).

A hard coded link will no longer work if the physical location (or name) of either text or map changes. Especially in environments where knowledge sources are produced and maintained by different people or organisations, hard coded linking creates undesired dependencies.

Building general applications for editing and searching combined text-map repositories will be very hard if not impossible if the format of texts and maps would vary from organisation to organisation. Then it would be difficult to predict where a link will be put and what it would look like. Consequently we need standardisation to prevent these problems. The W3C has proposed and adopted standards to cope i.a. with the changing names and location issue: the Universal Resource Identifier (URI) and Universal Resource Name (URN). Their XML and RDF/OWL standards allow for machine readable definitions of document and knowledge structures. Together these can be used to tackle the problem of different formats of texts and maps. Two interchange standards, MetaLex for legal texts and GML for geospatial data, are specified using XML schema and both also have an RDF data model. We have adopted these W3C and interchange standards to specify a general solution for coupling legal texts and maps and build a prototype viewer exploiting these links.

3   Spatial Aspects of Law

As already stated in the introduction legislation always has a spatial component, mediated through the concept of jurisdiction. Jurisdiction refers both to a power or competence to legislate, to apply or interpret legislation, or to take administrative decisions based on legislation and the territory within which this power can be exercised. This power can be delegated, by means of legislation, to a dependent legislator which has jurisdiction in a territory that is included in the territory of the delegating legislator. Legislation of the EU, the state, the province or region, and the municipality is in a sense linked to a space to which it applies. Usually this link to a geospatial object is an implicit one. The fact that Dutch legislation applies in Dutch territory and that across the border in the east German legislation applies is taken for granted.

The concept of space plays a central role in the domain of spatial planning. Therefore in such regulations space is indicated explicitly. Even more space is so dominant in this field that a legislative culture has grown in which the drafters of spatial regulations focus on drawing good maps, while in the regular legislative drafting processes legislative drafters focus on written norms, the definition of the things the norms are about, and the way those norms are described following certain rules, regulations, and traditions.

When we want to realise a service that should provide access to spatial regulations we can benefit from the fact that for that type of regulations the major conditions on applicability can be naturally displayed on maps. Wilson et al. [4] state that in spatial regulations, restrictions and rights are naturally associated with geospatial objects and consequently they are more easy to find compared to other regulations because:

Presentation: Everybody knows and understands maps; It is a familiar user interface paradigm.

Territorial jurisdiction: Because legislators tend to have jurisdiction in a specific area, and hierarchical relations between legislators are mirrored in spatial inclusion relationships, it is easy to find all legislators that could possibly have something to say about the usage of a specific parcel of land. Compare this with deciding which legislators could have jurisdiction over a sales transaction. Jurisdiction could be decided by factors like: Nationality of involved parties, location of relevant events and the use of mediating technologies like computers.

Adjacency: Occasionally regulations have indirect effects on other things one would not have thought about. A nice feature of the spatial planning domain is that these indirect effects are almost always mediated by adjacency in space, which means that rules and restrictions can be naturally grouped by region.

A number of advantages for eGovernment follow naturally from the use of map data:

Transparency of metadata: Retrievability of documents is highly dependent on whether the user understands the meta data attached to the legal documents. Choosing good descriptive keywords for a document is perceived as hard and subjective, while attaching coordinates to a document is considered easy and objective.

Citizen-centered organisation of data: GIS is a successful user interface for managing the territorial jurisdictions the user intends to interrogate. The GIS interface makes it natural to view legal information from different territorial jurisdictions in the same user interface, and makes it possible to activate different layers of distinctive (coloured) representations of restrictions at he same time in one interface to answer queries of the form "in which location will I have the least problems with legal constraints or permits when I perform this or that action?". This is a tangible improvement over the classical government-centered organisation of data, which typically requires text search in local regulations at municipal websites (cf. [4]).

Using maps has the great advantage that maps provide both the legislative drafters and the addressees (users) of the regulation with a means of communication that is easy to understand, but maps are not necessarily the only access method to a legal source containing a spatial regulation. There are circumstances in which it is more appropriate to access specific geospatial objects on maps through text retrieval. This is for example the case when one want to find similar cases to the one at hand. In that case one does not want to search for geospatial objects with the same properties from the point of view of the law of interest, regardless of where they are, in order to see whether relevant case law is attached to them.

4   Standards for Spatial Regulations

In order to link regulations to maps we need to be able to deal with different spatial concepts such as jurisdiction, application area etc. As stated before all regulations have a spatial component through jurisdiction. This is an attribute or feature connected to the authority that enacted the regulation or applies and upholds it. But there are other ways norms can have a relation to geographical data. One close to jurisdiction is what we call application area: A designated area to which the particular norm set applies, e.g. an area in which the police is allowed to search on suspicion. Furthermore, regulations refer to concepts or objects that have spatial extensions. As long as these objects are fixed in space at a certain time and are stored somewhere, access to them on a map from the regulation, or access to the regulations through a representation of these objects on a map is possible and useful.

Certain spatial regulations are connected to the concept (destination) of 'residential'; for a certain period of time the city council designates a particular area in its jurisdiction as a 'residential' area. Then the norms related to 'residential function' apply to that area. In a similar way norms may apply to objects belonging to a certain class with a spatial extension at a certain time, e.g. norms about petrol stations. If we have a database with all petrol stations in a community, it may be handy to either see where these norms apply in the area, or when we select such a station, see what are the regulations that apply to it. The general solution to solve this link between norms in text and spatial objects represented on a map is through intermediary concepts. The 'destination' concepts of the DURP project (IMRO 2006, see below) are an example of such intermediary concepts. Of course the level of granularity of both text and map objects should be useful. For texts this is typically the 'article' level if we talk about regulations. For spatial plans, the 'parcel' is a typical one on the map.

For linking texts containing (spatial) regulations and maps we can use three different standards. For the Legal Atlas application we have used the following 3:

1. MetaLex
2. GML
3. IMRO2006

We also use a mechanism for handling Links. An ontology is used to connect the three standards and our link mechanism. This ontology is an abstract one and helps us to create a flexible mechanism (preventing hard coded links) that enables linking legal sources described in one standard (MetaLex) with Maps (GML) and the spatial objects upon those maps and their catagories (IMRO2006).

The following paragraphs will shortly describe each of these standards, the link mechanism and the LegalAtlas Ontology.

4.1 MetaLex

For the text side of spatial and other regulations we chose MetaLex/NL, an interchange format for legal documents initiated by the Leibniz Center for Law. An interesting question was whether spatial regulations could be captured in MetaLex XML, which claims to be jurisdiction and domain-independent. The answer is yes. Translating spatial regulations to MetaLex is straightforward and linking geospatial references to MetaLex XML is not at all hard if RDF is used as a common platform. Any MetaLex XML structure can be translated with a standard XSL transformation to RDF conforming to the MetaLex Web Ontology Language (OWL) schema.

When legal documents are tagged in MetaLex, it provides a sophisticated version management system, for time related versions and language versions. MetaLex has a method for referencing concepts that are defined outside the scope of the document, it has citation elements to provide a way for complex citations to other regulations and parts of regulations. Finally very important for the coupling of legal texts and maps is the concept of *area of applicability* that has been added to the version 1.3 of MetaLex and up.[3] This gives the possibility to link every part of a regulation to an area of applicability concept that has an extension on the map. For a more detailed description of the functionalities of MetaLex see [1][2].

```
<Article id="idArtikel2">
- <IndexDesignation id="idIndexDesignation2">
- <Category id="idCategory2">
  <TextVersion xml:lang="en">Article</TextVersion>
  </Category>
- <Index id="idIndex2">
  <TextVersion xml:lang="en">2</TextVersion>
  </Index>
```

---

[3] http://www.metalex.eu

```
    </IndexDesignation>
-  <Title id="idTitle2">
-  <TextVersion xml:lang="en">
    <Reference xlink:type="simple"
      xlink:href="...legal2#idWerken1">Work</Reference>
    </TextVersion>
    </Title>
-  <Sentence id="idSentence2">
    <TextVersion xml:lang="en">The areas marked with this designation on
    the map are meant for building offices...
    </TextVersion>
    </Sentence>
  </Article>
```

This example shows one article translated from a Dutch spatial plan concerning the designation 'work' ('Werken' in Dutch) in MetaLex.

4.2   GML

The XML standard for exchanging geographical objects is the Geographical Modeling Language (GML), maintained by the Open Geospatial Consortium.[4] GML is used to describe geographical structures. It is a model that gives a hierarchy for geographical concepts like lines, dots, areas and polygons. The state of a feature is defined by a set of properties, where each property can be thought of as a $\{name, type, value\}$ triple, and GML can also be easily rendered as an RDF data model, which is also triple- based. This is the so-called *profile 3* serialization of GML. By treating both GML and MetaLex as RDF data, the integration problem is reduced to a matter of defining an OWL schema for the objects mediating between the legislative text and the geospatial object.

GML does not provide any references to other domains, but because it is an open XML- schema it makes it highly usable for the integration with other XML-schema.

4.3   IMRO2006

The third source of information that functions as the domain specific knowledge is in this case IMRO2006 (Information Model Spatial Planning). This is an XML model for spatial planning in the Netherlands. This model has been created to provide the municipalities with a standardized definition of all the terms that are needed for describing spatial plans.

But the scope of IMRO2006 is not limited to municipal spatial plans as can be seen in figure 1; e.g. national and regional plans are also covered. The heart of the IMRO2006 model is about the different designations an area can be assigned to. The different designation are a product of the legacy categories and a consensus of the many different municipalities in the Netherlands. The most

---

[4] http://www.opengeospatial.org

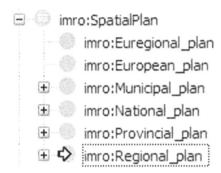

Fig. 1. The different spatial plans that could fit into IMRO2006

important categories can be seen in figure 2. There is a consensus in DURP that the GIS aspect of spatial regulations should be represented in GML, using the existing IMRO scheme. That is why IMRO uses GML to describe the geographical objects, and therefore when a spatial regulation is a valid IMRO documents it is completely compatible with GML. IMRO (and the GIS field in general) is a good example of the thesis that a shared abstract data model (which is the philosophy RDF is based on) can be as valuable or even more valuable than a shared syntax, such as XML.

Typical of spatial planning regulations is that legal concepts such as a 'residential destination' are defined in an *extensional* way on the map, by explicitly pointing out to which geospatial objects it applies, as opposed to normal legal concepts, which usually acquire meaning from an intensional definition. No interpretation is necessary or possible: a geospatial object only acquires the classification 'residential' by explicit assignment by the competent authorities.

```
<gml:featureMember>
 <SingleDesignation
xml:base="http://www.ravi.nl/imro2006"
gml:id="localid135">
  <identification>135</identification>
  <typePlanobject>
   designationarea; singledesignation
  </typePlanobject>
  <articlenumber>18</articlenumber>
  <designationfunctionInfo>
   <DesignationFunctionElement>
    <functionlevel>primary function
    </functionlevel>
    <designationfunction>housing; not stacked
    </designationfunction>
   </DesignationFunctionElement>
  </designationFunctionInfo>
```

Fig. 2. The most important Designation for spatial regulation in the Netherlands

```
<designationfunctionInfo>
 <DesignationFunctieElement>
  <functionlevel>primary function
  </functionlevel>
  <designationfunction>
   housing; house with garden
  </designationfunction>
 </DesignationFunctieElement>
</designationfunctionInfo>
<DesignationMainGroup>housing
</DesignationMainGroup>
<geometry>
 <gml:Polygon>
  ...
 </gml:Polygon>
</geometry>
<labelInfo>
 <Label>
  <text>H</text>
  <position>
   <Labelposition>
    <position>
     <gml:Point>
      ...
     </gml:Point>
    </position>
   </Labelposition>
  </position>
 </Label>
</labelInfo>
<name>Housing</name>
<linkToText>
 v_NL.IMRO.-.htm#housing
</linkToText>
<planarea xlink:href="#localidNL.IMRO.-"/>
</SingleDesignation>
</gml:featureMember>
```

The example above shows part of an IMRO2006 spatial plan. The GML information that is left out concerns the polygon and the point from the previous GML example.

4.4   Links

Now that the 3 standards, MetaLex, GML and IMRO2006, have a common RDF-based format they have to be able to connect with each other. There are many links imaginable between these 3 domains. Legal texts link to certain areas on the map, and the other way around. This means that identifiers, URI from the text, link to an URI from the map. With URI's as identifiers, let's have a look at how relations between identifiers work. A relation is a triple that relates an identifier to another identifier, where the third argument specifies what kind of relations it is about. This fits the RDF definition perfectly. There are many types of relations in the domain of legal texts and maps. The two trivial links that are important to make relations between maps and legal texts:

text to map: The area of applicability. This is the area where the particular piece of text enforces its content.
map to text: link to text. This is a link to a specific piece of text that has something to say about it.

Although these links are trivial, it is easy to fall in the trap of quick implementation. When these links refer directly from one identifier to the other, and one document gets a new version, the link to the old version stays in existence and does not automatically link to the new version. As was mentioned earlier about static hyper links, the solution is to introduce an intermediate concept. Lets take an article with an area of applicability, that refers to a region in general on which it applies instead of multiple links to multiple polygons on a map. In this case when another map is used with different polygons, the regulation can still stay unchanged. Only the new polygons have to be related to the region of that unchanged regulation. The other way around is similar.

4.5   Ontologies

While it is in principle possible to let people decide for themselves how spatial information is attached to MetaLex regulations – there are no technical impediments after all – we have designed two extensions in the form of an OWL schema for this purpose:

The LegalAtlas ontology: This ontology defines the general mechanism by which a source of law refers to either a class of geographical objects, specific geographical objects, a class of geospatial normative objects, or a specific geospatial normative object.
The IMRO ontology: This ontology implements the IMRO information model for spatial planning of the government of the Netherlands in OWL and uses the concepts of the LegalAtlas ontology.

5   LegalAtlas

LegalAtlas[5] is an open source application that is a showcase of all that has been discussed in the previous sections. It is a tool that provides a new way

---

[5] http://www.legal-atlas.org/

of searching through legal documents. Besides following hyper links to different regulations, it is also possible to browse through geospatial information in the form of maps. By browsing through the maps the regulations that apply to a certain area on the map can be found. LegalAtlas shows that when the data is available in the right formats it is possible to provide a new way of searching for valid regulations applying to a certain area. It is build around the Semantic Web standards and around MetaLex, GML and IMRO2006 and their appropriate ontologies.

LegalAtlas has been implemented in Java, based on an RDF store (Jena[6]) and an OWL DL reasoner (Pellet[7]). The LegalAtlas has two basic functionalities: it can show to which geospatial objects a concept in the legal source applies, and it can show which (parts of) legal sources apply to some selected geospatial object. The selection process is performed by SPARQL queries based on the classes of the LegalAtlas ontology. The LegalAtlas map viewer is based upon the open source GML viewer OpenJump[8].

To describe LegalAtlas we will summarize the user interface and possible actions. The interface has two main screens:

Map Browser: This browser provides the normal map browser functionalities like, move, zoom and selecting single or multiple items.

Regulation Browser: This browser is a simple html viewer, it is possible to follow links and to get a list of actions in the right mouse menu. There are also some filters available that can be used on the textual regulations.

The regulation browser is closely linked to the map browser. All effects of an action take place immediately, which gives a good and fast browsing experience. Between the two main browsers a variety of actions are possible that have an effect on both the selection of the geographical and the textual objects. To show what the actions are of both browser we will give an short description of all the actions for both browsers. The possible actions in the map browser, see figure 3, are:

Zoom: Zooming in and out on the map, to provide a better view on the details or provide a better overview.

Move: Moving around on the map, to access the area you want to interact with.

Select objects: Any number of objects can be selected with the mouse cursor. The selected items will automatically update the list of regulation that the items are linked with.

The regulation browser, see figure 4, is a bit more complicated, because MetaLex supports more functionalities. The following actions are available in the regulation browser:

---

[6] http://jena.sourceforge.com
[7] http://www.mindswap.org/2003/pellet/
[8] http://www.vividsolutions.com

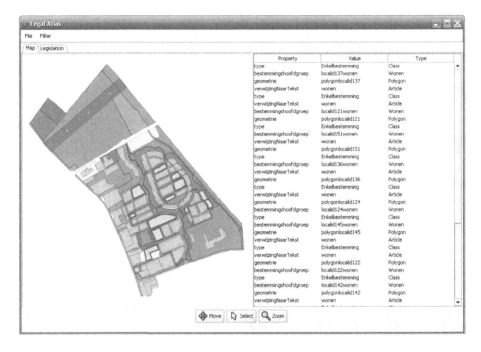

Fig. 3. Map view

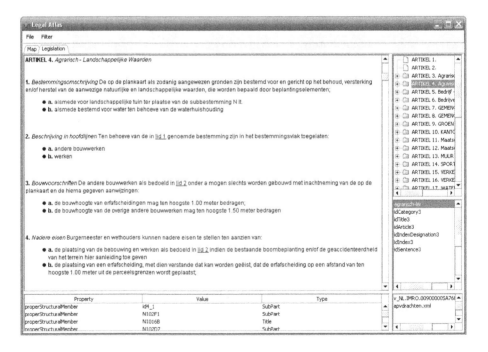

Fig. 4. Text view

Select applicable Regulation: The selected items on the map create a list
of applicable regulations. Selecting different items shows different (parts of)
regulations.

Follow Link to Regulation: This works as a hyper link. The target of the
link will be displayed in the regulation browser.

Show Region: Highlight all the map objects that are linked with this piece of
the regulations. (Little globe graphic)

Select Concept: Highlight all the map objects that are an instance of the
target of the link. (Right mouse click menu)

Change Language: Change the text to a different language, if it is available.

Change Validity Date: Shows the valid regulation for that moment in time.
Default is the current date.

Select Any Regulation: Switch to any regulation that is currently in the
data base.

The next sections will explain how the more interesting functions and actions
work in LegalAtlas. Only the actions that interact with the RDF repository will
be explained.

5.1 RDF Repository

Browsing through linked maps and regulations is only possible when this information is available in the right form. Therefore this data needs to be molded
into the right form. The functionalities in LegalAtlas are defined according to the
ontologies of the 4 domains; MetaLex, IMRO, GML and the LegalAtlas specific
ontology.

LegalAtlas runs on data as long as it validates with the appropriate schemas.
There are 2 different types of data sources that can be loaded into LegalAtlas; MetaLex and IMRO XML documents. The MetaLex documents have to

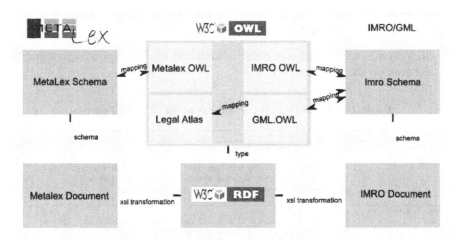

Fig. 5. This figure shows how the ontology and the data are related

validate with the MetaLex schema and the IMRO documents have to validate with the IMRO schema. The xsl-transformations that come with the LegalAtlas distributions translate the data source files into RDF-triples. These triples not only represent the information that is contained within the documents, they also consist of the relations between the data and their ontologies. In other words, the objects from the documents are of a type that is modelled in one of the ontologies.

In figure 5 the dependancies and relations between the ontologies and the different representations of the legal source (MetaLex and IMRO documents).

The RDF repository is used for many actions and functions within LegalAtlas. The following sections will go into that in more detail.

5.2 SPARQL Queries

A number of functions in LegalAtlas make it possible to browse from the map to the regulations and vice versa. These functions are represented in the interface and a SPARQL query is mapped with each function. These SPARQL function are designed specifically for the RDF repository and give back the expected result of the action.

The map browser only has one action that is mapped with a SPARQL query. Every time an area is added to or removed from the selection, the list of the regulations that are linked to the selection is updated.

1. Select or deselect 1 or more polygons and refresh the list of the corresponding legal documents that relate with the selected polygons.

For the regulation browser there are 2 possible actions that interact with the RDF repository. The MetaLex region attribute of a structural element of a regulation refers refers to its region on which it applies. This region is a intermediate concept that can then be related with a GML description of that area. Another possibility is to to have an in-line MetaLex reference in the text that refers to a concept that has a geographical description.

1. Click on a MetaLex region in order to highlight the corresponding geographical area on the map browser.

Table 1. Interaction and the corresponding SPARQL queries

Action	SPARQL query

Change selection on the map browser	SELECT ?verwijzingnaartekst
	WHERE{
	    {bpimro:GML_ID imro:verwijzingNaarTekst ?verwijzingnaartekst}
	    UNION{?verwijzingnaartekst metalex:region ?region .
	    ?region legalatlas:spatial_extension ?polygon .
	    bpimro:GML_ID imro:geometrie ?polygon}}

Click on a metalex region	SELECT ?gmlitems WHERE { metalex:REGION_ID legalatlas:spatial_extension ?geom . ?gmlitems imro:geometrie ?geom}
Click on a metalex reference	SELECT ?gmlitems WHERE { ?instance rdf:type metalex:REFERENCE_ID . ?gmlitems imro:bestemmingshoofdgroep ?instance}

Table 2. Functions that provides a some metadata about the selected objects

Function	SPARQL query
Give metadata for the MetaLex objects	SELECT ?property ?value ?type WHERE { bpimro:METALEX_ID ?property ?value . ?value rdf:type ?type .}
Give metadata for the GML objects	SELECT ?property ?value ?type WHERE { bpimro:GML_ID ?property ?value . ?value rdf:type ?type .}

Table 3. Functions that provide a better interface and the corresponding SPARQL queries

Function	SPARQL query
Give the polygon the right colour	SELECT ?hoofdgroepType WHERE { bpimro: GML_ID imro:bestemmingshoofdgroep ?hoofdgroepInstance . {bpimro:GML_ID imro:verwijzingNaarTekst ?verwijzingnaartekst}

2. Click on a MetaLex reference in order to highlight the corresponding geographical area on the map browser.

These 3 actions provide the interaction with the spatial regulation and the user within LegalAtlas. Table 1 shows what SPARQL query is triggered when an action is initiated in LegalAtlas. In this table the prefixes are omitted for better readability.

During the process of building LegalAtlas and modelling the content it was helpful to see what relations existed for a certain object. Therefore we also created a generic SPARQL query that just represented all outgoing relations of a selected object. This general purpose query was split in 2, one for the GML relations and one for the MetaLex relations. We split these in 2 so we could specify additional metadata for each browser. The SPARQL queries for the metadata can be found in table 2.

We also used a SPARQL query for finding out what the designation is of a certain polygon. We did this to colour the area in the appropriate colour in the map browser. This makes the map browser alot more user friendly. The SPARQL query we used for this functionality can be found in table 3.

6 Conclusion

We have not found grounds for treating spatial regulations different from other (normal?) regulations. The specific focus on spatial objects, the traditional ways of working in the field and the intuitive interaction accessing spatial regulations through a map based interface requires a specific map-oriented visualisation of these regulations. We can reuse the standards and tools that are being developed (or will be developed in the future) for the regular legal domain, but extensions will be needed in order to be able to address the specific spatial features of norms, and the way citizens use these as an effective selection criterion for searching legislation.

The MetaLex standard was developed to provide a generic solution for exchanging legal sources. In order to cover spatial regulations only minimal extensions to this standard were needed. LegalAtlas was developed as a demonstrator for how MetaLex could be integrated with GIS standards using a compact ontology to bridge between these different domains. In the DURP project the LegalAtlas application was intended to be used as a facilitator for the uptake of the new IMRO standards. The LegalAtlas application also demonstrates how future user friendly legal services and eGovernment applications for both citizens and civil servants could be created. If LegalAtlas actually provides useful support for tasks within the domain of spatial regulations is currently tested. We expect that such services will further improve government effectiveness. The Ministry of VROM shares this vision and is considering distributing our viewer for all future spatial plans through Geonovum[9].

We developed a compact ontology that helped us in creating a solution that is general enough to link any regulation or other legal source to a map (and vice versa). This opens up the possibility of what in the Netherlands is sometimes called a "What is allowed where"-map ("WatMagWaarKaart"). The idea is to have a WEB service that allows citizens and companies to investigate the rights, obligations, permissions etc. of a particular area or parcel on a map. There is even a new law enacted (Wet Kenbaarheid Publiekrechtelijke Beperkingen, WKPB), that should lead to enable people to get full insight in the legal status of immovable property. The use of Semantic WEB technology including ontologies like the one developed for this purpose is essential, since all (legal) data necessary to offer these kinds of services is and should be maintained at the source, i.e. distributed all over the Netherlands. Semantic WEB technologies are specifically aimed at supporting linking of heterogeneous sources and dynamically integrating those sources into services.

In order to get to the point that we will be able to deliver such services we still need to add a standardized geo-link to all legal sources, like the 'area of applicability' we introduced for MetaLex. The Dutch Ministry of Internal Affairs experiments with the use of the zip-code for linking decentral regulations to geospatial information. While this seems to provide us with a practical and fast solution it has some obvious disadvantages. First of all, the zip-codes are not publicly owned in the Netherlands as they are owned and maintained by the privatised postal services. They are also not stable; when a new area is being developed and (more) houses are built, new zip-codes emerge. Even more not all areas have (detailed) zip-codes, e.g. large agricultural ones. Finally, they are specific for the Netherlands and cannot be used to refer to areas outside the country.

Another omission that has hindered service scenarios like the one described in this paper is the unfulfillment of the requirement that all legal sources need to be structured and referenced in the same way and their identities should be unique and easily obtainable. The Dutch government is working on that, an agreement

---

[9] http://www.geonovum.nl

has been reached about the unique identification of legislation and published case law by the Judiciary Council.

Thirdly, all databases that are part of this network need to be kept up-to-date and a version management system needs to be in place. While this is already the case for legislation it is not for all other legal sources and GIS databases. International and national committees have started to define standards for management and exchange of geospatial data. The intention is to create a basis for eGovernment and eCommerce by standardisation of spatial datasets (e.g. cadastral information, road maps, power lines, etc.) and direct access to "data at the source" without local duplications. A large number of initiatives is mentioned in [5]. It also includes a comprehensive set of references to relevant technical documents. In the US (NSDI), Australia (ASDI), and Canada (CGDI) there are similar initiatives for making geospatial data collected and created by governments more accessible to business and public (cf. [6]).

We at the Leibniz Center for Law work on computational models of law. We believe that any serious large scale information system that application of legal knowledge and legal reasoning can best be build as within a service oriented architecture using Semantic WEB technologies. The latter are specifically aimed at supporting linking of heterogeneous sources and dynamically integrating those sources into services. The dynamics of legal sources having different authorities as owner, i.e. legal pluralism, the temporal aspects such as versioning, but also more basic things like dynamics in business needs and policy require a solution that balances adaptivity with stability. The Semantic WEB technologies provide a basis for such solution. In this paper we described a compact ontology written in one of the Semantic WEB standards, i.e. OWL, that helped us in creating a solution that is general enough to link any regulation or other legal source to a map (and vice versa). In other projects such as the 6th Framework sponsored Estrella project[10] we work on developing other ontologies, such as the LKIF-core ontology (a core ontology for legal knowledge). The use of ontologies is frequently limited to improving the precision-recall ratio for search and retrieval. These 'ontologies' often are not real ontologies but rather lexical taxonomies like Wordnet [7]. Little or no use is made of the ontological properties and the sufficient and necessary conditions for instances to fit a certain class. In the LegalAtlas example we try to go beyond this limited usage and we created an ontology for the purpose of supporting reasoning. And although the complexity of the reasoning in this example is still very limited the value of using ontologies in supporting reasoning is demonstrated by this example.

Acknowledgements

We thank the Dutch Ministry of Housing, Spatial Planning, and the Environment (VROM) for partially funding the research discussed here. The map of the example spatial plan used in the LegalAtlas demo and in gures in this paper are taken from: Codeervoorbeeldenboek PRBP2006 VROM. Sonsbeek Adviseurs

---

[10] http://www.estrellaproject.org

and Bureau Vijn. We also like to acknowledge the programming work of LegalAtlas by Sander Bisschops and Minze Tolsma.

References

1. Boer, A., Hoekstra, R., Winkels, R., van Engers, T., Willaert, F.: ^{META}Lex: Legislation in XML. In: Bench-Capon, T., Daskalopulu, A., Winkels, R. (eds.) Legal Knowledge and Information Systems (Jurix 2002), pp. 1–10. IOS Press, Amsterdam (2002)
2. Boer, A., Winkels, R.: What's in an interchange standard for legislative xml? i Quaderni 2(41), 32–43 (2005)
3. Winkels, R., Boer, A., Hupkes, E.: Legal atlas: Access to legal sources through maps. In: ICAIL, pp. 27–36. ACM, New York (2007)
4. Wilson, F., Peters, R.: Mapping the law: Knowledge support for business development enquiry. In: Proceedings of the eChallenges 2004 Conference, IST Programme, Vienna (2004)
5. Dorninger, P., Kippes, W., Jansa, J.: Technical push on 3d data standards for cultural heritage management. In: Schrenk, M. (ed.) Proceedings of the 10th International Conference on Information and Communication Technology in Urban Planning and Spatial Development and impacts of ICT on Physical Space, Wien, Austria, CORP (2004)
6. Hall, M., van Orshoven, J.: Spatial data infrastructures in australia, canada, and the united states. Commissioned by the eu, in the framework of the inspire initiative, KU Leuven (SADL+ICRI) (2003)
7. Fellbaum, C.: Wordnet: An Electronic Lexical Database. Bradford Books (1998)

Supporting the Construction of Spanish Legal Ontologies with Text2Onto

Johanna Völker[1], Sergi Fernandez Langa, and York Sure[2]

[1] Institute AIFB, Universität Karlsruhe (TH), Germany
voelker@aifb.uni-karlsruhe.de
[2] SAP Research Center CEC Karlsruhe, Germany
york.sure@sap.com

Abstract. The IST project SEKT (*Semantically Enabled Knowledge Technologies*) aims at developing semantic technologies by integrating knowledge management, text mining, and human language technology. Tools and methodologies implemented in the SEKT project are employed and optimized in three case studies, one of them being concerned with intelligent integrated decision support for legal professionals. The main goal of this case study is to offer decision support to newly appointed judges in Spain by means of iFAQ, an intelligent Frequently Asked Questions system based on a complex ontology of the legal domain. Building this ontology is a tedious and time-consuming task requiring profound knowledge of legal documents and language. Therefore, any kind of automatic support can significantly increase the efficiency of the knowledge acquisition process. In this paper we present Text2Onto, an open-source tool for ontology learning, and our experiments with legal case study data. The previously existing English version of Text2Onto has been adapted to support the linguistic analysis of Spanish texts, including language-specific algorithms for the extraction of ontological concepts, instances and relations. Text2Onto greatly facilitated the automatic generation of the initial version of the Spanish legal ontology from a given collection of Spanish documents. In further iterative steps which included a mixture of learning and manual effort the ontology has been refined and applied to the real-world case study.

1 Introduction

The EU IST integrated project Semantic Knowledge Technologies[1] (SEKT) developed and exploited semantic knowledge technologies. Core to the SEKT project was the creation of synergies by combining the three core research areas ontology management, machine learning and natural language processing. The SEKT technologies were applied in three case studies. One of them is the case study on "Intelligent Integrated Decision Support for Legal Professionals" which aims at supporting newly appointed judges in Spain by means of iFAQ, an intelligent Frequently Asked Questions. iFAQ relies on several complex ontologies of the legal domain, among them the Ontology of Professional Judicial Knowledge (OPJK) consisting of about 100 classes and more than 500 instances.

---

[1] http://www.sekt-project.com/

P. Casanovas et al. (Eds.): Computable Models of the Law, LNAI 4884, pp. 105–112, 2008.

Building and maintaining this ontology is a tedious and time-consuming task requiring profound knowledge of legal documents and language. Thus, any kind of automatic support can significantly increase the efficiency of the knowledge acquisition process. In recent years, several tools and frameworks have therefore been developed to learn ontologies in the legal domain, covering a multitude of languages such as French [7], Portuguese [9], German [10], and Italian [8]. However, there are few reports on experiences with adapting existing, multi-purpose ontology learning tools and frameworks to the requirements of a different language or domain.

In this paper we present a new version of the open-source software Text2Onto[2], a framework for ontology learning from open-domain unstructured text. Moreover, we report on the experiences we made when adapting the previously existing English version of Text2Onto to support the syntactic and semantic analysis of Spanish documents, including language-specific algorithms for the extraction of ontological concepts, instances and relations.

2   Text2Onto

Text2Onto [3] is a framework for ontology learning and data-driven ontology evolution which can be used to automatically generate ontologies from textual resources. It features both a graphical user interface and an API which makes it easy to integrate Text2Onto into Java-based applications. Finally, one of the most important advantages of Text2Onto is its flexibility with respect to interchangeability of algorithms and linguistic processing components. In the past we have already adapted the original, English version of Text2Onto to the German language – and in the SEKT project, we built on these experiences for developing a version that provides complete support for ontology learning from Spanish texts.

2.1   Linguistic Preprocessing

All of the algorithms being part of the Text2Onto framework largely rely on a combination of machine learning and natural language processing techniques in order to extract ontology entities and relationships from open-domain unstructured text. Since the necessary linguistic analysis is done by means of GATE [5] it is very flexible with respect to the set of linguistic components used, i.e. the underlying GATE application can be freely configured by replacing existing components or adding new ones such as a deep parser if required. Another benefit of using GATE is the seamless integration of JAPE which provides finite state transduction over annotations based on regular expressions.

- **Tokenizer:** splits text into individual tokens, e.g., words and punctuation symbols.
- **Sentence splitter:** detects sentence boundaries.
- **Part-of-Speech tagger:** assigns a syntactic category to each token.
- **Lemmatizer:** reduces each token to its lemma, i.e. base form.
- **JAPE transducer:** for shallow parsing: identifies chunks of tokens which constitute e.g. noun phrases or verb phrases.

---

[2] http://ontoware.org/projects/text2onto/

Linguistic preprocessing in Text2Onto starts by tokenization and sentence splitting. The resulting annotation set serves as an input for a POS tagger which in the following assigns appropriate syntactic categories to all tokens. Finally, lemmatizing or stemming (depending on the availability of the regarding processing components for the current language) is done by a morphological analyzer or a stemmer respectively.

In order to improve the quality of the linguistic analysis particularly for Spanish text, we replaced some of the standard GATE components by external resources. The Tree-Tagger[3] is a POS tagger and lemmatizer developed by the University of Stuttgart which can be adapted to a multitude of languages by means of language-specific parameter files. The following screenshot 1 shows the output of the TreeTagger's command line interface. Our GATE wrapper transforms this output into GATE annotations which are available at all subsequent stages in the linguistic processing pipeline.

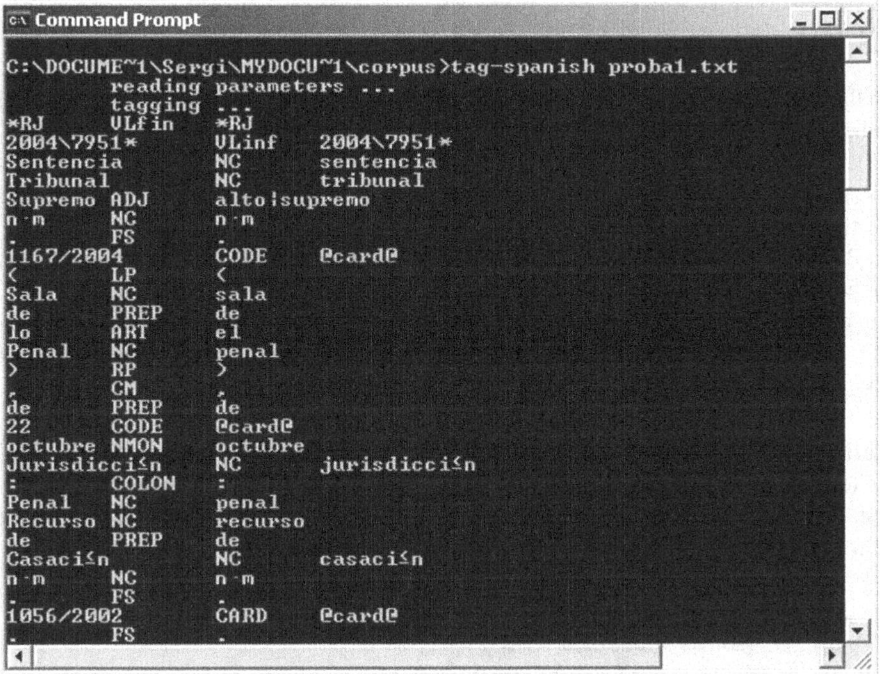
```
Command Prompt
C:\DOCUME~1\Sergi\MYDOCU~1\corpus>tag-spanish proba1.txt
        reading parameters ...
        tagging ...
*RJ     VLfin   *RJ
2004\7951*      VLinf   2004\7951*
Sentencia       NC      sentencia
Tribunal        NC      tribunal
Supremo ADJ     alto|supremo
n·m     NC      n·m
.       FS      .
1167/2004       CODE    @card@
(       LP      (
Sala    NC      sala
de      PREP    de
lo      ART     el
Penal   NC      penal
)       RP      )
,       CM      ,
de      PREP    de
22      CODE    @card@
octubre NMON    octubre
Jurisdicci≤n    NC      jurisdicci≤n
:       COLON   :
Penal   NC      penal
Recurso NC      recurso
de      PREP    de
Casaci≤n        NC      casaci≤n
n·m     NC      n·m
.       FS      .
1056/2002       CARD    @card@
.       FS      .
```

Fig. 1. TreeTagger

After the basic linguistic preprocessing is done, an additional JAPE transducer is run over the annotated corpus in order to match a set of particular JAPE patterns for shallow parsing. These JAPE patterns have to take into account the specific structure of Spanish noun phrases, verb phrases and prepositional phrases. For example, Spanish other than English noun phrases may contain adjectives before as well as after the head of the phrase, and a prepositional complement.

---
[3] http://www.ims.uni-stuttgart.de/projekte/corplex/TreeTagger/

In the following sections we describe the core algorithms provided by the Text2Onto framework and their required adaption to the Spanish language.

2.2 Concepts and Instances

The extraction of concepts and instances relies on a set of JAPE patterns for identifying common and proper noun phrases. Each noun phrase is assigned a relevance value before being mapped to a new or existing class in the ontology. The relevance values with respect to the particular domain are computed by means of statistical measures such as average TFIDF or entropy. While these measures are in principle language independent, the structure of Spanish noun phrases requires specific treatment. We therefore had to develop a number of new JAPE patterns for shallow parsing to be matched during the linguistic preprocessing phase.

2.3 SubclassOf Relations

In the previous version of Text2Onto mainly three algorithms were used for extracting subclassOf relationships from English text. As described in the following sections all of them had to be adapted to the requirements of the Spanish language. Figure 2 shows the results of the Spanish concept classification which is described in the following sections.

Please note that the screenshot as well as most of the examples in this chapter were created from a corpus of web documents about ontologies and the semantic web, since the terminology of this domain is more easily accessible to non Spanish speaking people. Evaluation results and detailed examples relating to original legal case study data are given in [1].

Patterns. The pattern-based concept classification algorithm relies upon a number of lexico-syntactic patterns indicating hyponymy relationships. SubclassOf relations are generated based on this evidence and annotated with a confidence value that corresponds to the normalized frequency of pattern occurrences.

$$NP_{superclass} \quad como \quad (por\ ejemplo)? \quad NP_{subclass}$$
$$NP_{subclass} \quad (son|es|eran|era) \quad NP_{superclass}$$
$$NP_{subclass} \quad (y|o) \quad (otros|otras|dem\acute{a}s) \quad NP_{superclass}$$
$$NP_{superclass} \quad (incluiendo|especialmente) \quad NP_{subclass}$$
$$NP_{superclass} \quad tal|tales \quad como \quad NP_{subclass}$$

WordNet. For any given pair of classes the WordNet-based concept classification aims to find evidence for a hyponymy relationship between the corresponding terms by querying WordNet.

Since the standard version of WordNet [6] provided by Princeton University has been developed particulary for the English language, it is unsuitable for processing Spanish texts. We therefore integrated a Spanish version of WordNet[4] which is developed and

---

[4] http://www.lsi.upc.edu/~nlp/

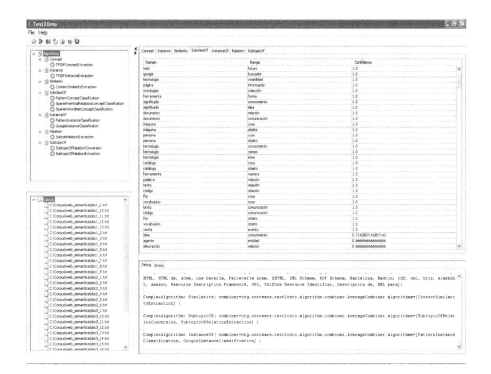

Fig. 2. Graphical User Interface of Text2Onto

maintained by the natural language processing group of the Technical University of Catalonia (UPC).

Vertical Relations Heuristic. The vertical relations heuristic generates subclassOf relations from composite noun phrases, assuming that the class denoted by the whole phrase is subsumed by the class which is represented by its head. For example, from the noun phrase "buscador semántico"[5] ("semantic search engine") the algorithm would conclude that the class *BuscadorSemantico* is subsumed by *Buscador*.

2.4   InstanceOf Relations

Basically three algorithms are available for learning instanceOf relationships from Spanish texts.

Patterns. The pattern-based extraction of instanceOf relationships is very similar to the concept-based concept classification described in Section 2.3. Some of the lexico-syntactic patterns used for detecting concept instantiations are listed below.

---

[5] Please note that the alternative spelling "buscador semantico" (without accentuation) will not be recognized as denoting the same class by the current version of Text2Onto. This bug still needs to be fixed.

$PNP_{instance}$, NP_{class} ,
$PNP_{instance}$ (NP_{class})
NP_{class} como   (por ejemplo)? $PNP_{instance}$
$PNP_{instance}$ (son|es|eran|era) NP_{class}
$PNP_{instance}$ (y|o)   (otros|otras|demás) NP_{class}
NP_{class} (incluyendo|especialmente) $PNP_{instance}$
NP_{class} tal|tales   como $PNP_{instance}$

Google. In line with [2] we implemented an approach to obtaining evidence for instanceOf relations by online pattern matching. For each instance lexically represented by $PNP_{instance}$ the algorithm poses a number of Google queries similar to those described in the previous section.

"como $PNP_{instance}$"
"$PNP_{instance}$ es un"
"$PNP_{instance}$ es una"

The results returned for each of these queries are then analysed in order to determine possible fillers for the open position in the regarding pattern. For the first query template, for example, the filler must be a noun phrase directly preceding the phrase matched by the query.

Context-Based Similarity. The assumption underlying the context-based instance classification is that each instance belongs to the class which is semantically most similar.

In order to compute the semantic similarity Text2Onto exploits the distributional hypothesis by Harris which basically states that the senses of two words, i.e. concepts, are similar to the degree the words share lexical context. Lexical context in its most simple form is a vector consisting of all the words which co-occur (e.g. in the same sentence or token window) with the words representing the class or instance of interest. For the new version of Text2Onto we implemented a more sophisticated context extraction based on both lexical and syntactic features.

The context vector of each instance is compared to all context vectors of concepts in the ontology by means of the cosine measure. If the similarity of the context vectors is above a certain threshold the instance classification algorithms assumes the instance to instantiate that particular concept. Further details regarding different types of context features and similarity measures for instance classification are given by [4].

2.5   Non-taxonomic Relations

For the extraction of non-taxonomic relationships Text2Onto relies upon subcategorization frames, i.e. predicate argument structures consisting of verbs and prepositional or nominal complements, enriched by ontological knowledge and statistical information. Confidence values for each relationship are computed based on the number of instantiations of that particular frame found in the corpus. Example: *incluir(Ontología, Definición)*.

3   Lessons Learned and Conclusion

Adapting Text2Onto to the requirements of the Spanish language confronted us with a number of unexpected technical challenges. Some of them had to do with syntactic and semantic particularities of the Spanish language, e.g. with respect to prepositional complements in noun phrases. Others were related to character encoding, compatibility of different versions of WordNet and the adaptation of new linguistic processing components.

On the other hand, we found that a flexible and extensible language processing framework such as GATE is of great use if multilinguality is required by the application. This flexibility made it possible to integrate specialized linguistic components for the Spanish language into Text2Onto without much effort, and significantly speeded up the implementation process.

Acknowledgements

Research reported in this paper has been partially financed by the European Union in the IST-2003-506826 project SEKT (http://www.sekt-project.com).

References

1. Casanovas, P., Vallbé, J.-J., Casellas, N., Poblet, M., Blázquez, J.C.M., Benjamins, V., Cobo, J.-M.L.: D10.4.1 SEKT legal case study: After analysis, with the collaboration of Z. Huang, and J. Völker (2006)
2. Cimiano, P., Ladwig, G., Staab, S.: Gimme the context: Context-driven automatic semantic annotation with c-pankow. In: Ellis, A., Hagino, T. (eds.) Proceedings of the 14th World Wide Web Conference, Chiba, Japan, MAY 2005, pp. 332–341. ACM Press, New York (2005)
3. Cimiano, P., Völker, J.: Text2onto - a framework for ontology learning and data-driven change discovery. In: Montoyo, A., Muñoz, R., Métais, E. (eds.) NLDB 2005. LNCS, vol. 3513, pp. 227–238. Springer, Heidelberg (2005)
4. Cimiano, P., Völker, J.: Towards large-scale, open-domain and ontology-based named entity classification. In: Proceedings of the International Conference on Recent Advances in Natural Language Processing (RANLP 2005), pp. 166–172 (September 2005)
5. Cunningham, H., Maynard, D., Bontcheva, K., Tablan, V.: GATE: A framework and graphical development environment for robust NLP tools and applications. In: Proceedings of the 40th Annual Meeting of the ACL (2002)
6. Fellbaum: WordNet: An Electronic Lexical Database (Language, Speech, and Communication). The MIT Press, Cambridge (1998)
7. Lame, G.: Using NLP techniques to identify legal ontology components: Concepts and relations. Artificial Intelligence and Law 12(4), 379–396 (2004)

8. Lenci, A., Montemagni, S., Pirrelli, V., Venturi, G.: NLP-based ontology learning from legal texts. a case study. In: Pompeu Casanovas, E.F.M.T.S., Biasiotti, M.A. (eds.) Proceedings of the Workshop on Legal Ontologies and Artificial Intelligence Techniques at ICAIL 2007, pp. 113–129 (June 2007)
9. Saias, J., Quaresma, P.: A methodology to create legal ontologies in a logic programming based web information retrieval system. Artif. Intell. Law 12(4), 397–417 (2004)
10. Walter, S., Pinkal, M.: Computational linguistic support for legal ontology construction. In: ICAIL 2005: Proceedings of the 10th international conference on Artificial intelligence and law, pp. 242–243. ACM, New York (2005)

Dynamic Aspects of OPJK Legal Ontology

Zhisheng Huang[1], Stefan Schlobach[1], Frank van Harmelen[1], Núria Casellas[2],
and Pompeu Casanovas[2]

[1] Vrije Universiteit Amsterdam, The Netherlands
{huang,schlobac,frankh}@cs.vu.nl
[2] Universitat Autonoma de Barcelona, Spain
{nuria.casellas,pompeu.casanovas}@uab.es

Abstract. The OPJK (Ontology of Professional Judicial Knowledge) is
a legal ontology developed to map questions of junior judges to a set of
stored frequently asked questions. In this paper, we investigate dynamic
and temporal aspects of one of the SEKT legal ontologies, by subjecting
the ontology OPJK to MORE, a multi-version ontologies reasoning System. MORE is based on a temporal logic approach. We show how the
temporal logic approach can be used to obtain a better understanding
of dynamic and temporal evolution of legal ontologies.

1 Introduction

SEKT is a European project on developing Semantically Enabled Knowledge
Technologies[1] [11]. The aim of SEKT is to develop and exploit semantically-
based knowledge technologies in order to support document management,
content management, and knowledge management in knowledge intensive work-
places. Specifically, SEKT aims at designing appropriate utilities to users in
three main areas: digital libraries, the engineering industry, and the legal domain, providing them with quick access to the right pieces of information at the
right time.

The SEKT Legal Case Study was based on the development of the prototype IURISERVICE[2], a web-based application that will provide access to Spanish judges to a repository of judicial practical experience, stored in the form of
question-answer pairs. A set of surveys provided the quantitative and qualitative
data necessary not only to assess the context of users (newly recruited judges in
Spain) and their specific needs with regard to the technology under development,
but also to gather the practical experience needed for the repository [7]. In particular, these data gave insight on institutional, organisational, and individual
constraints that could either facilitate the introduction of SEKT technologies

---

[1] http://www.sekt-project.com/
[2] The prototype is developed by the Institute of Law and Technology (IDT-UAB, Intelligent Software Components S.A. (iSOCO), with the collaboration of the Spanish
Judicial School and the Spanish General Council of the Judiciary (SEC2001-2581-
C02-01).

P. Casanovas et al. (Eds.): Computable Models of the Law, LNAI 4884, pp. 113–129, 2008.
© Springer-Verlag Berlin Heidelberg 2008

within the judicial units [6]. IURISERVICE was designed as a result of those surveys, and will offer decision-making support through the use of ontologies. Also from the data regarding the practical judicial experience with everyday problems (faced by newly recruited judges in the judicial unit), obtained during fieldwork, the Ontology of Professional Judicial Knowledge (OPJK) is built [2,3,4,5,9].

Ontologies are the backbone of the Semantic Web, as they allow to share vocabulary in a semantically sound way. With the rise of the Semantic Web, the need to create ontologies has become more prominent, and even highly sensitive applications depend on ontologies, which in turn have to be of the highest possible quality. Unfortunately, building such a high-quality ontology is a very time-consuming process that often requires highly qualified professionals and domain experts over a significant time span.

Often, building an ontology can take years, and many different versions are produced (these versions are often called the *version space*). In this process, keeping track of modeling decisions and changes is an extremely difficult task, for the success of which a dedicated versioning system is necessary. Versioning systems are known from Software Engineering, but they are restricted to keeping track of syntactic changes. In the development of ontologies the more significant changes are often semantic. Ontology modellers not only need to keep track of modelling decisions and changes over time, especially in distributed scenarios, but also they should be able to get a general insight to the changes that the ontology has suffered during all the modelling process. Were some versions of the ontology more stable than others? Are certain concepts or certain parts of the ontology more stable than others? How have the modelling decisions affected concepts over time?

Answers to these questions could provide developers with a) information regarding when and where a unstable (or less stable) version was made in the past and which one was its previous version, b) knowledge about the type of changes: additions or deletions, c) insight to the stability or unstability of concepts. This latter information is of particular interest as could reflect the ongoing discussion among developers and domain experts regarding a certain concept and, thus, the modelling difficulty that a particular concept has offered. The stability measure could reflect the most difficult discussions present during the design of the ontology.

A versioning system can be used to analyze various properties of the version space. Consider a situation, where people work on the same ontology, who disagree on a particular relation between two classes. In the development process, the disputed relation will probably not remain stable, i.e. in some versions the relation will hold, in others it will not, according to who edited the latest version. Such an unstable situation can be very damaging for the overall quality of an ontology, and it is important to detect unstable relations. Part of the task of a versioning system should be to detect instable relations or similar "problems" in the version space. For this purpose, we developed the system MORE, a Multi-versions Ontology Reasoning system, at the Vrije Universiteit Amsterdam as part of the SEKT project [13,14].

The framework of MORE is based on a temporal logic approach. Namely, we consider multi-versions of an ontology as a sequence of ontologies which are connected with each other via change operations. In this paper, we present the work of an investigation on the dynamic and temporal evolution of the SEKT legal ontologies, by subjecting the ontology OPJK to MORE. We show how the temporal logic approach can be used to obtain a better understanding of dynamic and temporal evolution of legal ontologies.

This paper is organized as follows: Section 2 overviews the system MORE. Section 3 presents the notion of effect space for investigating dynamic and temporal aspects of ontologies. Section 4 discusses the OPJK versioning. Section 5 presents the analysis of the OPJK ontology with the system MORE. Section 6 discusses further work, and concludes the paper.

2   MORE: A Multi-version Ontology Reasoning System

MORE is a multi-version ontology reasoning system, which is based on a temporal logic approach [13,14]. Under this approach, multi-versions of an ontology are considered as a sequence of ontologies which are connected to each other via change operations. Each of these ontologies has a unique name. Thus, a version space S over an ontology set Os is a set of ontology pairs, namely, $S \subseteq Os \times Os$. We use version spaces as a semantic model for our temporal logic, restricting our investigation to version spaces that present a linear sequence of ontologies. A linear version space S on an ontology set Os is denoted as a finite sequence S of ontologies as $S = (o_1, o_2, \cdots, o_n)$. An ordering $<$ with respect to a version space S is introduced as $o < o'$ iff o occurs prior to o' in the sequence S. We use $ontology(S)$ to denote the ontology set $Os = \{o_1, \cdots, o_n\}$ of the version space S.

A temporal logic has been developed in MORE for Multi-version Reasoning [13,14]. The Language $\mathcal{L}+$ of the temporal logic **LTLm** is defined as an extension to the ontology language $\mathcal{L}$ with Boolean operators and the backward temporal operators, which include the previous version operator **Prev**ϕ which denotes that the property ϕ holds in the previous version (with respect to the current version in the version space), the always-in-the-past operator **H**ϕ which denotes that the property ϕ always holds in any version before the current version, and the since operator ϕ**S**ψ which denotes that the property ϕ always holds (till the current version) since the property ψ held in a version before the current version. The sometimes-in-the-past operator **P**ϕ is defined in terms of the always-in-past operator as $\neg$**H**$\neg\phi$. In the temporal logic, the evaluation of a temporal formula ϕ on an ontology o (i.e., a version) in a version space S is defined as an entailment relation [18,19,17]:

$$S, o \models \phi$$

The semantics of the temporal operators is illustrated in Figure 1, where arrows denote the sequence relation of ontologies in the version space, and a formula under an ontology denotes that the formula holds on the ontology. For example, the first line in the figure shows that **Prev**ϕ holds on an ontology iff the formula ϕ hold on its previous ontology.

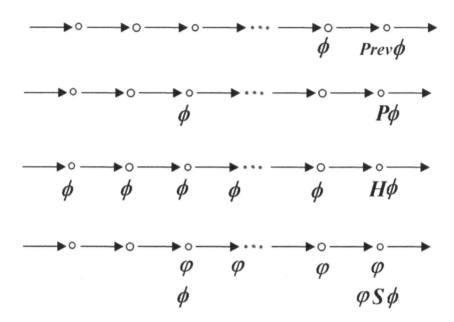

Fig. 1. Semantics of Temporal Operators

Many Description Logic Reasoners support so-called retrieval queries which return a set of concept names which satisfy a certain condition. For example, a children concept c' of a concept c, written $children(c, c')$, is defined as one which is subsumed by the concept c, and there exists no other named concepts between them.

Thus, the set of new/obsolete/invariant children concepts of a concept on an ontology o in the version space S is defined as follows:

$$new_{children}(S, o, c) =_{df} \{c' | S, o \models children(c, c') \wedge \neg \mathbf{Prev}\, children(c, c')\}.$$

$$obsolete_{children}(S, o, c) =_{df} \{c' | S, o \models \neg children(c, c') \wedge \mathbf{Prev}\, children(c, c')\}.$$

$$invariant_{children}(S, o, c) =_{df} \{c' | S, o \models children(c, c') \wedge \mathbf{Prev}\, children(c, c')\}.$$

We define the new/obsolete/invariant children concept relations of an ontology as a set of concept pairs as follows:

$$new_{children}(S, o) =_{df} \{\langle c, c' \rangle | S, o \models children(c, c') \wedge \neg \mathbf{Prev}\, children(c, c')\}.$$

$$obsolete_{children}(S, o) =_{df} \{\langle c, c' \rangle | S, o \models \neg children(c, c') \wedge \mathbf{Prev}\, children(c, c')\}.$$

$$invariant_{children}(S, o) =_{df} \{\langle c, c' \rangle | S, o \models children(c, c') \wedge \mathbf{Prev}\, children(c, c')\}.$$

The same definitions can be extended into the cases like parent concepts, ancestor concepts, descendant concepts.

We have implemented the prototype of MORE by using Prolog. MORE is powered by the XDIG interface [16], an extended DIG description logic interface for

Prolog[3]. MORE is designed to be a simple API for a general reasoner with multi-version ontologies. It supports extended DIG requests from other ontology applications or other ontology and metadata management systems and supports multiple ontology languages, including OWL and DIG [1][4]. This means that MORE can be used as an interface to any description logic reasoner as it supports the functionality of the underlying reasoner by just passing requests on and provides reasoning functionalities across versions if needed. Therefore, the implementation of MORE will be independent of those particular description logic reasoners.

3   Ontology Versioning and Effect Space

In order to measure ontology changes and their effect, we have to check the ramification of an ontology change on all possible semantic relations on the concept/role/individual relations of an ontology. We call the set of all possible semantic relations the *Effect Space* of a version space.

All of the ontology changes can be examined under an effect space which covers all of the possible changes and their ramification on the semantic relation with respect to concepts, roles, and individuals. In this paper, we consider effect spaces which are characterized by the new/obsolete/invariant relations on concepts, roles, and individuals on a version space, which are supported by the description logic based language DIG[5]. Thus, an effect space consists of new/obsolete/invariant concept relations with respect to children/parents/ancestors/descendants relations, new/obsolete/invariant role relations with respect to rchildren/rparents/rancestors/rdescendant relations[6], and new/obsolete/invariant instance relations between concepts and individuals, as they are defined in DIG.

Suppose that an ontology o in a version space S consists of about n_c concepts, n_r roles, and n_i individuals. The number of possible concept relations on new/obsolete/invariant with respect to children/parents/ancestors/descendants aspects is $N_C(S, o) = n_c \times n_c \times 3 \times 4$. The number of possible instance relations between a concept and an individual is $N_I(S, o) = n_c \times n_i \times 3$. The number of possible role relations is $N_R(S, o) = n_r \times n_r \times 3 \times 4$. Therefore, the effect space has $N(S, o) = N_C(S, o) + N_R(S, o) + N_I(S, o)$ possible semantic relations at each version. Suppose that version space S consists of n_v version ontologies and each ontology has almost the same numbers of concept/role/individual, then the number of possible semantic relations in a linear version space is $N(S, o) \times n_v$.

---

[3] http://wasp.cs.vu.nl/sekt/dig

[4] http://dl.kr.org/dig/

[5] Note that all ontology files which are specified by using OWL DL can be converted into ones with the DIG data format by using XDIG. Thus, MORE supports any OWL DL ontology.

[6] rchildren means the children role relation. Namely, a role r is a children role of r' iff r is a subrole of r' and there exists no other role between r and r'.

4   OPJK Versioning

The OPJK ontology lies at the core of the IURISERVICE FAQ system, developed within the SEKT Legal Case Study. This ontology has been developed collaboratively by legal experts from the IDT-UAB team and software engineers from iSOCO in a distributed environment. This ontology is used to link the input question in natural language to the repository of stored frequently asked questions (FAQs, which also contain their corresponding answers provided by experienced judges from the Judicial School). To this aim, OPJK conceptualises the most relevant terms for the judicial profession (extracted from the data gathered during the ethnographic survey focused on judicial practical problems). As this is highly specialised knowledge, enriching the current ontology and producing a new version of OPJK is a complex task. In practice, a team of legal experts meet in regular intervals to decide on the relevant changes to the OPJK, based on the analysis of the data [8,10].

The OPJK version space has been built to reflect the construction process; the legal experts met regularly to discuss the formalisation of the concepts, instances and properties extracted from the list of questions that contain practical judicial knowledge. During each meeting, a group of questions is discussed and either concepts, instances or properties were added to the ontology or current ontology terms were modified or deleted. Also, when ontology engineers intervened other changes were added. Therefore, the version space is not only a direct mirror of the creation process, it also constitutes an opportunity to analyse the epistemic process in a systematic way.

For this experiment, each version in this study represents the changes added to OPJK suggested by the discussion of one of the above-mentioned questions at a time. OPJK v1.1 of this study was already a mature version which included 97 concepts, 118 roles and 484 individuals. In general, the added modifications of each of the versions presented will be of low significance, however it will make explicit the changes that specific decisions cause in the ontology. 27 versions have been collected. Therefore, the OPJK version space can be analysed with two different motivations: first, as a well-constructed (and documented) version space of a complex ontology, and secondly, as a formalisation of an epistemological process, in which knowledge elicitation is made explicit in ontological changes.

In this paper we will only address the first issue, and leave the latter for future work. More concretely, we will study measures on the temporal dimension of the version space to detect peculiarities in the process as a whole and in the evolution of particular conceptualisations over time. Is the process a stable, monotonic one, where information is added to new versions? Or is there constantly information retracted, suggesting a more ad hoc process? It is important for developers to have a general insight to the construction process, especially when the construction is distributed (several ontology modellers and ontology engineers interact in a distributed environment) and to be able to detect which versions produced more changes (and the type of changes) to the ontology and also to detect which concepts have been stable or unstable during the process in order to be aware of the future changes that might affect them. In the following we will try to give an insight to OPJK version space in order to offer some insight to these issues.

5 Analysis of OPJK

In this section, we will measure the OPJK ontology change and their effects from the following different levels:

- Version level: We examine ontology changes on a single version, so that their effects can be displayed in a timeline of a version space, from which we provide a dynamic view on ontology changes.
- Concept/Role level: We examine ontology changes on a single concept or role to see its dynamic properties, from which we can obtain the picture of concepts and roles for their stability, difference, and the monotonicity.
- Logical Property Level: We examine a logical property to see its temporal aspects in a version space.

5.1 Ontology Change Measure on the Version Level

In this section, we measure ontology changes and their effect on the version level, so that the difference on changes can be presented as a timeline on a version space. Moreover, we measure ontology changes with respect to the following criteria respectively:

- **stability:** how stable are the semantic relations when an ontology is subjected to a change which leads to a new version. Namely, for each version i, we compare the effect space at version i with the effect space of its previous version. The intersection of the effect spaces at those two versions is considered as the stable part of the current version i.
- **difference:** what are the differences, more exactly what are new semantic relations when an ontology has been changed.
- **monotonicity:** whether or not some semantic relations which hold in the previous version no longer hold in the current version?

The relation among these three properties are illustrated in Figure 2.

Stability Measure. We measure the stability of an ontology as the sets of its invariant concept/role/individual relations when compared with its previous version.

Figure 3 shows the timeline differences of the cardinality of the stability sets in the version space OPJK[7]. A way to normalize the stability is to divide them by the corresponding relation numbers, i.e. N_C, N_R, N_I, and N, in a single version o of the effect space in a version space S.

---

[7] So far the OPJK ontology has flat role relations. Therefore, we do not count any role relation of the OPJK ontology in this document.

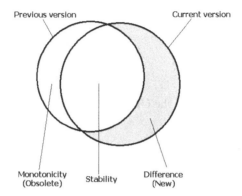

Fig. 2. The relation among stabilty, difference, and monotonocity

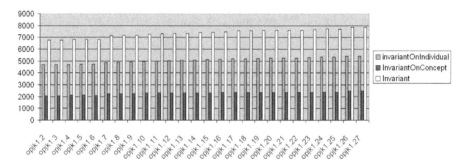

Fig. 3. Timeline of the stability of the OPJK ontology (opjk1.1-opjk1.27)

The timeline of the normalized stability in the effect spaces is shown in Figure 4. From the timeline graphs, we can observe a high degree of stability, although stability changed with time. The last few versions (opjk1.25 and opjk1.27) are less stable than their previous versions. Analysing the logs, the changes introduced to these versions were multiple (concept, role and instance level) and important (in quantity) with respect to the other versions. Version opjk1.25 included the addition of three concepts and some instances, although some other instances were deleted and moved to different concepts. Version opjk1.27, included four new concepts, some changes in position in the concept hierarchy, new roles and instances.

Difference Measure. The new and obsolete concept/role/individual relations show the difference of an ontology from its previous version. They can be used to measure how big a change has been done on the ontology. We are particularly interested in the difference measure by new concept/role/individual relations.

Figure 5 represents the timeline of the new relation of the OPJK ontology which shows the amount of different concept/individual relations of all ontologies in the version space OPJK. From the timeline figures we know that most of

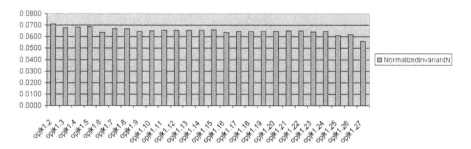

Fig. 4. Timeline of the normalized stability of the OPJK ontology (opjk1.1-opjk1.27)

the changes on the OPJK ontology are small, as expected. And that opjk1.25 and opjk1.25 include significant changes in relation to the version space. However, it is interesting to discover that the changes added to opjk1.5, shown as new relations on opjk1.6, are also significantly big, and were not equally represented in the previous analysis. The biggest change occurs on opjk1.6, the second biggest change occurs on opjk1.27, and the third one occurs on opjk1.25. Opjk1.6 includes mainly the addition of 6 concepts and several instances. However, most of the additions only affect one existing concept (`Organization`) that has been populated with concepts and instances, which could explain the stability measures obtained.

Similarly we introduce the normalized difference measure. Figure 6 is the timeline of the normalized new relation of the OPJK ontology. The result shows the biggest change occurs on opjk1.6, the second biggest change occurs on opjk1.27, and the third biggest change occurs on opjk1.25.

Monotonicity Measure. The obsolete concept/role/individual relations show that some semantic relations on an ontology which hold in the previous version do not hold in the current version any more. Therefore, it can be considered as a kind of measure for the monotonicity/non-monotonicity of an ontology change. By the monotonicity we mean that the change does not make any previously held property obsolete, otherwise it is called a non-monotonic change. Figure 5 is the timeline of the new relations of the OPJK ontology which shows the amount of different concept/individual relations of all ontologies in the version space OPJK. This measure on the OPJK version space shows clearly that the addition of individuals occurs more often than the addition of new concepts or roles; which in turn shows a high level of agreement on the ontology developed up to that moment. The above-mentioned changes on version opjk1.5, which are shown as new relations on opjk1.6, are significantly big. Figure 7 is the timeline of the obsolete relation of the OPJK ontology, which shows that only opjk1.6, opjk1.23, and opjk1.25 have some obsolete concept relations and obsolete individual relations, and opjk1.16 has only some obsolete individual relations.

Figure 8 is the timeline of the normalized obsolete relation of the OPJK ontology. Those timelines show that opjk1.25 has the biggest nonmonotonicity

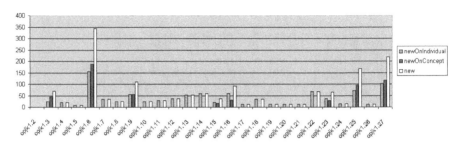

Fig. 5. Timeline of the new relation of the OPJK ontology

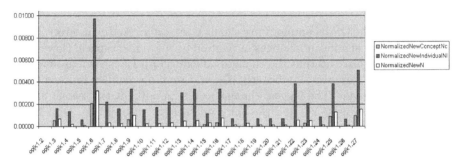

Fig. 6. Timeline of the normalized new relation of the OPJK ontology (opjk1.1-opjk1.27)

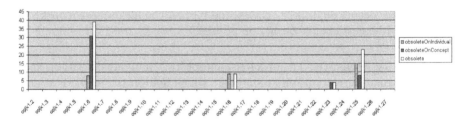

Fig. 7. Timeline of the obsolete relation of the OPJK ontologies

effect and opjk1.23 has the second biggest nonmonotonicity effect with respect to the individual relation.

5.2 Ontology Change Measure on the Concept Level

In this subsection we measure ontology changes and their effects on individual concepts, so that we can determine which concepts/roles in the ontology are more stable than others, by which we can find the core of an ontology.

Stability Measure. We measure the stability of a concept in a version space as the amount of its invariant relations when compared with its previous version.

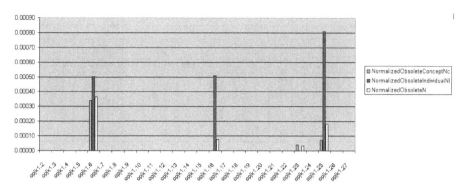

Fig. 8. Timeline of the normalized obsolete of the OPJK ontologies w.r.t. the effect space (opjk1.1-opjk1.27)

Figure 9 shows the stability of the concepts in the OPJK ontology. Similarly we normalize the concept stability measure by dividing the invariance cardinarity by the maximal invariance cardinarity. The normalized concept stability is shown in Figure 10. Both the concept stability and the normalized concept stability produce the same results. For example, the 5 concepts that have shown more concept stability over time are Happening, Legal_Abstraction, Event, Procedural_Phase and Agent.

The individual stability, more unstable than concept stability, is shown in Figure 11. In this case, the 5 concepts that have shown more individual stability over time are Situation, Document, Procedural_Document, Event and Organization.

Figure 12 shows the top 30 most stable concepts from opjk1.1 till opjk1.15 in the OPJK ontology. As an example, the 10 more stable concepts are: Object, Happening, Abstraction, Legal_Abstraction, Document, Event, Procedural_Document, Procedural_Phase, Civil_Procedure_Phase and Phase. The analysis of the different list of concepts not only provides an idea of the concepts that have been more stable over time, but also of the parts or branches of the ontology that are most stable. Within the above-mentioned list, we discover that Happening is superconcept of Event, which is a superconcept of Phase. The latter is at the same time a superconcept of Procedural_Phase that has Civil_Procedure_Phase as its subconcept. Also, if we analyse the log of the 27 versions, only one instance has been added to one of those concepts.

Difference Measure. Similar to the difference measure in the ontology level, we measure the difference on the concept level in terms of the new concept relations and new instances relations.

Figure 13 shows the concept difference in the OPJK ontology. The normalized concept difference is shown in Figure 14. Both include the same concepts. The concepts that have experienced more changes are: Public_Administration (modified deeply in version opjk1.5), Parent (modified in version opjk1.9), Organization (sublcasses added in version opjk1.11), Adulthood (modified in version opjk1.3), Competence, Guarantees (modified in version 1.14), Court

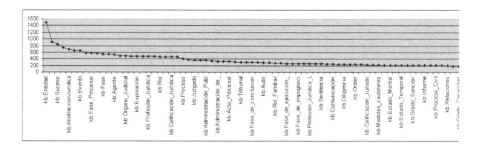

Fig. 9. The concept stability of the OPJK ontology (opjk1.1-opjk1.15)

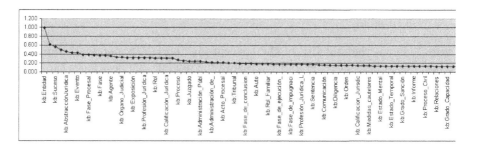

Fig. 10. The normalized concept stability of the OPJK ontology (opjk1.1-opjk1.15)

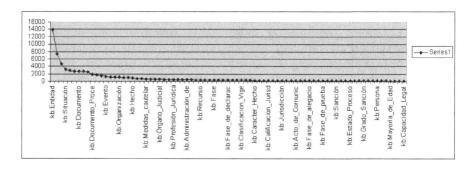

Fig. 11. The individual stability of the OPJK ontology (opjk1.1-opjk1.15)

(version opjk1.5), `Group` (version opjk1.5), `Family_Role` (modified in versions opjk1.9, 1.17 and 1.24), `Police_Forces` (version opjk1.5), `Tribunal` (version opjk1.5) and `Happening` (version opjk1.9).

5.3   Ontology Change Measure in the Logical Property Level

A logical property can be examined with respect to its temporal aspects. If a property ϕ on the ontology is changed once, it is never changed back again in any sequel version as illustrated in Figure 15. We consider that kind of change

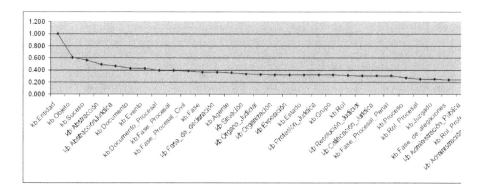

Fig. 12. Top 30 most stable concepts in the OPJK ontology

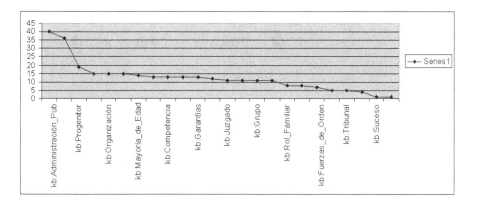

Fig. 13. The concept difference of the OPJK ontology (opjk1.1-opjk1.15)

as a stable one, because it occurs only once in the whole version space. The corresponding temporal query of stable change on a property can be expressed as

$$\neg\phi\mathbf{SH}\phi.$$

The query on the existence of only two changes with respect to a property ϕ, as shown in Figure 16, can be expressed as

$$\neg\phi\mathbf{SPrev}(\phi\mathbf{SH}\neg\phi).$$

We examined the OPJK ontology for its changes with respect to the temporal aspects. It shows that all concept relations on the OPJK ontology from opjk1.1 till opjk1.27 are stable. That can be confirmed by examining the obsolete concept relations only in opjk1.6, opjk1.23, and opjk1.25, because the timeline presentation of the obsolete concept in Figure 7 shows that obsolete concept relations occur only in those versions. More examples of temporal queries on the OPJK

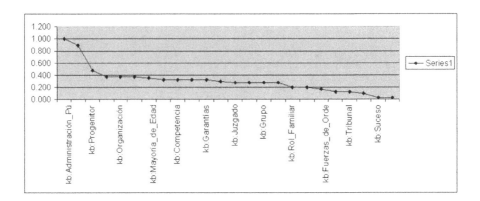

Fig. 14. The normalized concept difference of the OPJK ontology (opjk1.1-opjk1.15)

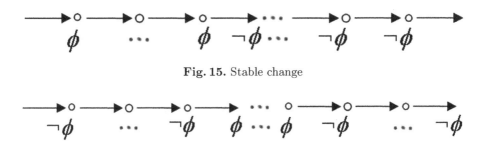

Fig. 15. Stable change

Fig. 16. Only two changes with respect to ϕ

ontology can be found in our SEKT deliverables [13,12,15], which are available at the website of MORE[8].

6   Discussion and Conclusions

OPJK is an ontology of professional judicial knowledge that has been developed for the IURISERVICE prototype, to support the semantic matching of natural language input questions from junior Spanish judges with FAQs stored in a database (question-answer pairs). The OPJK ontology has been developed by legal experts and judges over a significant time. To obtain further knowledge regarding the construction process and the implications of decisions a number of successive versions have been recorded in the so-called version space. This space contains valuable information on the knowledge elicitation process, but is also an interesting test case for the use of semantic versioning techniques in the creation of ontologies.

---

[8] http://wasp.cs.vu.nl/sekt/more

The MORE tool offered semantic versioning support for OPJK ontology development based on a combination of change detection and Linear Temporal Logic. The experiments we conducted were three-fold. First, we studied the properties of the overall version space of the OPJK ontology by checking for stability, novelty and monotonicity of the process. More concretely, we studied at which time semantic relations between classes were added, deleted or remained the same. This indicated that there are different types of changes in the OPJK version space, the smaller ones with only minor, often cosmetic, variations, and the more substantial ones, in which many new semantic relations change. The second study was a more detailed analysis on concept level. Here, we identified the most stable concepts in the OPJK ontologies, those with the most frequent changes, and the most commonly effected concepts. Finally, we provided an initial case study on the stability of change in the OPJK version space using the power of the temporal logic which underlies MORE.

From the findings of the stability and the difference measure on the version level, we may conclude that a version that only included changes related only to a concept and its individuals and subconcepts could be considered more stable as a whole than a version that included less changes in quantity but that affected different concepts of the ontology (figures 3-7). Furthermore, we may also conclude that more individuals have been added than concepts, most additions were at the instance level, i.e., at A-box, not at the terminology level, i.e., at T-box (figure 8). This could indicate a high level of agreement on the developed concepts, roles and individuals until that time (opjk1.1). Finally, although he stability level of OPJK has increased over time (figure 3), the results provided above in this paper, also show that some recent OPJK versions are less stable than their previous ones, as shown in Figure 4. The property indicates that some conceptual restructuring occurs on OPJK. In these last 3 versions, due to insight provided by the data (competency questions) the modellers modified modelling decisions made in previous versions; in particular they modified subclass relationships and added children to certain classes. For example, in version 1.25, a new subclass of `FamilyRole` is created and instances belonging to another class have been moved to it. Also a new subclass is added in version 1.27 and existing instances are reassigned to it.

From the analysis of the stability on the concept level, we could not only observe that some concepts were more stable than others, but also that there are some areas of stability within the OPJK ontology. The areas of concepts, roles and individuals related to `Phase` are of particular stability. Also, we may detect that some of the so-called top" or upper" ontological concepts appear as particularly stable (`Object`, `Happening`, `Abstraction`, `Event`). Finally, the analysis of the concept difference measure brings about two different types of changes: concepts that have suffered modification (addition of subconcepts and individuals, for example, `Public_Administration`) and concepts that have just been added to the ontology (`Parent`, `Guarantees`).

Regarding the implications for ontology development, ontology modelers may rely on the ontology stability measurement to analyze the modeling decisions that involved a great level of instability. In the same way, if the amount of instability inflicted by adding or modifying subclasses could be assessed, different modeling options could be sought to promote stability. Also, it is interesting to see how certain knowledge experts discussions are mirrored in the stability measures of the ontology or in the measurements on the concept level. As an example, the list of the most stable concepts in OPJK did not include `Act` or `Fact`. Although they had been in the ontology since the initial versions, they did not appear within the most 30 stable concepts. Modelers analysed that those concepts had been the focus of an ongoing discussion within the modeling team and, this analysis might support the view that a final agreement on the final conceptualization has not yet been made. Future work would then involve the revision of the modeling decisions that involve those concepts in order to reach an agreement.

Acknowledgements. The authors thank the anonymous reviewers for constructive comments on the earlier draft of the paper. The work reported in this paper was partially supported by the EU-funded SEKT project(IST-506826).

References

1. Bechhofer, S., Möller, R., Crowther, P.: The DIG description logic interface. In: International Workshop on Description Logics (DL2003), Rome (September 2003)
2. Benjamins, V.R., Contreras, J., Blázquez, M., Rodrigo, L., Casanovas, P., Poblet, M.: The sekt use legal case components: ontology and architecture. In: Gordon, T.B. (ed.) Legal Knowledge and Information Systems, pp. 69–77. IOS Press, Amsterdam (2004)
3. Blazquez, M., Pena-Ortiz, R., Contreras, J., Benjamins, R., Casanovas, P., Vallbe, J.J., Casellas, N.: Legal case study: Prototype. Project Deliverable D10.3.1, SEKT (2005)
4. Casanovas, P., Casellas, N., Vallbe, J.-J., Poblet, M., Benjamins, V.R., Blazquez, M., Pena-Ortiz, R., Contreras, J.: Semantic web: A legal case study. In: Davies, J., Studer, R., Warren, P. (eds.) Semantic Web Technologies: Trends and Research in Ontology-based Systems, pp. 259–280. John Wiley & Sons, Chichester (2006)
5. Casanovas, P., Casellas, N., Vallbe, J.J., Poblet, M., Blazquez, M., Contreras, J., Benjamins, V.R., Lopez-Cobo, J.-M.: After analysis: First steps of the iuriservice implementation at the spanish judicial school. Project Deliverable D10.4.1, SEKT (2007)
6. Casanovas, P., Poblet, M., Casellas, N., Contreras, J., Benjamins, V.R., Blazquez, M.: Supporting newly-appointed judges: a legal knowledge management case study. Journal of Knowledge Management 9(5), 7–27 (2005)
7. Casanovas, P., Poblet, M., Casellas, N., Vallbé, J.-J., Ramos, F., Benjamins, V.R., Blázquez, M., Contreras, J., Gorronogoitia, J.: Legal scenario. case study intelligent integrated decision support for legal professionals. Project Deliverable D10.2.1, SEKT (2005)
8. Casanovas, P., Casellas, N., Tempich, C., Vrandecic, D., Benjamins, V.R.: Opjk and diligent: ontology modeling in a distributed environment, vol. 15, pp. 171–186 (2007)

9. Benjamins, V.R., Casanovas, P., Contreras, J., López-Cobo, J.M., Lemu, L.: Iuris-ervice: An intelligent frequently asked questions system to assist newly appointed judges. In: Benjamins, V.R., et al. (eds.) Law and the Semantic Web, pp. 205–522. Springer, London (2005)

10. Casellas, N., Blazquez, M., Kiryakov, A., Casanovas, P., Poblet, M., Benjamins, R.: OPJK into PROTON: Legal domain ontology integration into an upper-level ontology. In: Meersman, A.R., et al. (eds.) OTM-WS 2005. LNCS, vol. 3762, pp. 846–855. Springer, Heidelberg (2005)

11. Davies, J., Studer, R., Warren, P. (eds.): Semantic Web Technologies: Trends and Research in Ontology-based Systems. John Wiley and Sons, Ltd., Chichester (2006)

12. Huang, Z., Schlobach, S., van Harmelen, F., Klein, M., Casellas, N., Casanovas, P.: Reasoning with multiversion ontologies: Evaluation. Project Deliverable D3.5.2, SEKT (2006)

13. Huang, Z., Stuckenschmidt, H.: Reasoning with multiversion ontologies. Project Deliverable D3.5.1, SEKT (2005)

14. Huang, Z., Stuckenschmidt, H.: Reasoning with multiversion ontologies: a temporal logic approach. In: Proceedings of the 2005 International Semantic Web Conference (2005)

15. Huang, Z., ten Teije, A., van Harmelen, F.: MORE2: An extended reasoning and management system for multi-version ontologies. Project Deliverable D3.5.3, SEKT (2007)

16. Huang, Z., Visser, C.: Extended DIG description logic interface support for PRO-LOG. Deliverable D3.4.1.2, SEKT (2004)

17. Rescher, N., Urquhart, A.: Temporal Logic. Springer, Heidelberg (1971)

18. van Benthem, J.: The Logic of Time, 2nd edn. Kluwer Academic Press, Dordrecht (1991)

19. van Benthem, J.: Temporal logic. In: Handbook of Logic in Artificial Intelligence and Logic Programming, vol. 4, pp. 241–350. Clarendon Press, Oxford (1995)

Improvements in Recall and Precision in Wolters Kluwer Spain Legal Search Engine

Angel Sancho Ferrer[1], Jose Manuel Mateo Rivero[2],
and Alejandro Mesas García[3]

Wolters Kluwer Spain,
Research & Development Department
C/ Collado Mediano, 9. Las Rozas. Madrid. Spain
{asancho,jmmateo,amesas}@wke.es

Abstract. In this paper we describe the search technology in production in Wolters Kluwer Spain for the legal research market. This technology improves the "Google like" experience by increasing both the total number of retrieved documents *(recall)* and the quality of the very best ones *(precision)* while maintaining the ease of entering a natural language query. We propose a hybrid approach, both in the working methodology and in the codification of the legal knowledge -subject matter expert and librarian-, through new layers of semantic analysis and algorithms. We improve the traditional tf-idf vector space model by creating a mixed document indexing schema of terms and concepts as well as a proprietary ranking algorithm trained by a hybrid genetic algorithm. These calculations also improve the quality of keyword-in-context.

Keywords: Legal Knowledge Representation, Information Retrieval, Semantic Indexation, Hybrid Methodologies, Genetic Algorithms, Machine Learning.

1 Introduction

Legal professionals have critical information needs, both in the quality of the retrieved results and in the value of their invested time. The traditional approach to create better legal research tools has been to provide more sophisticated options, such as search operators and advanced query forms that expose editorially created metadata and taxonomies. The Google™ search engine, however, has demonstrated that quick and simple queries can retrieve quality results.

Legal publishers have historically created value in print-based research by collecting and enriching public domain content. In the print paradigm, publishers create knowledge-management aids, such as tables of contents, topical indexes and other metadata, to indentify potentially relevant documents. Publishers prepare abstracts to help customers determine whether a document is relevant. Concordances provide convenient paths to accessing related information.

Changes in technology, however, imply the need for legal publishers to increase their value by optimizing the legal-research experience online. This optimization challenges publishers because of the need to increase scalability and efficiency of their publishing processes. But along with these challenges are opportunities to apply

P. Casanovas et al. (Eds.): Computable Models of the Law, LNAI 4884, pp. 130–145, 2008.
© Springer-Verlag Berlin Heidelberg 2008

R&D in computer science to solve legal professionals' critical research needs. Success with any new tool or media requires learning it's new capabilities (i.e. physical volume is no longer a limitation, the knowledge can be put both in metadata and in algorithms). Optimizing the online legal research experience requires learning to think outside of the traditional print-publishing paradigm (i.e. from content designed for print-based browsing stored on large bookshelves to a small, interactive computer screen).

Wolters Kluwer Spain views the search engine as one of the main tools for optimizing the online legal research experience [20]. Applying the standard of the Google search engine of retrieving relevant results with simple queries in both the consumer and legal research markets, Wolters Kluwer Spain believes that codifying and managing legal knowledge in a new way enables legal professionals to run simple queries and retrieve highly relevant results.

Our R&D project has produced a Semantic Search Engine to optimize our customers' research experience in its legal research products. This paper explains these developments in the following order: Section 2 summarizes a study of our users' research behaviour; Section 3 presents and explains the innovative features of the Semantic Legal Searcher, including the semantic processing of the index and the customers research queries, relevance review of the vector space model, the use of a Hybrid Genetic Algorithm to "train" the relevancy-ranking algorithm applied to search results, and an improved presentation of keyword-in-context; Sections 4 and 5 conclude the paper with the conclusions and an overview of future R&D efforts.

2   User's Search Behaviour

The first step to solving the "search" problem is more fundamental than applying algorithms and technology, semantic or otherwise. We simply must understand how people formulate queries in search engines to solve their research problems. Based on our experience, it is unreasonable to expect researchers to learn how to configure Boolean search options, select metadata fields for their queries, and fine-tune search strategies according to their research problems.

We thus conclude that the search engine must adapt to the user's behaviour. Let's examine the data.

2.1   Users Have Difficulty Formulating Queries

There are several challenges that end users face during query formulation.

– First, users frequently do not understand the logic of query transformation applied by a search system. Users type in a string of characters or words in a search form and then submit a query. In order to improve the effectiveness of search and increase the number of query "hits", the search system might transform the user's query, for example, by finding all forms of search term with a particular stem as well as applying "wildcards" to a search string. The problem with this logical processing of the search parameters is that the user may not understand why certain results are returned, and why other "obvious" hits are not returned [18].
– Users often do not translate what they know about the problem they need to solve into a successful query. We can identify two main reasons: (1) Users inadequately

represent their problem with too few search terms; with an inadequate number of search terms, search engines cannot provide relevant results; (2) When customers do provide a good query, they often use terms that do not match the vocabulary used in the document collection being researched.

The report "What's Wrong with Internet Searching", confirmed that nearly all users (experts and novice) have difficulty formulating good searches even when they have all needed information. In the real world, users enter a library or a shop and express their requirements in too few, too many or imprecise terms. Alternatively, they confine their research to browsing the materials being offered [16]. Users are not accustomed to creating an artificial text string that matches their requirements.

Users may not understand that searching is an iterative process, often requiring refinement of search queries and winnowing of search results. Moreover, users may equate a negative response to their search query with the non-existence of valid results [18].

Numbers of Terms Per Query
On average, our users describe their information needs with 2.84 terms per query, which is similar to the 2.4 average number of terms identified by other studies [10], [12], [19]. This number of terms per query confirms that users have problems in formulating good queries, because it is almost impossible to describe a real problem in just two or three terms. It's also significant that one out of four queries consist of only one term.

Table 1. Wolters Kluwer Spain typology of queries

Number of terms in the query	Percentage of queries
One term	25.4%
Two terms	36.9%
Three terms	22.8%
Four of more terms	14.9%

Users' queries consist of so few query terms probably because users find that long queries often produce zero results. Users probably prefer losing time navigating through long answer sets resulting from too few query terms rather than entering long queries that yield no results.

2.2 Full Text Search and Boolean Operators

Users don't know the best type of search for their needs. It's probed that users may not understand the different functions of the different input boxes.

Search logs from Wolters Kluwer Spain reveal that almost 81% of the users prefer running full-text search instead of other different type of search: topical taxonomy, table of contents, metadata filtering.

Software engineers have invested a lot of effort in making search engines more powerful. One of the key features that have created is Boolean operators. Today many search systems offer extensions of the "core" Boolean operators of "and, "or" and "not." Customers often can use sophisticated variations, such as numerical and grammatical proximity to create complex and powerful queries. The problem, as different studies reveal, is that fewer than 5% of all queries contain any Boolean operators. Moreover, those users who do use them make mistakes [19] [13].

So the question arises: why invest effort in creating more complex Boolean operators if users do not use them or use them incorrectly? For this reason, we believe that search systems evolution is to adapt to the user rather than sophisticating each option and requiring the user more time to learn and to think about the query.

3   The Semantic Legal Search Engine

This section presents the Semantic Legal Search Engine developed by Wolters Kluwer Spain for its legal research products.

The quality of the results presented by a search engine is measured by the total number of good retrieved documents *(recall)* and the quality of the very firsts ones *(precision)*. Trying to increase recall typically introduces more bad hits into the hitlist.

The Semantic Legal Search Engine is built using Apache Lucene [9] [14], which offers state-of-the-art engineering. Wolters Kluwer Spain decided to use open-source technology rather than a commercially available search engine because of its flexibility. While we find commercially available search engines excellent for normal search scenarios, we needed to modify the API's beyond what commercial search engines expose. Lucene's programming model is easy to extend. We have been able to modify the source code even for critical components as the index formats[1] or specific performance scenarios. Excellent support is available in the Lucene community and the source code analysis itself serves as the best diagnose methodology.

On top of Lucene we have developed functionalities to improve the recall and precision: a semantic processing of the documents and the queries, a relevance review algorithm trained with a Hybrid Genetic Algorithm, and an improved presentation of keyword-in-context.

3.1   Natural Language: Semantic Indexation

Most of the legal knowledge is unstructured, so a document parsing and token analysis process is needed. A search engine deals with "terms" as its dictionary symbol, and the easy way to do it is tokenizing the document into words. But humans deal with concepts, and a single concept often does not map to a single word: it can be more than one word and one word can mean more than one concept.

The main issues of words versus concepts in Natural Language are related to compounds expressions and synonym-based expansion.

Compound Expressions
Most of the concepts human have in their minds don't match with just one word but are compound expressions (i.e. Value Added Tax, duty free shop, gas station, United States of America, Supreme Court ...).

---

[1] As said, for instance in [2], "the index has become a significant area of innovation for all search companies, where much of a search engine's secret sauce is applied. Noting statistical patterns and algorithmic potentials".

At first, searching for compound terms as literal expressions seems a likely solution. But this approach presents two problems:

- Most users do not know how to create "literal" searches; even if they know how, they don't use them.
- Literal searches can exclude relevant documents. Just consider the following example: if you search for "despido improcedente" as a literal those documents that discuss "despido considerado improcedente" or "despido calificado como improcedente" won´t be retrieved even though they are discussing the same concept.

Semantic Expansion

Most of the concepts that appear in a query have a synonym. For the user a document that talks about "gas station" and "petrol station" are the same, but not for a traditional search engine. If user searches only for "gas station", what will happen with those documents that refer to "petrol station"? Many relevant documents will not be retrieved.

Again, very few users realize the need to explicitly add all desired synonyms to the search. Moreover, even if a user tries to enter synonyms, the following two problems arise:

- The user may not know how to use Boolean operators to add synonyms to the search;
- The user will not remember all the combination of words that can represent a concept.

Even if the user solves these two problems, yet another challenge arises. With the use of the correct Boolean operators, the resulting query might be so complex that the search engine could not execute the query.

To solve these previous problems, we have developed methodologies to create semantic knowledge and new algorithms.

3.1.1 Methodology to Create Semantic Knowledge

Tests of machine learning and natural-language techniques have not proved successful for editorially created products. These techniques can help to create suggestions, but today it is not possible through technology alone to map one term (and its synonyms) in a document to a set of terms in another (i.e. "mobbing", and "acoso moral en el trabajo" never will be mapped by a machine learning algorithm as the same concept). Subject-matter experts are required for identifying the concepts, compound expressions and synonyms that a search system must address.

Another challenge is that legal queries and documents consist of more than just legal terminology. Half of the concepts represented in a user's query of legal content often does not use legal terminology but instead represents factual issues (i.e. house, car accident).

Also, documents and users' queries can contain typographical errors. For example, there are Supreme Court cases talking about "desaucio" (incorrectly omitting the "h" – deshaucio), or talking about "garage" (incorrectly using two "g's" – garaje). So documents that refer to "desaucio" or to "garage" instead of "desahucio" or "garaje" will not be retrieved when users enter these words with correct spelling. The system will have indexed the documents using the incorrect spelling.

3.1.2 Semantic Analysis' Algorithms

To enable users to conduct research using Natural Language, we need to find a way for the search system to identify concepts and compound expressions but not search for stop words. Also, it's necessary to search for all the synonyms of the different concepts user has written.

All the former semantic knowledge must be managed by algorithms in different scenarios. Although synonym injection is a known technique to add phantom terms in a certain position, it has drawbacks. It is not easy to create query parsers and high-lighters with compound expressions and to be able to inform the user of the semantic applied, and it's not easy to combine different levels of semantic analysis in the in-dexation and search.

We thus have developed two analyzers: one for the indexation and the other for the query, although both must share the same semantic knowledge and basic language processing.

This semantic processing has also additional benefits over the synonym injection in the indexation and over the query cooking in the search:

- For the performance of the search, as the Boolean logic of the queries is solved at indexation time and not at runtime.
- To get better relevance, because the number of terms is reduced. As an average, each term in the query is expanded to four terms in the index, and this will increase the relevance calculations.

The query interface presented another challenge. It must be possible for the system to inform users of what it understands as they enter search terms. In other words, while the user is typing search terms, the interface must inform about the editorially created synonyms that will be included in the search.

Let's consider an example. Suppose that a customer submits the following search: *"consideración de las stock options como rentas irregulares"*. The query parser will recognize that "consideración", "de las" and "como" are stop words; and that "stock options" and "rentas irregulares" are different concepts and will inform of the syno-nyms that will be transparently included in the search (Fig. 1).

3.2 Relevance Review of the Result List

The most powerful quality of a search engine is effective relevance ranking. Simply put, relevancy is the measure of how well the retrieved document answers the query. It arranges documents based on the mathematical measurement of similarity between the query and the content of each record. This is valuable because users are still only willing to look at the first few tens of results [19].

Lucene's and all commercial search engines relevance-ranking algorithms are based in the Vector Space Model (VSM) [1]. In this model, both the user's query and each document in the collection are represented by vectors of terms. Each term is a dimension that is weighted to represent its importance in the query (Inverse Docu-ment Frequency and user assigned) and in each document (Term Frequency).

The VSM model has different classical improvements, including (1) document nor-malization to overcome the problem of excess document length negatively influencing relevancy ranking; (2) adjusting relevance ranking according to the authority of each document; (3) weighted document-zone scoring; and (4) proximity of query terms.

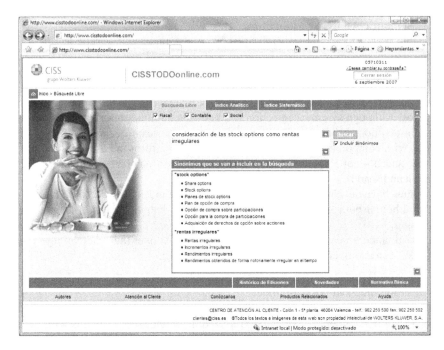

Fig. 1. Screenshot query interface: natural language recognition and synonym expansion

The relevancy-score ranking involves calculating a mathematical similarity between the vector of the query and the vector for each of the documents. For instance, Lucene has the following implementation:

$$score\,(q,d) = \sum_{t\ in\ q} tf\,(t\ \ in\ \ d)\,*\,idf\,(t)\,*\,getBosst\,(t.field\ \ in\ \ d)\,*\,lengthNorm\,(t.field\ \ in\ \ d\,*\,cord\,(q,d)\,*\,queryNorm\,(q) \tag{1}$$

The Google search engine has made a pragmatic and valuable contribution to the field of information retrieval through its use of hypertext links to documents to rank them computing the authority of the hypertext link and the relative weight, or value, of each part of the document to describe the document's content. The PageRank is based on a graph analysis of links to a document. The Fancy Hits feature examines text of links external to the document but which point to it and analyzes the semantics of HTML tags (such as titles or headings).

In a product controlled by a publisher these same concepts of determining the authority and finding good descriptors to documents can be done through subject matter experts instead of guessing patterns from big numbers of links and making adjustment to filter the spam of uncontrolled authored pages.

For example, there are 80 laws that have the words "Código Civil" in the title, so externally provided metadata are needed to know which one is the good one.

However, the difficulty of evaluating search engines, even by experts, is a recognized fact because the nature of the problem is subjective [3] [7].

3.2.1 Proprietary Relevance Algorithms

With a commercial search engine, the way to improve the users experience is *"query cooking"* [8]. This permits the use of semantic capabilities (identifying literal expressions, and inserting synonyms with the OR operator) and relevance (zones, proximity, weights), but this has limits in semantic and performance as commented in 3.1, but also in the relevance that all the previous parameters could achieve, as explained in 3.2.2.

To explain our algorithm of Relevance Review we start with the intuitive concepts that led us to think that with the out-of-the-box algorithms we were not going to be able to achieve certain goals: for instance, given a query consisting of three terms is better a document with the three terms together in the title than a document with lots of hits scattered through the document.

One key idea we use is the concept of "cluster" or "density zones". Our algorithm considers how query terms are distributed within a document, including the attribution of different weights to query terms according to the document zone in which they occur. These clusters' scores can be related to:

- Number of different query terms.
- Total number of query terms.
- The inside proximity between terms, as well as the total size of the cluster.
- Proximity to the previous cluster of query terms to avoid "jump errors" or "horizon effect".
- Whether the query terms occur in the document in the same order as they appeared in the query.
- If query terms appear at the beginning of a document, as it usually means that they are slightly more important.

Each cluster of query terms in a document receives a score, and the interrelation between clusters provides another level of scoring, combined with all the others factors.

This is somehow similar in concept with the kind of "evaluation function" of a chess position that looks for clues of patterns and pieces. As experts generate new knowledge about document-scoring, Wolters Kluwer Spain will "codify" that knowledge into its evaluation function (Fig. 2).

Cut-offs of the Solutions Space

The former described function is very heavy computationally, and in a commercial application one of the most important qualities is the performance as commented in 3.4. In addition our applications have big volumes of information (more than 3,000,000 documents). For this reason before designing the relevance algorithm we did stress tests and confirmed that time response was not satisfactory.

Traditional cut-off techniques [3] [17] are to order the documents in the inverted index not by "doc id" but by "term frequency" or its "doc value" (i.e. PageRank), with the hypothesis that the best ones will have that attribute. Based in the performance of Lucene, we have organized the relevance algorithm in two phases: a first selection of candidates is done using the standard (slightly improved) ranking criteria. The hypothesis is that in the first 400 hits are our "best 10", so then a relevance

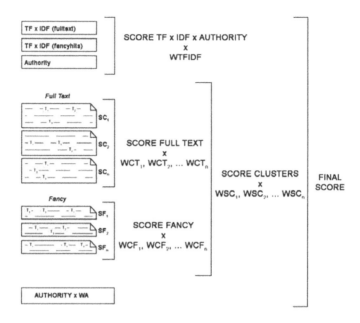

Where WTFIDF, WCT, WCF, WSC ... are different weighs to evaluate the relevance factors

Fig. 2. Relevance score ranking algorithm

review of those documents is done analyzing them in more detail. This way we do present very good response time without sacrificing relevance.

3.2.2 Training the Relevance Review with a Hybrid Genetic Algorithm

The relevance algorithm described before obtained very good results, but each improvement in the quality was more difficult. Until a certain point in which no good prediction could be done of how the tuning of one aspect will impact the others. As Google's creators said "The ranking function has many parameters like the type-weights and the type-prox-weights. Figuring out the right values for these parameters is something of a black art" [3]. This limited not only new tunings, but including new potential parameters found by the expert analysis of documents, and a limitation in our capacity to do different relevance profiles.

At this point we decided that a Genetic Algorithm (GA) approach was worth testing. GA [11] is a stochastic general method to find exact or approximate solutions to optimization and search problems. It proceeds in an iterative manner by generating new populations of individuals from the old ones. Every individual is the encoded (binary, real, etc.) version of a tentative solution. The canonical algorithm applies stochastic operators such as selection, crossover, and mutation on an initially random population in order to compute a new population. The pseudo-code of a GA can be seen in Fig. 3.

```
t:=0;
Initialize:              P(0) := {a₁(0),...,a_μ} ∈ I^μ;
Evaluate:                P(0) : {Φ(a₁(0)),...,Φ(a_μ(0)))};
While ( P(t) ≠ true ) do
        Select :         P(t): sΘ_z(P(t));
        Recombine        P'(t) := ⊗_Θ_c (P(t));
        Mutate:          P''(t) := mΘ_m(P'(t));
        Evaluate         P'''(t) := {Φ(a₁ᵐm(t)),...,Φ(a_z^m(t))};
        Replace:         P(t+1) := {r_Φ_c (P''(t)∪Q)};
        t := t+1;
endWhile;
```

Fig. 3. Pseudo-code of a Genetic Algorithm

Genetic algorithms may be crossed with various problem-specific search techniques to form a *hybrid algorithm* [5] [6] [15] that exploits the global perspective of the GA and the convergence of the problem-specific technique. Hybrid Genetic Algorithms is a population-based approach for heuristic search in optimization problems. They execute orders of magnitude faster than traditional Genetic Algorithms for some problem domains.

In our approximation we have used the heuristic to define "feasible" individuals, doing a best guess based on the experience of the previous subject matter expert works.

Individual representations. We have designed a structure of individuals in which we have a unique chromosome composed by 25 genes that represented the parameters used for the algorithm of relevance review, described in 3.2.1.

Fitness function. In our fitness function we have used one hundred queries, exactly like have been formulated by users, without any metadata, and we manually identified which laws should have been retrieved for that query. Then the objective is to maximize the number of identified norms at the firsts positions in the result list:

$$fitness = \sum_{i=0}^{|Q|} \left[g(q_i, d_i) + (\alpha - f(q_i, d_i)) * h(q_i, d_i) \right] \qquad (2)$$

$$g(q_i, d_i) = \begin{cases} w_{1-2} & \text{if} \quad f(q_i, d_i) <= 2 \\ w_{3-5} & \text{if} \quad 2 < f(q_i, d_i) <= 5 \\ w_{6-11} & \text{if} \quad 5 < f(q_i, d_i) <= 10 \\ w_{11-30} & \text{if} \quad 10 < f(q_i, d_i) <= 30 \\ 0 & \text{if} \quad 30 < f(q_i, d_i) \end{cases} \qquad (3)$$

$$h(q_i, d_i) = \begin{cases} lengPag - f(q_i, d_i) & \text{if} \quad f(q_i, d_i) <= 10 \\ 0 & \text{if} \quad 10 < f(q_i, d_i) \end{cases} \qquad (4)$$

Where $f(q_i, d_i)$ is the position of document d_i after evaluating the relevance algorithm with query q_i; n is the total number of test cases; $lengPag = 10$ (pondering that appears in the first result page [19]); $\alpha = lengPag*3$ (empirical upper limit of user navigation, as they see as much as three result pages per query).

The parameters w_{1-2}, w_{3-5}, w_{6-10} and w_{10-30} are established empirically and modified during the training process, to simulate two strategies:

- The first strategy increases the recall, defining recall in this context such as the number of document in α positions: $w_{1-2} > w_{3-5} > w_{6-10} > w_{10-30}$
- A second phase of precision in which to get good score the documents must be in the two first positions: $w_{1-2} \gg w_{3-5} \gg w_{6-10} \gg w_{10-30}$

Crossover. We sort the individuals, and crossover one of the best ones with other chosen randomly.

Selection. It has been used the roulette wheel algorithm. Also some of the best individuals of the previous generation are saved directly.

3.2.3 Precision Comparison

To verify the precision we have used a set of 250 real user's queries that look for concretes norms. We have selected a very demanding scenario of the most objective kind of search: a norm that should be retrieved in the first two positions.

We present a comparison between other commercial search engines and our Semantic Search Engine (Fig. 4). In addition we compare the relevance review after training it with the Hybrid Genetic Algorithm.

To use a good guess as a starting point has been very valuable. The human-tuned algorithm was stuck to a percentage of 74% that was difficult to overcome. Now we have reached an 88% of quality.

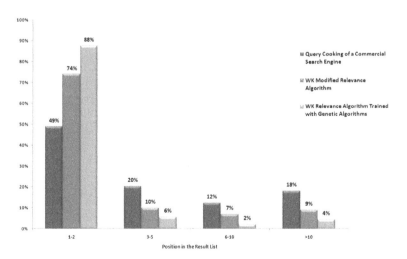

Fig. 4. Relevance algorithms improvements

Another problem we found with the commercial search engine is the "other documents of the result list". Although the 49% reflects that it can retrieve the wanted one in good positions, the surrounding ones are sometimes unpredictable while with the trained pattern they are related and as good as the known.

3.3 Keyword-in-Context

The figure of the snippets in the result list [1] [21] has not received a lot of attention, but it is an important part of the search workflow that consumes a lot of time to user. KWIC serves several purposes:

- Help to decide if a document is worthwhile reading and save the cost of clicking and waiting until is rendered. A clue of how is related to the problem described in the formulated query.
- Also, we've found that in some cases it can provide another level of response, as the answer can be contained in one sentence of the document, so the engine should be able to extract it.
- In documents without editorially created static summaries there is no other way to provide a glimpse of its contents.

The classical solution is to provide a number of words before and after some terms of the query extracted from a certain limited amount of text. The improvements we have developed are:

- The fragments are readable, usually complete sentences, not cut in a random word. We use our XML structure to obtain better fragments.
- There is no limit in the size of the document. Long laws and court cases are completely analyzed.
- The best fragment is extracted, not some parts from the start. Also assuring that all the words of the query are shown if more than one fragment is needed.
- To achieve all the former, we use the same cluster information calculated for the relevance review that has done a complete analysis of the best zones of the document (clusters) as one factor of the ranking.

For instance, in Fig. 5 can be seen that the extracted text is a sentence that seems like a human written answer, simply because the system has been able to identify a fragment of the document that is the best candidate, and fragmented it with certain logic instead of the usual ellipsis.

Fig. 5. Best fragment of the document extracted for the result list

3.4 Performance

This is a non functional requirement, but it impacts so directly in the user's experience that has been a cornerstone for each feature developed.

– It permits more trials. In the same amount of time the user can try several alternative queries.
– It equals user's time, and this means money for a professional.
– Users don't lose their track, so get an immersion on the problem.

3.5 Other Functionalities

It's interesting to mention other developments done related with search, to illustrate the kind of value propositions that a search engine can provide in the different tasks of the search workflow.

– Quick Facts. Sometimes the answer does not need a "set of documents" but a simple figure or date. For instance to the query "minimum salary in 2006" the system can present, before the result list, the required figure "the minimum salary in 2006 is: 590.23€".
– Clustering the results by Metadata (also known as parametric indexing, faceted queries or supervised clustering). A tool of associative retrieval that permits to refine the results, in a kind of dialog using one metadata as the dimension.
– Did You Mean? Around a 12.5% of the queries contain misspellings that lead to spoiled searches.

4 Conclusions

In this paper we've presented a Semantic Legal Search Engine that proposes a solution for the information retrieval problem in the legal domain, as a result of a pragmatic approach.

The benefits over the standard solutions are:

– Improvement in the recall: as an average 90% more documents are presented in the result list with the semantic expansion of the query.
– Improvement in the precision: the first two positions of the results have a 75% more of quality documents.
– Reduction of the time needed for the user to learn and to think "a good query" because of the natural language analysis of the concepts, and thus the possibility to create more complex queries.
– Reduction of the time needed to select documents of the result list because of how the keyword-in-context looks for more text and with better algorithms.

This means not only time saving but more quality of the outcome of the overall research process, as some good documents would otherwise probably never have been found by the user (not even retrieved or hidden in the result list).

These benefits have been achieved by developing the following techniques of Software Engineering and Artificial Intelligence:

- A way to codify legal knowledge both in semantic structures and in algorithms. This permits at the same time an improvement in the recall, the precision and the performance. It can be seen as layers on top of the components of a search engine: In the dictionary terms we use concepts instead of simple tokenization; in the relevance model, instead of vectors of term frequencies we analyze sets of clusters.
- A two phased relevance algorithm that uses classic relevance concepts (frequency, document authority, weighted zones) for a cut-off and applies a second in-depth review of the documents to identify new of potential arguments of relevance.
- A hybrid genetic algorithm used to train the relevance ranking algorithm.
- An improved keyword-in-context, that uses the score calculations to increase the quality of the selected text (more readable, and more representative of why that document has been considered relevant), and that doesn't sacrifice the performance of the system.

These Knowledge Engineering developments need a mixed profiled team of software engineers and legal experts. None expertise separately can solve the problem because the research continuously influences the work of the other profile. For this kind of hybrid AI solutions, the overall intelligence of the systems increases in phases, alternating the leadership by experts and by engineers, as the improvements in the relevance ranking illustrates so well. This kind of teams has challenges and also core value.

Buckland, in 1992 talked about the *expert human search assistant* [4], but a sufficient population of human expert search intermediaries is unaffordable. The challenge, therefore, is to provide automatically the kind of expert prompting that an expert human search intermediary would provide.

We cannot teach users how to search, but we can learn from them and adapt our data structures and algorithms to codify that legal knowledge.

5 Future Works

One surprising thing has been to realize how much can still be done in search, despite its both innocent and hyped look. The more we work, the more ideas seem to appear as a result of the new steps achieved. We think that there are two reasons for this.

- Search is a basic need of any intelligent activity: for any decision, the rules, the previous experiences and the advices of experts (law, sentences, authors, document analysts), must be obtained to know the alternatives and to evaluate them, and the heuristics to process that in the most efficient way.
- In Wolters Kluwer we have the advantage of having control on all the pieces in the production of knowledge tools: from the structure of the documents, to expert resources to add additional metadata and semantic knowledge where needed, and the source code to create new algorithms. All this resources are very strongly focused because of the feedback that provides a real market with very concrete and valuable problems to solve.

Our future research ranges from semantic improvements to create better queries (suggesting terms that can refine or expand with ontologies), all kinds of Associative

Retrieval features (suggesting documents not retrieved in the first page or even in the whole set but that can be also relevant), or collaborative filtering techniques to personalize to one user and to learn from the behavior of the expert community we have.

What we foresee as the future of the research tools is more post-processing work of the search engine as we are able to add more knowledge codified in semantic structures and in algorithms that understand the documents and the research process.

Acknowledgments

This project was initiated at the Centro de Estudios Garrigues under the responsibility of Angel Bizcarrondo and Antonio Ortega, and was finished thanks to the strong support of Rosalina Díaz Valcárcel, Chief Publishing Officer of Wolters Kluwer Spain.

Part of this research has been carried out in the context of the PROFIT projects: FIT-150500-2002-135 and FIT-350100-2007-161, funded by the Spanish Ministry of Industry, Tourism and Trade.

References

1. Baeza-Yates, R., Ribeiro-Neto, B.: Modern Information Retrieval. Addison Wesley, Reading (1999)
2. Battelle, J.: The Search: How Google and Its Rivals Rewrote the Rules of Business and Transformed Our Culture. Penguin Group (2005)
3. Brin, S., Page, L.: The Anatomy of a Large Hypertextual Web Search Engine. Computer Science Department, Stanford University (1998)
4. Buckland, M., Chen, A., Chen, H.M., Youngin, K., Lam, B., Larson, R., Norgard, B., Purat, J.: Mapping Entry Vocabulary to Unfamiliar Metadata Vocabularies (1999)
5. Cano, J.R., Herrera, F., Lozano, M. (2006). A Study on the Combination of Evolutionary Algorithms and Stratified Strategies for Training Set Selection in Data Mining. Advances in Soft Computing Series, pp. 271–284. Springer, Heidelberg (2005)
6. Cotta, C., Alba, E., Troya, J.M., Schoenauer, M.: Utilising Dynastically Optimal Forma Recombination in Hybrid Genetic Algorithms. In: Bäck, T., Eiben, A.E., Schoenauer, M., Schwefel, H.-P. (eds.) Parallel Problem Solving from Nature V. LNCS, pp. 305–314. Springer, Berlin (1998)
7. Cutting, D.:
   http://www.theserverside.com/tt/talks/videos/DougCutting/int
   erview.tss?bandwidth=dsl
8. Gelotte, K.: The Art and Science of Query Cooking. In: New Idea Engineering (2005), http://www.ideaeng.com/pub/entsrch/v2n7/article01.html
9. Gospodnetic, O., Hatcher, E.: Lucene in Action. Manning Publications (2005)
10. Hallerman, D.: Search engine marketing: search users and usage. In: eMarketer (2005)
11. Holland, J.H.: Adaptation in Natural and Artificial Systems. The University of Michigan Press, Ann Arbor (1975)
12. Jansen, B.J., Spink, A.: How are we searching the World Wide Web? A comparison of nine search engine transaction logs. In: Information Processing and Management (2004)
13. Jansen, B.J., Spink, A., Bateman, J., Saracevic, T.: Real life information retrieval: a study of user queries on the Web. SIGIR Forum 32(1), 5–17 (1998)

14. Lucene: `http://lucene.apache.org/`
15. Peláez, J.I., Mesas, A., Pelta, D.: Soft Computing based decision support system: Two Prototypes for Combinatorial optimization problems. In: Fourth Conference of the European Society for Fuzzy Logic and Tecnology (EUSFLAT 2005) (2005)
16. Pollock, A., Hockley, A.: What's Wrong with Internet Searching. At: Designing for the Web: Empirical Studies (october 30, 1996 Microsoft Corporate Headquarters)
17. Singitham,, Pavan Kumar, C., Mahabhashyam, M.S., Raghavan, P.: Efficiency-quality tradeoffs for vector score aggregation. In: Proc. VLDB, pp. 624–635 (2004)
18. Sisson, D.: Assumptions About User Search Behaviour (1998-2002)
19. Spink, A., Wolfram, D., Jansen, B.J., Saracevic, T.: Searching the Web: the public and their queries. Journal of the American society for information science and technology (February 1, 2001)
20. Susskind, R.: Transforming the Law. Essays on Technology, Justice and the Legal marketplace. Oxford University Press, Oxford (2000)
21. Turpin, A., Tsegay, Y., Hawking, D., Williams, H.E.: Fast generation of result snippets in web search. In: Proc. SIGIR, pp. 127–134. ACM Press, New York (2007)

Three Senses of "Argument"

Adam Z. Wyner, Trevor J.M. Bench-Capon, and Katie Atkinson

Department of Computer Science
University of Liverpool
Ashton Building, Ashton Street
Liverpool, L69 3BX, United Kingdom
{azwyner,tbc,katie}@csc.liv.ac.uk
http://www.csc.liv.ac.uk

Abstract. In AI approaches to argumentation, different senses of argument are often conflated. We propose a three-level distinction between arguments, cases, and debates. This allows us to modularise issues into separate levels and identify systematic relations between levels. Arguments, comprised of rules, facts, and a claim, are the basic units; they instantiate argument schemes; they have no sub-arguments. Cases are sets of arguments supporting a claim. Debates are sets of arguments in an attack relation; they include cases for and against a particular claim. Critical questions, which are characteristic of the particular argument schemes, are used to determine the attack relation between arguments. In a debate, rankings on arguments or argument relations are given as components based on features of argument schemes. Our analysis clarifies the role and contribution of distinct approaches in the construction of rational debate. It identifies the source of properties used for evaluating the status of arguments in Argumentation Frameworks.

Keywords: argumentation, argument, case, debate.

1   Introduction

A central concern of AI and Law is the modelling of legal argument. In AI we find a number of approaches to argumentation and argument. Some approaches represent arguments as trees or graphs (e.g. [1], [2], and [3]), some are highly concerned with the structure of arguments (e.g. [4]) and the way arguments support one another (e.g. [5]). From informal logic we have the notion of argument schemes (e.g. [6]), while much of the more formal work has taken place in the context of abstract argumentation frameworks (e.g. [7]). With this variety of approaches it is important to determine the relations between them, and in particular to avoid conflation of distinct ideas. To this end we will, in this paper, explore three different senses of the word "argument", all of which are represented in the previous work mentioned above, in order to give a clear characterisation of what may be intended by argument, and to identify the appropriate role of various senses in argumentation as a whole.

The Oxford English Dictionary lists seven senses of the word "argument", of which three will concern us in this paper. We begin by giving the definitions

P. Casanovas et al. (Eds.): Computable Models of the Law, LNAI 4884, pp. 146–161, 2008.
© Springer-Verlag Berlin Heidelberg 2008

below: although these are senses 3a, 4 and 5 in the OED, we will introduce our own numbering for clarity. In Sense 1 an argument is a self-contained entity, a reason for a conclusion.

> **Sense 1:** "3. a. A statement or fact advanced for the purpose of influencing the mind; a reason urged in support of a proposition."

Thus we can see an argument in Sense 1 as a pair <reason, conclusion>, which makes no reference to any other arguments. This is quite a common use in AI and elsewhere: Toulmin's scheme [8], as originally presented, was "stand alone" in the sense that it made no reference to the grounds on which the reasons were believed, nor the uses to which the claim might be put. The arguments based on the many schemes found in [6] share this feature. Most common of all in AI are arguments of the form "Q because P" representing the application of a single (defeasible) rule. In law this is akin to a single point made within a case.

In the second sense, reference is made to where the reasons come from:

> **Sense 2:** "4. A connected series of statements or reasons intended to establish a position (and, hence, to refute the opposite); a process of reasoning; argumentation."

In Sense 2 we move beyond a single step of reasoning, giving grounds for the reasons advanced for the conclusion. An argument in Sense 2 may be seen as a chain of reasons, reasons for reasons. In AI this can appear as a proof tree, as with the typical "how" explanation of a rule based expert system, and is a commonly used notion of argument in work such as [9] when an "argument" has sub-arguments: e.g. "P → Q" and "Q → R" are sub-arguments of the argument "P, P → Q, Q → R, so R" where "→" is some kind of, possibly defeasible, implication. In law this may be seen as the whole case to be presented for a particular party.

The third sense relates arguments in the previous senses:

> **Sense 3:** "5. a. Statement of the reasons for and against a proposition; discussion of a question; debate."

In Sense 3 we have the possibility of conflict: we have reasons *against* as well as *for*, the proposition, and we may have multiple arguments in the preceding two senses on both sides. In AI this corresponds more to an argumentation framework in the sense introduced by [7]. In law it corresponds to the whole of a suit with all the arguments for both parties and perhaps also the adjudication of a judge.[1]

---

[1] In AI sometimes "argumentation" is used rather that "argument": in fact no distinction between these terms is reflected in the definitions given in the OED. There are senses of "argumentation" corresponding to each of the senses of "argument" discussed above. Differences seem to be in connotation: "argumentation" is sometimes used pejoratively, and sometimes is intended to convey a sense of process, the putting forward of arguments.

In this paper, we shall distinguish between these three senses of argument. In the following we will refer to Sense 1 as an *argument*: we shall always here mean an argument which cannot be divided into sub-arguments. For Sense 2, a collection of arguments advocating a particular point of view, we shall use the term *case*. This picks up on phrases such as "the case for the prosecution", but should not be confused with the whole of a case as mentioned above.[2] Rather, for a collection of arguments for and against a point of view, we shall use the term *debate*.

In distinguishing the three senses, we also relate them. Arguments are *parts* of cases, and a case is a *part* of a debate. Furthermore, changes in one of the parts may induce a change in another, as we shall see.

Before proceeding further, we should mention, for purposes of comparison, Prakken's well-known four layer model of argumentation [10]. He distinguishes a *logic* layer, which is concerned with arguments and is where questions such as whether the argument is sound can be posed. Prakken, however, does not distinguish between Senses 1 and 2, and so both arguments and cases may emerge from the logic layer. Next there is a *dialectical* layer, which examines conflicts between the arguments/cases identified in the logic layer. This layer corresponds to what we are terming debate, and it is intended to resolve conflicts between the arguments/cases identified. Next there is a *procedural* layer, which controls the conduct of the dispute, how arguments can be introduced and challenged. Finally, there is a *strategic* layer: while the procedural layer controls what it is possible or legal to do, the strategic layer determines what it is advisable to do. In what follows we will be concerned only with the logical and dialectical layers.

In Section 2, we present arguments as the basic unit. However, arguments do have parts, which are specified by the argument schemes which they instantiate; for instance, all arguments have claims, which are the propositions that hold if the arguments succeed. The key notion is that arguments do not have other arguments as parts. In Section 3, critical questions are presented as a means to establish attack relations between arguments; given an argument scheme and a critical question associated with it, an affirmative answer to the question implies that another argument attacks the argument and indicates how the first attacks the second. Given arguments and attack relations, we move to the level of debates in Section 4, where sets of arguments are provided for and against a particular claim. Different sets of arguments are derived from different attack relations; in turn, the attack relations depend on the critical questions and the argument schemes that have been instantiated. In Section 5, we discuss abduction in Argumentation Frameworks. We present cases in Section 6 in terms of admissible sets in an Argumentation Framework, for a case is a set of arguments that collectively supports a particular claim in a debate. We discuss the role of evaluation metrics such as preference or value rankings in Section 7; the rankings use properties that come from particular argument schemes and have consequences for properties of sets of arguments at the level of the Argumentation Framework.

---

[2] In AI and Law, a *case* usually refers to all aspects of what we have previously referred to as a *suit* and which has been brought before a court.

2   Arguments

In order to generate some arguments, we will need some facts and some means of inferring conclusions from those facts. We will use as a starting point a very simple knowledge base, KB1, comprising four defeasible rules and three facts, from which we can generate a standard form of argument: *P and if P then Q, so Q*. The facts F1-F3 and rules R1-R4 of KB1 are:

R1: $P \rightarrow Q$
R2: $Q \rightarrow R$
R3: $S \rightarrow \neg Q$
R4: $T \rightarrow \neg R$
F1: P
F2: S
F3: T

We begin by forming arguments by applying the available rules to the available facts. Each of the facts is the antecedent of a rule, and so we get three arguments:

A1: F1, R1 so Q
A2: F2, R3 so $\neg$Q
A3: F3, R4 so $\neg$R

Note that A1 and A2 have conflicting claims. This is not unusual: it simply means that we have a reason to believe Q, and a reason to disbelieve Q: we are not saying that the claims of all the arguments are true, only that we have a reason to think they may be. We expect such conflicts to appear in the logic level of argumentation: it is the role of the dialectical layer to resolve them. In our terms, such conflicts open up the possibility of debate. Of course, it needs to be ensured at the dialectical level that arguments with conflicting claims are not co-tenable.

Since we have obtained Q using A1 and Q is itself the antecedent of rule R2, perhaps we can add an argument:

A4: Q, R2, so R

Alternatively we might want to reflect that Q was derived as the conclusion of A1 and so include A1 as a sub-argument.

C1: A1, R2, so R.

Note that C1 is, in our terms, a case and not an argument: it contains A1 as a sub-argument. It is a chain of arguments for R, and so what we call a case. A difference between these approaches emerges if we add another rule and fact to KB1 to get KB2:

R5: U → Q
F4: U

Now we have a second argument for Q:

A5: F4, R5, so Q

Now A4 still applies, so we get no extra argument for R, but using the approach with sub-arguments we would get a second case for R:

C2: A5, R2, so R

Although the production of such cases is very natural in AI, in which the chaining of rules is standard practice, and although these cases (i.e. arguments with sub-arguments) have been termed arguments in a number of common approaches ([4] and [9]), we will restrict ourselves for the time being to strict arguments in Sense 1. This gives us the ability to individuate arguments and cases: we have one argument for R, but two distinct cases for it.

We see arguments in Sense 1 as the *instantiation of an argument scheme*. So far, we have used two argument schemes, AS1 and (implicitly) AS2:

AS1: Defeasible Modus Ponens
    Data: Type: Fact | Conjunction of Facts
    Warrant: Type: Rule with Data as antecedent
    Claim: Type: Fact, namely the consequent of Warrant.

AS2: Argument by Assertion
    Data: Type: Fact
    Claim: Type: Fact, namely Data

Now A1-5 are all instantiations of AS1: instantiating AS2 gives us four more arguments:

A6: P, so P
A7: S, so S
A8: T, so T
A9: U, so U

While in Sense 1, arguments do not have sub-arguments, arguments nonetheless have *parts*, as indicated by the argument schemes. Among the parts of an argument we have Data, Warrant, and Claim, and other argument schemes may have other parts.

We have now identified all the arguments that can be generated from KB2. All these arguments are sound in that they are valid instantiations of our permitted argument schemes. Our argument schemes do not allow the production of cases such as C1 and C2: that would require a scheme which allowed an argument to act as Data like a Fact. We do not want to allow this, since our conception of argument (Sense 1) does not permit arguments to be parts of one other.

3   Critical Questions

Having identified the arguments, we will now wish to identify relations between them. In particular we need to identify which arguments attack one another. As noted above, A1 and A2 are in mutual conflict because the claim of one negates the claim of the other; we say that A1 *attacks* A2 (and vice versa). This is but one way that one argument can attack another. In order to make our identification of attacks systematic, we will draw on the notion of *critical questions*, taken from informal logic. In [6] each argument scheme is associated with a characteristic set of critical questions. Let us suppose an argument A which instantiates a scheme and with respect to which we ask a critical question. An affirmative answer to the question implies an argument which is the instantiation of some scheme and which is in some conflict with our initial argument A. As we remark below, there are several ways the conflict can arise. So what are the critical questions in our example?

For AS2, the only possibility is that we deny the premise and conclusion, which are of course, the same for this scheme. Thus:

AS2CQ1: Have we reason to believe the premise/claim is false?

If there is an argument A which instantiates AS2 and the answer to this question is "Yes", then there will be another argument B which instantiates some argument scheme in conflict with A. Thus, we have two arguments A and B which we say attack one another, for they make claims which are in conflict.

For AS1 we would expect to have three critical questions corresponding to the standard kinds of attack found in the literature, namely premise defeat, undercut and rebut. AS1, however, cannot be undercut by an argument instantiation, since the claim is always a fact, not a rule, and so we cannot infer that a rule is inapplicable. Accordingly we modify AS1 to AS3:

AS3: Defeasible Modus Ponens with undercut
    Data: Type: Fact | Conjunction of Facts
    Warrant: Type: Rule with Data as antecedent
    Claim: Type: Fact | Rule, namely the consequent of Warrant

This gives the following three critical questions.

AS3CQ1: Have we reason to believe the data is false?
AS3CQ2: Have we reason to believe the warrant does not apply?
AS3CQ3: Have we reason to believe the claim is false?

Thus if an argument has a claim which is the negation of the data, or the warrant, or the claim of an instantiation of AS3, it will correspondingly attack that instantiation. Note that AS3CQ3 gives rise to a symmetric attack, the others to asymmetric attacks. The use of these critical questions thus allows us to determine which of our arguments are in conflict.

Having discussed arguments and their relationships, we can move the discussion to the level of debates, for which we will use argumentation frameworks. There we consider the arguments only in terms of the relationships we have determined hold between them, namely attack. After having discussed debates, we return to discuss the cases, which we define as part of a debate.

4   Argumentation Frameworks and Debates

For our dialectical layer we will use Dung's Argumentation Framework (AF), introduced in [7]. In an AF, we have arguments in attack relations. We recall some key notions of that framework.

Definition 1. *Definition 1 An argument system is a pair $AF = <X,A>$ in which X is a set of arguments and A is the attack relationship for AF. Unless otherwise stated, X is assumed to be finite, and A comprises a set of ordered pairs of distinct arguments. A pair $<x, y>$ is referred to as "x attacks (or is an attacker of) y" or "y is attacked by x".*

For R, S subsets of arguments in the system AF we say that:

a. *$s \in S$ is attacked by R if there is some $r \in R$ such that $<r, s> \in A$.*
b. *$x \in X$ is acceptable with respect to S if for every $y \in X$ that attacks x there is some $z \in S$ that attacks y. In this case, we say that z defends x.*
c. *S is conflict-free if no argument in S is attacked by any other argument in S.*
d. *A conflict-free set S is admissible if every argument in S is acceptable with respect to S.*
e. *S is a preferred extension if it is a maximal (with respect to set inclusion) admissible set.*
f. *S is a stable extension if S is conflict free and every argument y, $\neg(y \in S)$, is attacked by some $x \in S$.*
g. *S is a complete extension if S is a subset of X, S is admissible, and each argument which is defended by S is in S.*
h. *S is a grounded extension if it is the least (wrt set inclusion) complete extension.*
i. *An argument x is credulously accepted if there is some preferred extension containing it; x is sceptically accepted if it is a member of every preferred extension.*

Dung specifically states that arguments are abstract, and that attack is the only relation between them. This in part motivates our desire to exclude cases, in which arguments are related to other arguments, from the dialectical layer.

As discussed above, we can use our argument schemes and critical questions to identify the sets X and A. So, what is the argumentation framework, AF2, corresponding to KB2?

X is the set of all arguments generated in the previous section: {A1, A2, A3, A4, A5, A6, A7, A8, A9}.

Using AS3CQ3, we can see A1 and A2 are in conflict, since the claim of one is the negation of the claim of the other. Next AS3CQ1 shows that A2 must attack A4, since the claim of A2 negates a premise of A4. Applying these two principles gives us the attack relation: {<A1,A2>, <A2,A1>, <A2,A4>, <A3,A4>, <A4,A3>, <A2,A5>, <A5,A2>}. A graphical representation of AF2 is given in Figure 1: here, to help understanding of the diagram, we label arguments with their claim as well as their name, even though strictly the claims are abstracted away with the rest of the structure when we form an AF.

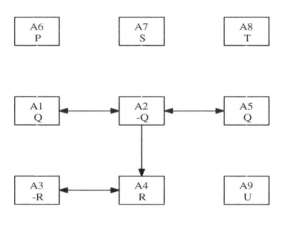

Fig. 1. AF2

The grounded extension is {A6,A7,A8,A9}, which is rather disappointing since it contains only the given facts. We have a number of preferred extensions, giving possible sets of inferences from these facts:

{A1,A3,A5,A6,A7,A8,A9}
{A1,A4,A5,A6,A7,A8,A9}
{A2,A3,A6,A7,A8,A9}

These extensions allow us, therefore, to accept any of the arguments credulously, but only the arguments from assertion sceptically. This is, of course, not very useful, and so we often want to use some notion of priority between arguments. This is often based on a notion of priority between the rules on which they are based. For example we might say R5 > R3 > R1. The effect of this is to break the symmetry of the attack relation between arguments with the same conclusion: thus from KB1, A2 would now defeat A1, but the additional rule, R5, in KB2 means that in AF2 the attacks <A1, A2> and <A2, A5> are both removed so that A2 is defeated. We would still then need to decide the priority between A3 and A4. Note again that we have to resort back to the logical level to identify the rules and their priorities.

To illustrate undercutting, suppose we extend KB2 to KB3 by adding:

R6: $U \rightarrow \neg R2$ (i.e. $U \rightarrow \neg (Q \rightarrow R)$)

Now we can extend AF2 to AF3 by adding an extra argument which instantiates AS3:

A10: F4, R6, so $\neg R2$

A10 attacks A4 (by undercut), but not vice versa, so <A10,A4> is added to the attack relation of AF3.

5   Another Argument Scheme

The above discussion used two argument schemes. There is, however, no reason to limit ourselves as to the sorts of arguments we can generate. For example, let us consider KB4, which is KB2 but with F1 and F4 replaced by F5, namely R. Using the argument schemes AS1-3, we can show arguments A2, A3, A7, A8 and A9 and, using argument by assertion,

A11: R, so R.

Suppose we now introduce an additional argument scheme:

AS4: Argument from Abduction
    Data: Type: Fact
    Warrant: Type: Rule with Data as consequent
    Claim: Type: Fact: the antecedent of Warrant

This enables us to produce the following arguments:

A12: F5, R2, so Q
A13: Q, R1, so P
A14: Q, R5, so U

Like any argument scheme, AS4 will need its characteristic critical questions. For this scheme we need to consider not only the usual notions of premise defeat, undercut and explanation, but also the possibility of there being a competing, perhaps better, explanation of the claim. It is part of the notion of arguing by abduction that the justification for abducing the antecedent is that it represents the best explanation of the consequent. Here P and U are competing explanations for Q. We assume that two abductive arguments conflict when they have the *same data*, since we cannot reuse the explanation. This is an important point: determining whether arguments attack one another depends crucially on the argument scheme which they instantiate.

We therefore have four critical questions:

AS4CQ1: Have we reason to believe the data is false?
AS4CQ2: Have we reason to believe the warrant does not apply?
AS4CQ3: Have we reason to believe the claim is false?
AS4CQ4: Is there another explanation of the data?

Thus, instantiations of AS4 are attacked by arguments with the same data as well as the attacks applicable to AS3.

Now we can organize this into an argument framework AF4. The set of arguments is now {A2, A3, A7, A8, A11, A12, A13, A14}. What of the attacks? A3 and A11 are in mutual conflict, as are A2 and A12. But now using AS4CQ4 we can see that A13 and A14 are in conflict. Additionally if A3 is accepted, by AS4CQ1 A12 must fail, since the abductive premise fails. Similarly A2 attacks A13 and A14, using AS4CQ1.

Thus attacks = {<A2, A12>, <A12, A2>, <A3, A11>, <A11, A3>, <A13, A14>, <A14, A13>, <A3, A12>, <A2, A13>, <A2, A14>}

We can show the resulting AF4 in Figure 2.

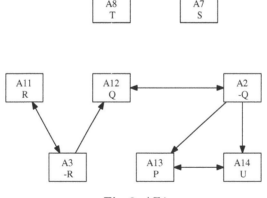

Fig. 2. AF4

Preferred extensions of AF4 are:

{A7, A8, A11, A12, A13}
{A7, A8, A11, A12, A14}
{A7, A8, A11, A2}
{A7, A8, A3, A2}

We will leave for later consideration how we might choose between these preferred extensions.

A further possibility is that we might think that there may be another explanation of the claim of an instantiation of AS4, even if we dont know what it is:

AS4CQ5: Might there be another explanation?

A positive answer to this critical question instantiates AS5. Note that AS5 is only used to critique an argument instantiating AS4. Suppose we refer to the critiqued argument as 'Arg':

AS5: Argument from Unknown Explanation
    Data: Type: Fact: Data of Arg
    Claim: Type: Fact: Negation of claim of Arg

Note that AS5 is not legitimate if we believe that our knowledge of possible explanations is complete. This gives two critical questions:

AS5CQ1: Do we have an independent reason to believe Claim?
AS5CQ2: Is our knowledge of the explanations for Claim complete?

Supposing our knowledge of possible explanations for, say U, were complete, we would have an argument for instantiating AS6CQ2:

A15: Our knowledge of explanation of U is complete.

Here, however, we will not make any such closed world assumption. Now, applying AS5 to KB4 gives A16-18.

A16: ¬Q since there may be an unknown explanation for R
A17: ¬P since there may be an unknown explanation for Q
A18: ¬U since there may be an unknown explanation for Q

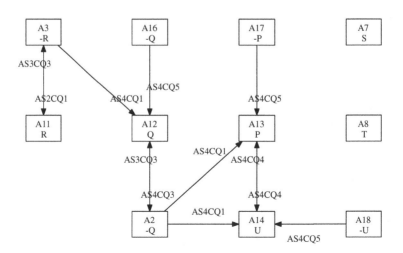

Fig. 3. AF4a

We can usefully label the arcs in the framework with the critical questions. If we add A16-A18 to AF4 we get AF4a as shown in Figure 3.

6   Cases

We now need to return to the notion of a case. Recall that we decided to admit only arguments without sub-arguments into our framework, thus precluding the possibility of representing support for an argument as sub-argument. Also we want to stay within Dung's original intentions, and so do not wish to include an additional relation to show support, as is done, for example, in [5]. We can, nevertheless, obtain a clear notion of support, and hence of arguments in Sense 2, by considering admissible sets. An admissible set is conflict free and able to defend itself against attackers. This means that a given argument in the admissible set which is attacked will have defenders in the admissible set. Moreover if these defenders have attackers, they too will have defenders in the admissible set. Thus the minimal admissible set containing a given argument will contain all the arguments needed to make that given argument part of an admissible set. It is in this way that we can express the notion of support while staying within Dung's framework, as originally specified. Consider, as an example, A13 in AF4 above. This argument appears in only one preferred extension: {A7, A8, A11, A12, A13}. A12 is needed to defend A13 against A2, and A11 is needed to defend A12 against A3. A7 and A8 are included only to make the extension maximal. Thus the minimal admissible set containing A13 is {A11, A12, A13}. Thus we can say that A13 is supported by A11 and A12, and that these three arguments form the case for the claim of A13, P. This would make the case something like "P is the best explanation of Q, which is the best explanation of R, which is known to hold." Had we adopted the sub-argument approach we would have had

C3: A11, A12, R1, so P

showing the connection between chains of arguments and admissible sets. Note, however, that on this notion of case, A2 is not supported by A7, which would, as being the datum required to infer ¬Q using A2, often be thought to be a sub-argument of A2. We argue that we should not see A7 as supporting A2, because this aspect of A2 is not in question, the only attack on A2 coming from A12, which is a rebuttal, not a premise defeat. In other words, A7 is accepted without question, and so its claim can be presumed in any argument that requires it, meaning that the argument stands in no need of support in this respect. Of course, if the logic level had in fact generated an argument with claim ¬S, we would have an argument attacking the datum of A2, but that argument would itself be attacked by A7. In that case A7 would be required to admit A2 into an admissible set, and so would be regarded as supporting it. We feel that this notion of support, which only calls in potential supporters if they are required, is sharper than notions which attempt to identify all potential supporters at the logical level and without regard to their supporting role in a debate.

7   Evaluation of Arguments

When discussing AF2 and AF4, we used the standard notion of evaluating the argumentation framework in which all arguments have equal weight, and all attackers succeed, and where we calculate the grounded, preferred or stable extensions, according to our semantic preferences. Yet, as noted earlier, we may have multiple preferred extensions which we want to differentiate; we want to have some principled reason to choose between them.

The usual method of distinguishing between multiple preferred extensions, and so provide a reason to choose between them, is to ascribe some property to the arguments representing their strength, and to require an attacker to be at least as strong as the attacked argument if the attack is to succeed. In virtue of these more fine-grained attacks, we can distinguish among previously undistinguished preferred extensions. For example [11] use preferences in this way, and [12] uses the notion of value (the social interest promoted by the acceptance of an argument) to determine the relative strength of pairs of arguments. But where do these properties come from?

The answer must be that they come from the argument schemes instantiated to produce the arguments in the framework. At the very least therefore the arguments can be ascribed the property of being instantiations of a particular argument scheme. This in turn means that we could apply a preference order to schemes: for example we might rate Argument from Assertion most highly, since this requires a known fact in the database, then Defeasible Modus Ponens, then Abduction. Or we could choose a different order if we desired. The general idea is that the arguments can be ascribed properties, these properties can be ranked, and this ranking is used in determining the status of arguments in the framework. Note that although the schemes determine which properties can be ascribed to the arguments, the ranking is produced independently, and that different rankings may be applied to the framework for different purposes or by different audiences.

If we use different argument schemes, we may be able to ascribe a wider range of properties. Three examples are:

- One well known argument scheme is Argument from Authority (e.g. [6]). In order to instantiate this scheme an authority must be identified. All arguments instantiating this scheme therefore will have the property of being endorsed by some particular authority. If we have several competing authorities, we can use a ranking of confidence in these authorities to determine the strength of arguments.
- In [13] an argument scheme for practical reasoning is proposed. In this scheme the social value promoted by acceptance of the argument has to be identified in order to instantiate the scheme. This allows arguments from this scheme to be labelled with these values, which in turn means that the resulting framework can be regarded as a Value-Based Argumentation Framework [12], and evaluated according to a particular audiences ranking of the values.
- Work on case-based reasoning in law such as [14], effectively identifies a set of argument schemes and critical questions tailored to reasoning with legal

precedents. Each of these argument schemes is related to the citation of a legal decision, and so comes with information such as the date of the case, the jurisdiction in which it was decided, and the level of court which made the decision. All of these things represent useful properties of argument which can feed into the evaluation of the status of arguments when they are formed into a framework.

Properties of arguments will not, however, suffice for AF4a. The use of Argument Scheme AS5 means that any abductive argument will have an attacker. If attacks always succeeded, this means that we simply could not use abductive arguments. The implication is that we need to provide some way for attacks to fail independent of any attack. One obvious strategy is to use the labels on the attacks. For example it might be that one considered that AS4CQ5 should not defeat the argument it attacks, unless that argument is attacked by some other argument and then succeeds whatever the relative merits of the other argument. Thus in AF4a, none of the abductive arguments will succeed, because they have independent attackers. But suppose we did not have the fact that S, so that A2 no longer can be made. Now if we accept A11 to defeat the other attacker of A12, we will accept A12. A13 and A14 are, however, still defeated since they mutually attack, as well as being attacked using AS4CQ5. This seems reasonable, since we do not have another explanation of R, but P and U are competing explanations for Q, and we have no reasons given for preferring one to another.

There are two important points to note here. First, the properties of arguments can play an important role in deciding the status of arguments in an argumentation framework, since they can form the basis for rational choice between competing preferred extensions. Second, the properties ascribed to arguments in the AF need to have their origin in the argument schemes which ground the arguments in the framework. The schemes used will thus determine the properties which are available at the framework level.

8   Conclusion

In this paper we have attempted to make clear distinctions between three senses in which "argument" may be used, and which can sometimes appear to be conflated in work on argumentation.

First we have the level of the atomic argument. For us this is an instantiation of an argument scheme, and cannot be divided into any constituent parts which are themselves arguments. There is a wide variety of argument schemes found in the literature: the choice of which schemes to use will depend on the nature of the application since different schemes are appropriate for legal, practical, scientific, mathematical and evidential reasoning. These schemes have associated with them critical questions, and various arguments will form the basis of these questions posed against other arguments. This provides a principled basis for deciding which arguments are in conflict, and whether the conflict is symmetric or not. Also the different critical questions permit attacks to be labeled according to the question being posed. Finally particular schemes will permit the ascription of properties to these arguments.

The above allows us to form the arguments into an argumentation framework, which represents the notion of argument as debate, sets of reasons for and against particular propositions. At this level it is possible to evaluate arguments to form a view as to which should be accepted and which should be rejected. Where suitable argument schemes have been used, properties of arguments and attacks can be used to inform the evaluation, according to rankings of these properties.

From debates, we can extract a case, a set of supporting arguments for a particular point of view, in terms of a minimal admissible set taken from the framework.

We believe that it is important to maintain a distinction between these three senses. Moreover we can see that our separation shows clearly the links between them. An argumentation framework is independent of the argument schemes used to form it. The properties of arguments do depend on the schemes used, and so some evaluations will be possible only if the arguments instantiate particular argument schemes. The notion of support is derived from the status of arguments in the framework level, rather than being identified at the logic level and thus is dependent on the method of evaluation for the framework. We believe that these important distinctions will play an important role in clarifying approaches to modelling legal argument.

Acknowledgments

During the writing of this paper, the first author was supported by the Estrella Project (The European project for Standardized Transparent Representations in order to Extend Legal Accessibility (Estrella, IST-2004-027655)).

References

1. Reed, C., Rowe, G.: Araucaria: Software for argument analysis, diagramming and representation. International Journal on Artificial Intelligence Tools 13(4), 961–980 (2004)
2. Hunter, A.: Making argumentation more believable. In: [15], pp. 269–274
3. Hunter, A.: Towards higher impact argumentation. In [15], pp. 275–280
4. Caminada, M., Amgoud, L.: An axiomatic account of formal argumentation. In: Veloso, M.M., Kambhampati, S. (eds.) Proceedings of The Twentieth National Conference on Artificial Intelligence, Pittsburgh, Pennsylvania, USA, July 9-13, 2005, pp. 608–613. AAAI Press/The MIT Press, Cambridge (2005)
5. Cayrol, C., Lagasquie-Schiex, M.C.: On the acceptability of arguments in bipolar argumentation frameworks. In: Godo, L. (ed.) ECSQARU 2005. LNCS (LNAI), vol. 3571, pp. 378–389. Springer, Heidelberg (2005)
6. Walton, D.: Argumentation Schemes for Presumptive Reasoning. Erlbaum, Mahwah (1996)
7. Dung, P.M.: On the acceptability of arguments and its fundamental role in nonmonotonic reasoning, logic programming and n-person games. Artificial Intelligence 77, 321–358 (1995)

8. Toulmin, S.: The Uses of Argument. Cambridge University Press, Cambridge (1958)
9. Pollock, J.: Defensible reasoning with variable degrees of justification. Artificial Intelligence 133, 233–282 (2001)
10. Prakken, H.: Logical Tools for Modelling Legal Argument. Kluwer Academic Publishers, Dordrecht (1997)
11. Amgoud, L., Cayrol, C.: Inferring from inconsistency in preference-based argumentation frameworks. J. Autom. Reason 29(2), 125–169 (2002)
12. Bench-Capon, T.J.M.: Persuasion in practical argument using value-based argumentation frameworks. J. Log. Comput. 13(3), 429–448 (2003)
13. Atkinson, K.: What Should We Do?: Computational Representation of Persuasive Argument in Practical Reasoning. PhD thesis, Department of Computer Science, University of Liverpool, Liverpool, United Kingdom (2005)
14. Ashley, K.: Modeling Legal Argument: Reasoning with Cases and Hypotheticals. Bradford Books/MIT Press, Cambridge (1990)
15. McGuinness, D.L., Ferguson, G.: Proceedings of the Nineteenth National Conference on Artificial Intelligence, Sixteenth Conference on Innovative Applica tions of Artificial Intelligence, July 25-29, 2004. AAAI Press / The MIT Press, San Jose (2004)

Constructing Legal Arguments with Rules in the Legal Knowledge Interchange Format (LKIF)

Thomas F. Gordon

Fraunhofer FOKUS
Berlin, Germany
thomas.gordon@fokus.fraunhofer.de

Abstract. The Legal Knowledge Interchange Format (LKIF), being developed in the European ESTRELLA project, defines a knowledge representation language for arguments, rules, ontologies, and cases in XML. In this article, the syntax and argumentation-theoretic semantics of the LKIF rule language is presented and illustrated with an example based on German family law. This example is then applied to show how LKIF rules can be used with the Carneades argumentation system to construct, evaluate and visualize arguments about a legal case.

1 Introduction

The Legal Knowledge Interchange Format (LKIF) is an XML application being developed in the European ESTRELLA project (IST-4-027655) with the goal of establishing an open, vendor-neutral standard for exchanging formal models of the law, suitable for use in legal knowledge-based systems. By the end of the ESTRELLA project, LKIF will enable four kinds of legal knowledge to be encoded in XML: arguments, rules, ontologies and cases. The focus of the present paper is the LKIF language for modeling legal rules.

Legal rules express norms and policy. These are not only norms or policies about how to act, but also about how to reason about the law when planning actions or determining the legal consequences of actions. For example, the definition of murder as the "unlawful killing of a human being with malice aforethought" expresses both the legal (and moral) norm against the intentional killing of another human being and the reasoning policy creating a presumption that an accused person has committed murder once it has been proven that he killed another human being intentionally. Such presumptions are not sufficient for proving guilt. Rather, a guilty verdict would be legally correct only at the end of a properly conducted legal trial. If during this trial the defendant is able to produce evidence of the killing having been done in self-defense, for example, a guilty verdict would be correct only if the prosecution meets its burden of persuading the court or jury that the killing was in fact not done in self-defense.

Thus, the semantics of the LKIF rules is based not on the model theory of first-order logic, but rather on the dialectical and argumentation-theoretic

P. Casanovas et al. (Eds.): Computable Models of the Law, LNAI 4884, pp. 162–184, 2008.

approach to semantics articulated by Ron Loui in "Process and Policy: Resource-Bounded Non-Demonstrative Reasoning" [25]. Essentially, legal rules are interpreted as policies for reasoning in resource-limited, decision-making processes. In argumentation theory, such reasoning policies are viewed as inference rules for presumptive reasoning, called *argumentation schemes* [37]. Arguments are instances of argumentation schemes, constructed by substituting variables of a scheme with terms of the object language. An *argument graph* is constructed from a set of arguments. A set of argumentation schemes defines a search space over argument graphs. Reasoning with argumentation schemes can be viewed as heuristic search in this space, looking for argument graphs in which some disputed claim is acceptable or not given the arguments in the graph. In dialogues, the parties take turns searching this space, looking for counterarguments. Turn-taking, termination conditions, resource limitations and other procedural parameters are determined by the applicable rules of the legal proceeding, i.e. by the argumentation protocol for the particular type of dialogue.

The rest of this article is organized as follows. First, we provide an informal introduction and overview of LKIF rules, including some examples. This is followed by the formal definition of its abstract syntax. Then we define the semantics of the rule language, by mapping rules to argumentation schemes, using the Carneades model of argument [19]. LKIF rules is then illustrated with a more lengthy legal example about support obligations, based on German family law. This example is also used to illustrate an XML syntax for interchanging rule bases in LKIF, presented in the following section. Then we show how the Carneades argumentation system can use LKIF rule bases to construct and visualize arguments about cases. Finally, we conclude with a brief dicussion of related work, summarize the main results and suggest some ideas for future work.

2   Informal Overview

For simplicity and readability, we will be using a concrete syntax based on Lisp *s-expressions* to represent rules. Variables will be represented as symbols beginning with a question mark, e.g. `?x` or `?y`. Other symbols, as well as numbers and strings, represent constants, e.g. `contract`, `23.1`, or `"Jane Doe"`.

An *atomic sentence* is a simple declarative sentence containing no logical operators (negation, conjunction or disjunction). For example, the sentence "The mother of Caroline is Ines." can be represented as (`mother Caroline Ines`).

If `P` is an atomic sentence, then (`not P`) is a *negated atomic sentence*. Sentences which are either atomic sentences or negated atomic sentences are called *literals*. The *complement* of the literal `P` is (`not P`), and the complement of (`not P`) is `P`.

Rules are *reified* in this language, with an identifier and a set of properties, enabling any kind of meta-data about rules to be represented, such as a rule's date of enactment, issuing governmental authority, legal source text, or its period of validity. We do not define these properties here. Our focus is on defining the syntax and semantics of these rules.

Rules have a *body* and a *head*. The terms 'head' and 'body' are from logic programming, where they mean the conclusions and antecedents of a rule, respectively, interpreted as Horn clauses. Unlike Horn clause logic, a rule in our system may have more than one conclusion, including, as will be explained shortly, negated conclusions.

Here is a first example, a simplified reconstruction of a rule from the Article Nine World of the Pleadings Game [17], meaning that all movable things except money are goods.

```
(rule §-9-105-h
   (if (and (movable ?c)
            (unless (money ?c)))
       (goods ?c)))
```

§-9-105-h is an identifier, naming the rule, which may be used as a term denoting the rule in other rules.

We use the term *condition* to cover both literals and the forms (unless P), called *exceptions*, and (assuming P), called *assumptions*, where P is a literal. The head of a rule consists of a list of *literals*. Notice that, unlike Horn clause logic, rules may have negative conclusions. Negated atomic sentences may also be used in the body of a rule, also in exceptions and assumptions. Exceptions and assumptions are allowed only within the body of rules. The example rule above illustrates the use of an exception.

Legal rules are defeasible generalizations. Showing that some exception applies is one way to defeat a rule, by *undercutting* [27] it. Intuitively, a rule applies if its conditions are met, *unless* some exception is satisfied . A party who wants to use some rule need not show that no exception applies. The burden of proof for exceptions is on those interesting in showing the rule does not apply. Assumptions on the other hand, as their name suggests, are assumed to hold until they have been called into question. After an assumption has been questioned, a party who wants to use the rule must prove the statement which had been assumed.

Another source of defeasibility is conflicting rules. Two rules conflict if one can be used to derive P and another (not P). To resolve these conflicts, we need to be able to reason (i.e. argue) about which rule has priority. To support reasoning about rule priorities, the rule language includes a built-in predicate over rules, prior, where (prior r1 r2) means that rule r1 has priority over rule r2. If two rules conflict, the arguments constructed using these rules are said to *rebut* each other, following Pollock [27].

The priority relationship on rules is not defined by the system. Rather, priority is a substantive issue to be reasoned (argued) about just like any other issue. One way to construct arguments about rule priorities is to apply the argumentation scheme for arguments from legal rules to meta-level rules, i.e. rules about rules, using information about properties of rules, such as their legal authority or date of enactment. The reification of rules and the built-in priority predicate make this possible. In knowledge bases for particular legal domains, rules can be prioritized both *extensionally*, by asserting facts about which rules have priority over which other rules, and *intensionally*, using meta-rules about priorities.

For example, assuming metadata about the enactment dates of rules has been modeled, the legal principle that later rules have priority of earlier rules, *lex posterior*, can be represented as:

```
(rule lex-posterior
    (if (and (enacted ?r1 ?d1)
             (enacted ?r2 ?d2)
             (later ?d2 ?d1))
        (prior ?r2 ?r1)))
```

Rules can be defeated in two other ways: by challenging their validity or by showing that some exclusionary condition applies. These are modeled with rules about validity and exclusion, using two further built-in predicates: (`valid <rule>`) and (`excluded <rule> <literal>`), where `<rule>` is a constant naming the rule, not its definition. The second argument of the `excluded` predicate is a compound term representing a literal. Thus, literals can also be reified in this system.

The `valid` and `excluded` relations, like the `prior` relation, are to be defined in models of legal domains. Rules can be used for this purpose. For example, the exception in the previous example about money not being goods, even though money is movable, could have been represented as an exclusionary rule as follows:

```
(rule §-9-105-h-i
    (if (money ?c)
        (excluded §-9-105-h (goods ?c))))
```

To illustrate the use of the validity property of rules, imagine a rule which states that rules which have been repealed are no longer valid:

```
(rule repeal
    (if (repealed ?r1)
        (not (valid ?r1))))
```

This rule also exemplifies the use of negation in the conclusion (head) of this rule.

3   Syntax

This section presents a formal definition of an s-expression syntax for rules, in Extended Backus-Naur Form (EBNF)[1]. This syntax is inspired by the Common Logic Interchange Format (CLIF) for first-order predicate logic, which is part of the draft ISO Common Logic standard.[2] While inspired by CLIF, no attempt is made to make this rule language conform to Common Logic standard.[3]

---

[1] EBNF is specified in the ISO/IEC 14977 standard.

[2] http://philebus.tamu.edu/cl/

[3] Common Logic is a family of concrete syntaxes for first-order predicate logic, with its model-theoretic semantics and classical, monotonic entailment relation. These semantics are sufficiently different as to not make it useful to attempt to make the syntax of our rule language fully compatible with CLIF.

The syntax uses the Unicode character set. White space, delimiters, characters, symbols, quoted strings, boolean values and numbers are lexical classes, not formally defined here. For simplicity and to facilitate the development of a prototype inference engine using the Scheme programming language, we will use Scheme's lexical structure, as defined in the R6RS standard, which is based on the Unicode character set.[4]

Variable and Constant Symbols

```
variable ::= symbol
constant-symbol ::= symbol
```

Variable and constant symbols are disjunct. As mentioned in the informal overview, variables begin with a question mark character. Symbols are case-sensitive. Constant symbols may include a *prefix* denoting a *namespace*. Some mechanism for binding prefixes to namespaces is presumed, rather than being defined here. The prefix of a constant symbol is the part of the constant symbol up to the first colon. The part of the constant symbol after the colon is the local identifier, within this namespace.

Here are some example variable and constant symbols:

```
?x
?agreement
contract-1
lkif:permission
event-calculus:event
```

Term

A term is either a constant or a compound term. A constant is either a variable, constant symbol, string, number, or boolean value. A compound term consists of a constant symbol and a list of terms.

```
constant ::= variable | constant-symbol
          | string | number | boolean
term ::= constant | | '´' term |
       '(' constant-symbol term* ')'
```

Quoted terms are used, as in Lisp, to denote lists. Here are some example terms:

```
?x
contract-1
"Falkensee, Germany"
12.345
#t
(father-of John)
'(red green)
```

---

[4] http://www.r6rs.org/

Literal

Literals are atomic sentences or negations of atomic sentences.

```
atom ::= constant-symbol
      | '(' constant-symbol term* ')'
literal ::= atom | '(' 'not' atom ')'
```

Notice that constant symbols can be used as atomic sentences. This provides a convenient syntax for a kind of propositional logic.

The following are examples of literals:

```
liable
(initiates event1 (possesses ?p ?o))
(holds (perfected ?s ?c) ?p)
(children Ines '(Dustin Caroline))
(not (children Tom '(Sally Susan)))
(applies UCC-§-306-1 (proceeds ?s ?p))
```

Rule

Since Horn clause logic is widely known from logic programming, it might be helpful to begin the presentation of the syntax of LKIF rules by noting that it can be viewed as a generalization of the syntax of Horn clause logic, in the following ways:

1. Rules are reified with names.
2. Rules may have multiple conclusions.
3. Negated atoms are permitted in both the body and head of rules.
4. Rule bodies may include exceptions and assumptions.
5. Both disjunctions and conjunctions are supported in the bodies of rules.

Here is the formal definition of the syntax of rules:

```
condition ::= literal
            | '(' 'unless' literal ')'
            | '(' 'assuming' literal ')'

conjunction ::= condition
              | '(' 'and' condition condition+ ')'

disjunction ::= '(' 'or' conjunction conjunction+ ')'

body ::= condition | conjunction | disjunction

head ::= literal
       | '(' 'and' literal literal+ ')'
```

```
rule ::= '(' 'rule' constant-symbol
             '(' 'if' body head ')' ')'
          | '(' 'rule' constant-symbol
             literal literal* ')'
```

The second rule form is convenient for rules with empty bodies. These should not be confused with Prolog 'facts', since they are also defeasible. Conditions which are not assumptions or exceptions are called *ordinary conditions*.

Here are a few examples of rules and facts, reconstructed from the Article Nine World of the Pleadings Game [17]:

```
(rule §-9-306-3-1
    (if (and (goods ?s ?c)
             (consideration ?s ?p)
             (collateral ?si ?c)
             (collateral ?si ?p)
             (holds (perfected ?si ?c) ?e)
             (unless (applies §-9-306-3-2
                              (perfected ?si ?p))))
        (holds (perfected ?si ?p) ?e)))

(rule §-9-306-2a
    (if (and (goods ?t ?c)
             (collateral ?s ?c))
        (not (terminates ?t
               (security-interest ?s)))))

(rule F1 (not (terminates T1
                 (security-interest S1))))
(rule F2 (collateral S1 C1))
```

Reserved Symbols

The following predicate symbols have special meaning in the semantics, as explained in Section 4, and are thus reserved: `prior`, `excluded`, `valid`, and `applies`.

4   Semantics

We now proceed to define the semantics of the LKIF rules language. Due to space limitations, knowledge of the Carneades model of argument [19] is presumed. A rule denotes a *set* of argumentation schemes, one for each conclusion of the rule, all of which are subclasses of a general scheme for arguments from legal rules.[5]

---

[5] We do not claim that argumentation schemes can be modeled as or reduced to rules. Here we go in the other direction: each rule is mapped to a set of argumentation schemes.

Applying a rule is a matter of instantiating one of these argumentation schemes to produce a particular argument. Reasoning with rules is viewed as a process of applying these schemes to produce arguments to put forward in dialogues.

The scheme for arguments from legal rules is based on the rule language we developed for the Pleadings Game [17], but has also been influenced by Verheij's reconstruction of Reason-Based Logic in terms of argumentation schemes [36]. The scheme can be defined informally as follows:

Premises
1. r is a legal rule with ordinary conditions $a_1, \ldots, a_n$ and conclusion c.
2. Each a_i in $a_1 \ldots a_n$ is presumably true.

Conclusion. c is presumably true.

Critical Questions
1. Does some exception of r apply?
2. Is some assumption of r not met?
3. Is r a valid legal rule?
4. Does some rule excluding r apply in this case?
5. Can some rule with priority over r be applied to reach an contradictory conclusion?

Our task now is use this scheme to define the semantics of the formal language of Section 3, by mapping rules in the language to schemes for arguments in Carneades. We begin by mapping rule conditions to argument premises.

Definition 1 (Condition to Premise). *Let p be a function mapping conditions of rules to argument premises, defined as follows:*

$$p(c) = \begin{cases} c & \text{if } c \text{ is a literal} \\ \bullet s & \text{if } c \text{ is } (\texttt{assuming } s) \\ \circ s & \text{if } c \text{ is } (\texttt{unless } s) \end{cases}$$

If a conclusion of a rule is an atomic sentence, $\texttt{s}$, then the rule is mapped to a scheme for arguments *pro* $\texttt{s}$. If a conclusion of the rule is a negated atomic sentence, $(\texttt{not } \texttt{s})$, then the rule is mapped to a scheme for arguments *con* $\texttt{s}$.

Definition 2 (Scheme for Arguments from Rules). *Let r be a rule, with conditions $a_1 \ldots a_n$ and conclusions $c_1 \ldots c_n$. Three premises, implicit in each rule, are made explicit here. The first, $\circ \upsilon$, where $\upsilon = (\texttt{not } (\texttt{valid } r))$, excepts r if it is an invalid rule. The second, $\circ \epsilon$, where $\epsilon = (\texttt{excluded } r\ c_i)$, excepts r if it is excluded with respect to c_i by some other rule. The third, $\circ \pi$, where $\pi = (\texttt{priority } r_2\ r)$, excludes r if another rule, r_2, exists of higher priority than r which is applicable and supports a contradictory conclusion.*

For each c_i in $c_1 \ldots c_n$ of r, r denotes an argumentation scheme of the following form, where d is 'pro' if c_i is an atomic sentence and 'con' if c_i is a negated atomic sentence:

$$\frac{p(a_1) \ldots p(a_n), \circ \upsilon, \circ \epsilon, \circ \pi}{d\ c_i}$$

To construct an argument from one of these argumentation schemes, the variables in the scheme need to be systematically renamed and then instantiated using a *substitution environment*, i.e. a mapping from variables to terms, constructed by unifying the conclusion of the argumentation scheme with some goal atomic statement, as in logic programming.

The `valid` and `excluded` relations used in the argumentation scheme are to be defined in the models of legal domains, as explained in Section 2. Rules can be used to define the priority relation, as in the Pleadings Game [17] and PRATOR [28]. Legal principles for resolving rule conflicts, such as *lex posterior*, can be modeled in this way, as illustrated in Section 2.

The `applies` predicate is a 'built-in', meta-level relation which cannot be defined directly in rules. It is defined as follows:

Definition 3 (Applies). *Let σ be a substitution environment and G be an argument graph. Let r be a rule and S be the set of argumentation schemes for r, with all of the variables in these schemes systematically renamed. There are two cases, for atomic literals and negated literals. The rule r applies to a literal* P *in the structure* ⟨σ, G⟩, *if there exists a* pro *argumentation scheme s in S, if* P *is atomic, or a* con *argumentation scheme, if* P *is negated, such that the conclusion of s is unifiable with* P *in σ, and every premise of s, with its variables substituted by their values in the σ, holds in G.*

Given a set of rules and an argument graph, this definition of the `applies` predicate enables some meta-level reasoning. It allows one to find rules which can be used to generate defensible pro and con arguments for some goal statement or to check whether a particular rule can be used to generate a defensible pro or con argument for some statement.

The semantics of negation is dialectical, not classical negation or negation-as-failure. Exceptions do not have the semantics of negation-as-failure. The closed-world assumption is not made. In Carneades, a negated sentence, (`not p`), is acceptable just when the complement of the proof standard assigned to `p` is satisfied, where the complement of a proof standard is constructed by reversing the roles of pro and con arguments in the standard. See [19] for details.

5   A German Family Law Example

Let's now illustrate LKIF rules using a small, toy legal domain, roughly based on German family law. The question addressed is whether or not a descendent of some person, typically a child or grandchild, is obligated to pay financial support to the ancestor.

§1601 BGB (Support Obligations). Relatives in direct lineage are obligated to support each other.

```
(rule §-1601-BGB
  (if (direct-lineage ?x ?y)
      (obligated-to-support ?x ?y)))
```

§1589 BGB (Direct Lineage). A relative is in direct lineage if he is a descendent or ancestor. For example, parents, grandparents and great grandparents are in direct lineage.

```
(rule §-1589-BGB
    (if (or (ancestor ?x ?y)
            (descendent ?x ?y))
        (direct-lineage ?x ?y)))
```

§ 1589 BGB illustrates the use of disjunction in the body of a rule.

§1741 BGB (Adoption). For the purpose of determining support obligations, an adopted child is a descendent of the adopting parents.

```
(rule §-1741-BGB
    (if (adopted-by ?x ?y)
        (ancestor ?x ?y)))
```

§1590 BGB (Relatives by Marriage). There is no obligation to support the relatives of a spouse (husband or wife), such as a mother-in-law or father-in-law.

```
(rule §-1590-BGB
    (if (relative-of-spouse ?x ?y)
        (not (obligated-to-support ?x ?y))))
```

§ 1590 BGB illustrates the use of negation in the head of a rule.

§1602 BGB (Neediness). Only needy persons are entitled to support by family members. A person is needy only if unable to support himself.

```
(rule §-1602a-BGB
    (if (not (needy ?x))
        (not (obligated-to-support ?y ?x))))

(rule §-1602b-BGB
    (if (not (able-to-support-himself ?x))
        (needy ?x)))

(rule §-1602c-BGB
    (if (able-to-support-himself ?x)
        (not (needy ?x))))
```

In § 1602 we see examples of negation in both the head and body. This example also illustrates that it is not always possible to represent a section of a piece of legislation as a single LKIF rule. Thus, although LKIF brings us closer to the ideal of "isomorphic modeling", this goal remains illusive, at least if one takes the view that each section of legal code always expresses a single rule.

§1603 BGB (Capacity to Provide Support). A person is not obligated to support relatives if he does not have the capacity to support others, taking into consideration his income and assets as well as his own reasonable living expenses.

```
(rule §-1603-BGB
    (if (not (capacity-to-provide-support ?x))
        (not (obligated-to-support ?x ?y))))
```

§1611a BGB (Neediness Caused By Immoral Behavior). A needy person is not entitled to support from family members if his neediness was caused by his own immoral behavior, such as gambling, alcoholism, drug abuse or an aversion to work.

```
(rule §-1611a-BGB
    (if (neediness-caused-by-own-immoral-behavior ?x)
        (excluded §-1601-BGB (obligated-to-support ?y ?x))))
```

Here we have interpreted § 1611a BGB to be an exclusionary rule. If one instead takes the view that it states conditions under which there is no obligation to provide support, independent of the general obligation to provide support stated in § 1601 BGB, then the following LKIF rule would be a more faithful representation:

```
(rule §-1611a-BGB
    (if (neediness-caused-by-own-immoral-behavior ?x)
        (not (obligated-to-support ?y ?x))))
```

§91 BSHG (Undue Hardship). A person is not entitled to support relatives if this would cause him undue hardship.

```
(rule §-91-BSHG
    (if (undue-hardship ?x (obligated-to-support ?x ?y))
        (excluded §-1601-BGB (obligated-to-support ?x ?y)))))
```

As with § 1611a BGB, we have interpreted § 91 BSHG as an exclusionary rule, mainly to illustrate how statements are reified in LKIF and can be quoted in other statements. Here the statement (obligated-to-support ?x ?y) is quoted in the statement (excluded s1601-BGB (obligated-to-support ?x ?y)).

6   XML Syntax

The Legal Knowledge Interchange Format (LKIF) defines two ways to representing arguments, rules, ontologies and cases in XML. One uses OWL, the Ontology Web Language [26], to define concepts and relations for the structure of arguments, rules and cases. Particular arguments, rules and cases are represented in

OWL as instances of these classes. LKIF ontologies are defined directly in OWL. This approach offers the advantage of uniformity. An entire knowledge base can be represented using a single, widely supported existing standard, OWL, and be developed, maintained and processed using existing OWL editors and other tools.

As it turns out, however, rules and arguments cannot be conveniently written or maintained using generic OWL editors, such as Protege [14] or TopBraid Composer [35], at least not without first extending them with 'plug-ins' for special purpose editors, along the lines of the Protege plug-in for the Semantic Web Rule Language [24].

Moreover, OWL is not, strictly speaking, an XML format. Rather, OWL is defined at a more abstract level. OWL documents can be 'serialized' using a variety of concrete syntaxes. Some of these are XML-based, for example using RDF/XML [7]. Other serializations of OWL, some based on the Notation 3 [8] language, aim to be compact and more readable and thus do not use XML. For this reason, implementing a translator for LKIF documents encoded in OWL requires the document to first be preprocessed into some canonical concrete syntax, using for example Jena [23], a Java library for the Semantic Web.

For these reasons, LKIF offers an alternative, more compact, XML syntax. This syntax is defined using the XML Schema Definition Language [13]. An eqivalent definition of the grammar using Relax NG [10], an ISO standard schema definition language (ISO/IEC 19757), is also available. One advantage of Relax NG is that it offers a compact, readable language for schema definitions, in addition to an XML language. Here is the Relax NG version of the compact syntax of LKIF Rules:

```
start = element lkif  Statement*, Rule*, ArgumentGraph*

Rule = element rule
    attribute id  xsd:ID ?,
    attribute strict  "no" | "yes" ?,
    (Literal+ | Implies)

Literal = Statement | Not
Statement = element s
    attribute id  xsd:ID ?,
    attribute src  xsd:anyURI | xsd:string ?,
    ((text* &  Statement*)*)?
Not = element not  Statement
Implies = (Head, Body) | (Body, Head)
Head = element head  Literal+
Body = element body  Or | Condition+
Or = element or  (Condition | And)+
And = element and  Condition+
Condition = Literal
```

```
| element if  attribute role  text ?, Literal
| element unless  attribute role  text ?, Literal
| element assuming  attribute role  text ?, Literal
```

The specification of the compact syntax for argument graphs has been omitted, since the focus of this article is LKIF's rule language.

The German family law example of the previous section can be represented in XML using the compact syntax as follows:

```xml
<?xml version="1.0" encoding="UTF-8"?>
<lkif>
  <rule id="s1601-BGB">
    <body><s>direct-lineage ?x ?y</s></body>
    <head><s>obligated-to-support ?x ?y</s></head>
  </rule>

  <rule id="s1589a-BGB">
    <body><s>ancestor ?x ?y</s></body>
    <head><s>direct-lineage ?x ?y</s></head>
  </rule>

  <rule id="s1589b-BGB">
    <body><s>descendent ?x ?y</s></body>
    <head><s>direct-lineage ?x ?y</s></head>
  </rule>

  <rule id="s1741-BGB">
    <body><s>adopted-by ?x ?y</s></body>
    <head><s>ancestor ?x ?y</s></head>
  </rule>

  <rule id="s1590-BGB">
    <body><s>relative-of-spouse ?x ?y</s></body>
    <head><not><s>obligated-to-support ?x ?y</s></not></head>
  </rule>

  <rule id="s1602a-BGB">
    <body><not><s>needy ?x</s></not></body>
    <head><not><s>obligated-to-support ?y ?x</s></not></head>
  </rule>

  <rule id="s1602b-BGB">
    <body><not><s>able-to-support-himself ?x</s></not></body>
    <head><s>needy ?x</s></head>
  </rule>
```

```
<rule id="s1602c-BGB">
  <body><s>able-to-support-himself ?x</s></body>
  <head><not><s>needy ?x</s></not></head>
</rule>

<rule id="s1603-BGB">
  <body><not><s>capacity-to-provide-support ?x</s></not></body>
  <head><not><s>obligated-to-support ?x ?y</s></not></head>
</rule>

<rule id="s1611a-BGB">
  <body>
    <s>neediness-caused-by-own-immoral-behavior ?x</s>
  </body>
  <head>
    <s>excluded s1601-BGB <s>obligated-to-support ?y ?x</s></s>
  </head>
</rule>

<rule id="s91-BSHG">
  <body>
    <s>undue-hardship ?x <s>obligated-to-support ?x ?y</s></s>
  </body>
  <head>
    <s>excluded s1601-BGB <s>obligated-to-support ?x ?y</s></s>
  </head>
</rule>
</lkif>
```

7 Reasoning with LKIF Rules Using Carneades

'Carneades' is the name of both a compuational model of argumentation [19] and
an implementation of this model in PLT Scheme [30]. The ESTRELLA Reference

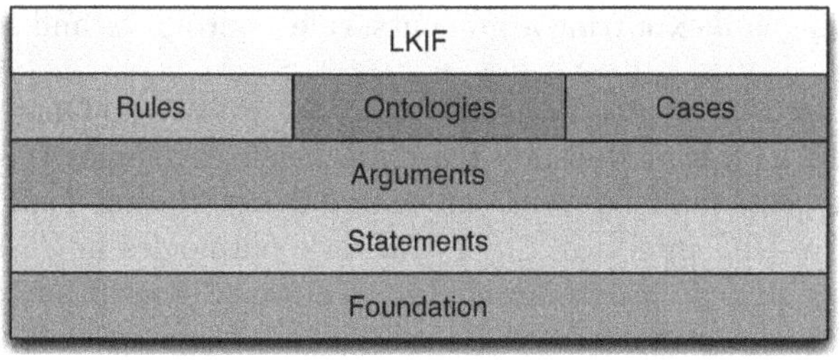

Fig. 1. Module Layers

Inference Engine for LKIF rules is being built using this implementation of Carneades.

This ESTRELLA platform, of which the LKIF rules inference engine is a part, has the layered architecture shown in Figure 1

Each layer consists of one more modules, where a module may make use of the services of another module in the same layer or any layer below it in the diagram. Conversely, no module depends on the services of any module in some higher layer.

Since the higher layers build upon the lower layers, we will describe the lowest layers first:

Foundation. The foundation layer consists of modules for configuring the system for a particular installation (`config`), managing possibly infinite sequences of data generated lazily (`stream`), and for heuristically searching problem spaces (`search`).

Statements. The statement layer provides a module for comparing and decomposing statements (`statement`), abstracting away syntactic details which are irrelevant for the higher layers, and a module implementing a unification algorithm (`unify`), needed for implementing inference engines for logics with variables ranging over compound terms, such as first-order logic and LKIF rules.

Arguments. The argument layer provides modules for constructing, evaluating and visualizing argument graphs, also called 'inference graphs' (`argument`, `argument-diagram`). It also provides modules (`argument-state`, `argument-search`) for applying argumentation schemes to search heuristically for argument graphs in which some goal statement is acceptable (i.e. presumably true) or not acceptable. An `argument-builtins` module provides an argument generator for common goal statements about arithmetic, strings, lists, dates and so on.

Rules. The rule layer implements LKIF rules. It provides a `rule` module for representing defeasible legal rules and generating arguments from sets of rules.

Ontologies. The ontologies layer provides a module for defining and reasoning with concepts, using Description Logic [4].

Cases. The cases layer provides a module for representing legal precedents and constructing arguments from these precedents using argumentation schemes for case-based reasoning.

LKIF. The Legal Knowledge Interchange Format (LKIF) layer provides a module for importing and exporting arguments, rules, ontologies and cases in XML.

Ontologies are represented in LKIF using the OWL Web Ontology Language [11]. The ESTRELLA module for reasoning with ontologies is still being designed. It may communicate with an external description logic reasoner, for example via the DIG interface [6], or translate ontologies into LKIF rules, using the description logic programming intersection of description logic and Horn clause logic [20]. However, since LKIF rules is more expressive than Horn clause logic, it may be possible to translate a larger subset of description logic into LKIF rules.

A first prototype of an implementation of case-based argumentation schemes, based on a reconstruction [38] of schemes modeled by HYPO [3] and CATO [1] has been completed and is currently being evaluated.

Some people have expressed surprise at the ordering of these layers, perhaps because of familiarity with Berners-Lee's vision of the Semantic Web [9], which has a similar architecture, consisting of the following layers, from bottom to top: Unicode and URI, XML, RDF, ontology, logic, proof and trust. Statements are expressed using the RDF layer. Rules are represented in the logic layer. One difference between these architectures is that the proof layer, which seems closely related to argument, is above the logic layer where rules reside in Berners-Lee's model, whereas rules are built on top of the argument layer in our system. Another difference is that ontologies form a foundation for logic and proof in Berners-Lee's model, whereas ontologies, rules and cases are all at the same layer in our system, as they are all interpreted as knowledge representation formalisms from which arguments can be constructed, using argumentation schemes appropriate for each type of knowledge.

We close this section with an example showing how to use Carneades to load the Germany family law example rule base, ask a query, and visualize the result. To simplify the example, let's suppose the rule base has been extended with rules defining 'direct-lineage' in terms of common-sense family relations, along with some facts about a case, for example that Gloria is needy and an ancestor of Dustin, but Dustin does not have the capacity to provide support. We will omit the definitions of family relations (such as ancestor, descendant, parent, grandparent, sibling, and relative) to keep this short. The facts of the case can be represented in LKIF rules as follows.

```
(rule* facts
   (ancestor Dustin Tom)
   (ancestor Tom Gloria)
   (needy Gloria)
   (not (capacity-to-provide-support Dustin)))
```

Let's suppose the XML file for the rule base is stored in a file named "family-support.xml". This file can be imported to create a rule base as follows:

```
(define family-support
   (add-rules empty-rulebase
              (import "family-support.xml")))
```

This code defines family-support to be the rule base created by importing the "family-support.xml" LKIF file. Now we can pose a query, about whether Dustin is obligated to support his grandmother, Gloria, as follows:

```
> (define s1 (initial-state '(obligated-to-support Dustin Gloria)
                            default-context))
> (define g1 (generate-arguments-from-rules family-support null))
> (define r (make-resource 50))
```

```
> (define results (find-best-arguments depth-first r 1 s1 (list g1
    builtins)))
> (define s2 (stream-car results))
> (view* (state-arguments s2)
         (state-context s2)
         (state-substitutions s2) #t)
```

We begin by defining a problem space. The first command defines the root, initial state of the problem state. The query, (obligated-to-support Dustin Gloria), is part of this initial state. The next command defines the transitions available between states in the search space. These transitions are induced by the rules available in the family-support rule base. Since the search space can be infinite, we use resources to limit the amount of searching done and assure the search process terminates. The (make-resource 50) constructs a resource with 50 units. Each state visited during the search for a solution consumes one unit of this resource. The find-best-arguments command in this example looks for arguments for Dustin being obligated to support Gloria, in this case using a (resource-limited) depth-first search strategy. A few others search strategies are also available, including breadth-first, and iterative-deepening. A stream of solution states is returned, where a stream is conceptually a sequence of states, where each member of the sequence is computed as needed. Thus, to backtrack and search for further solutions, one only needs to access subsequent members of the stream. If the search process fails, finding no state satisfying the query, the resulting stream will be empty. Each state in the search space contains an argument graph. The goal of the search, using the find-best-arguments command, is to find the best arguments pro and con the statement in the query, given the resources supplied and the number of turns to alternate between the roles of proponent and opponent of the statement. That is, in the terminology of the Carneades model of argument, we are interested in finding argument graphs which provide sufficient grounds, or reasons, for 'accepting' the statement of the query, presumptively, as true, when taking the perpective of the proponent, and finding extensions of these argument graphs which succeed in countering these arguments, making the statement of the query no longer acceptable, when taking the perspective of the opponent.

The (view* ...) command displays a diagram of an argument graph, using GraphView [12]. Figure 2 shows the argument graph found first by the find-best-arguments command above.

One way to look for a counterargument is to repeat the find-best-arguments command, but this time with 2 turns. In the second turn the system takes the perspective of the opponent.

```
> (define results
    (find-best-arguments depth-first r 2 s1
        (list g1 builtins)))
```

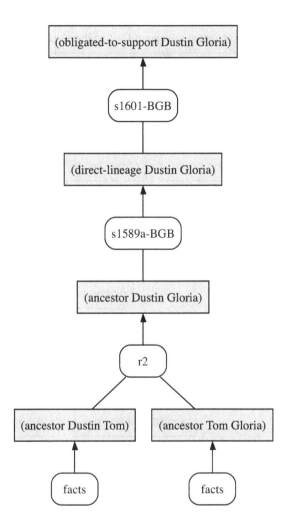

Fig. 2. An argument pro the obligation to support

In this example, a counterargument was found, as shown in Figure 3. Since **find-best-arguments** returns the best arguments for both sides, it would have returned the first argument graph again, shown in Figure 2, had it been unable to find a counterargument to this argument on the second turn.

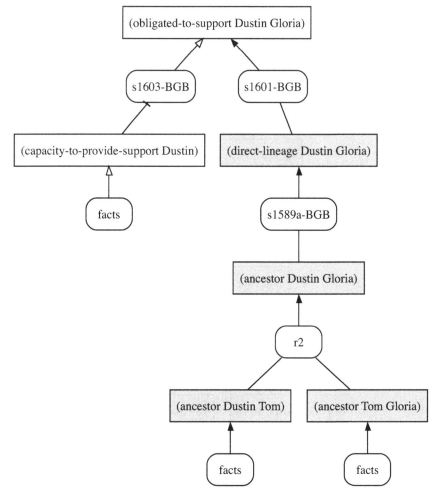

Fig. 3. A counterargument

8   Discussion

LKIF rules builds on the results of about 20 years of research in Artificial Intelligence and Law. Edwina Rissland, Kevin Ashley and Ronald Loui published a good summary of the field of Artificial Intelligence and Law, as of 2003, in a special issue of the Artificial Intelligence Journal [32]. A recent treatise on AI and Law is Giovanni Sartor's "Legal Reasoning: A Cognitive Approach to the Law" [34].

A major lesson from research on Artificial Intelligence and Law is that legal reasoning cannot be viewed, in general, as the application of some deductive logic, such as first-order predicate logic, to some theory of the facts and relevant legal domain. In fact, no one in the field ever seriously took the position that legal reasoning in its entirety could be viewed this way, although some critics have misunderstood or misrepresented the field by assuming this to be the case. As pointed out by Rissland et al. [32]:

> Contrary to some popular notions, law is *not* a matter of simply applying rules to facts via *modus ponens*, for instance, to arrive at a conclusion. *Mechanical jurisprudence*, as this model has been called, is somewhat of a strawman. It was soundly rejected by rule skeptics like the realists. As Gardner puts it, law is more "rule-guided" than "rule-governed."

The reference to Gardner here, is to Anne Gardner's thesis "An Artificial Intelligence Approach to Legal Reasoning" [15], one of the first books to be published in the field. Legal reasoning is not only deductive, because legal concepts cannot be defined by necessary and sufficient conditions. Better, one *can* define legal concepts this way, but such definitions are only hypotheses or theories which will not be blindly or "mechanically" followed, using deduction, when one tries to apply these concepts to decide legal issues in concrete cases. Legal concepts are, as the legal philosopher H.L.A. Hart put its, "open-textured" [22]. Whether or not a legal concept applies in a particular case requires the interpretation, or reinterpretation of the legal sources, such as statutes and case law, in light of such things as the history of prior precedent cases, the intention of the legislature, public policy, and evolving social values.

The process of determining whether the facts of a case can be "subsumed" under some legal concept is one of argumentation. Legal argumentation is a dialogue, guided by procedural norms, called "protocols". Which protocol applies depends on the particular type of dialogue and the task at hand.

Although argumentation has always been at the heart of work on modeling legal reasoning in the field of AI and Law, it wasn't until two papers on computational models of legal argumentation in a special issue of the International Journal of Man-Machine Studies on AI and Law [16,33] that argumentation became a hot topic in AI and Law and efforts began in earnest to use argumentation theory to integrate case-based, rule-based and other approaches to legal reasoning. The procedural aspects of argumentation, i.e. as a dialogue and not just a way of comparing pros and cons, began to come into focus [17]. A dialogical approach to integrating arguments from rules and cases was presented not much later [29].

The rule language developed here is much like the one the author developed for the Pleadings Game [17]: rules are reified and subject to exceptions; conflicts between rules can be resolved using other rules about rule priorities; the applicability of rules can be reasoned about and excluded by other rules; and the validity of rules can be questioned. The rule language of the Pleadings Game is similar to other systems developed independently at about the same time [21,28]. All of these systems viewed reasoning with legal rules as argumentation, but unlike in our semantics for LKIF rules, none of them interpreted legal rules as argumentation schemes. Rather, these prior systems either represented legal rules as sentences in a nonmonotonic logic [21,28] or compiled rules to a set of such sentences [17].[6] Verheij was the first to explicitly discuss the modeling of legal rules using argumentation schemes [36] but like the Pleadings Game interprets rules as abstractions of sets of formulas in a nonmonotonic logic, rather than interpreting rules as abstractions of arguments, i.e. as argumentation schemes in Walton's sense [37]. With the exception of the Pleadings Game, all of these prior systems model argumentation as deduction in a nomonotonic logic, i.e. as a defeasible consequence *relation*. In the Pleadings Game, argumentation was viewed procedurally, as dialogues regulated by protocols, but this was accomplished by building a procedural layer on top of a nonmonotonic logic. In LKIF, the relational interpretation of rules is abandoned entirely, in favor of a purely procedural view, and is thus more in line with modern argumentation theory in philosophy [37] and legal theory [2]. Argumentation cannot be reduced to logic.

The rule language presented here is syntactically similar to the rule languages of the Pleadings Game [17] and the PRATOR system [28]. Our main original contribution is the particular argumentation-theoretic semantics we have given these rules, by mapping them to argumentation schemes using the Carneades model of argument. This approach has at least two advantages:

1. The system can be extended with comparable models of other argumentation schemes. Argumentation schemes provide a unifying framework for building hybrid reasoners. The ESTRELLA platform will make use of this feature to support legal reasoning with ontologies, rules and cases, in an integrated way.
2. Despite the expressiveness of the rule language, which would result in an undecidable logic using the relational approach, since the semantics of LKIF rules is purely procedural, argumentation protocols can be defined for using these rules in legal proceedings which are guaranteed to terminate with procedurally just legal conclusions [31,2,5]

The ESTRELLA reference inference engine for LKIF rules has been fully implemented, in PLT Scheme [30]. Our work in the near future, together with our colleagues in the European ESTRELLA project, will focus on completing the modules for reasoning with cases and ontologies and validating LKIF in pilot applications, for example in the domain of European tax directives.

Our primary goal with LKIF rules has been to develop a knowledge representation formalism for legal rules which is theoretically well-founded, reflecting the state-of-the-art in AI and Law, and practically useful for building legal knowledge-based systems. Rule-based systems have been commercially successful, also for legal applications, but all of the products currently available on the market, to our knowledge, interpret rules either as formulas in propositional or first-order logic or as production rules. Legal rules are neither material implications nor procedures for updating variables in working memory, but rather schemes for constructing

---

[6] Technically speaking, the rules in PRATOR also may be viewed as domain-dependent inference rules, since they may not be used contrapositively, but nonetheless they are formulated as sentences in the object language.

legal arguments. There are many kinds of rules and correspondingly many kinds of formalisms for modeling rules. LKIF rules is designed to be better suited for modeling legal rules than existing alternatives on the market.

Acknowledgments

This is an extended version of a paper published at the International Conference on Artificial Intelligence and Law [18]. The work reported here was conducted as part of the European ESTRELLA project (IST-4-027655). I would like to thank Alexander Boer, Trevor Bench-Capon, Tom van Engers, Jonas Pattberg, Henry Prakken, Doug Walton, and Adam Wyner for fruitful discussions about topics related to this paper.

References

1. Aleven, V.: Teaching Case-Based Argumentation Through a Model and Examples. Ph.d., University of Pittsburgh (1997)
2. Alexy, R.: A Theory of Legal Argumentation. Oxford University Press, New York (1989)
3. Ashley, K.D.: Modeling Legal Argument: Reasoning with Cases and Hypotheticals. Artificial Intelligence and Legal Reasoning Series. MIT Press, Bradford Books (1990)
4. Baader, F., Calvanese, D., McGuinness, D., Nardi, D., Patel-Schneider, P. (eds.): The Description Logic Handbook – Theory, Implementation and Applications. Cambridge University Press, Cambridge (2003)
5. Bayles, M.D.: Procedural Justice; Allocating to Individuals. Kluwer Academic Publishers, Dordrecht (1990)
6. Bechhofer, S.: The DIG Description Logic interface: DIG 1.1. Technical report, D1 Implementation Group, University of Manchester (2003)
7. Beckett, D.: Rdf/xml syntax specification (revised) (February 2004), http://www.w3.org/TR/2004/REC-rdf-syntax-grammar-20040210/
8. Berners-Lee, T.: Notation 3 (1998), http://www.w3.org/DesignIssues/Notation3
9. Berners-Lee, T., Hendler, J., Lassila, O.: The semantic web. Scientific American 284(5), 34–43 (2001)
10. Clark, J.: Relax ng (September 2003), http://relaxng.org
11. Deborah, S.U.D.L.M., McGuinness, L. (Knowledge Systems Laboratory and van Harmelen, F.: OWL web ontology language overview, http://www.w3.org/TR/owl-features/
12. Ellson, J., Gansner, E., Koutsofios, L., North, S.C., Woodhull, G.: Graphviz — open source graph drawing tools. In: Mutzel, P., Jünger, M., Leipert, S. (eds.) GD 2001. LNCS, vol. 2265, pp. 483–484. Springer, Heidelberg (2002)
13. Fallside, D.C., Walmsley, P.: Xml schema part o: Primer, 2nd edn (2004), http://www.w3.org/TR/xmlschema-0/
14. S.C.: for Biomedical Informatics Research. The protege ontology editor and knowledge acquisition system (October 2007), http://protege.stanford.edu/
15. Gardner, A.: An Artificial Intelligence Approach to Legal Reasoning. MIT Press, Cambridge (1987)

16. Gordon, T.F.: An abductive theory of legal issues. International Journal of Man-Machine Studies 35, 95–118 (1991)
17. Gordon, T.F.: The Pleadings Game; An Artificial Intelligence Model of Procedural Justice. Springer, New York, Book version of 1993 Ph.D. Thesis; University of Darmstadt (1995)
18. Gordon, T.F.: Constructing arguments with a computational model of an argumentation scheme for legal rules. In: Proceedings of the Eleventh International Conference on Artificial Intelligence and Law, pp. 117–121 (2007)
19. Gordon, T.F., Prakken, H., Walton, D.: The Carneades model of argument and burden of proof. Artificial Intelligence 171(10-11), 875–896 (2007)
20. Grosof, B.N., Horrocks, I., Volz, R., Decker, S.: Description logic programs: Combining logic programs with description logics. In: Proceedings of the Twelfth International World Wide Web Conference (WWW 2003), Budapest, Hungary, May 2003, pp. 48–57. ACM, New York (2003)
21. Hage, J.C.: Monological reason-based logic. a low level integration of rule-based reasoning and case-based reasoning. In: Proceedings of the Fourth International Conference on Artificial Intelligence and Law, pp. 30–39. ACM, New York (1993)
22. Hart, H.L.A.: The Concept of Law. Clarendon Press, Oxford (1961)
23. Hewlett-Packard. Jena – a semantic web framework (October 2007), http://jena.sourceforge.net/
24. Horrocks, I., Patel-Schneider, P., Boley, H., Tabet, S., Grosof, B.N., Dean, M.: SWRL: A semantic web rule language combining OWL and RuleML, http://www.w3.org/Submission/SWRL/
25. Loui, R.P.: Process and policy: resource-bounded non-demonstrative reasoning. Computational Intelligence 14, 1–38 (1998)
26. McGuinness, D.L., van Harmelen, F.: OWL Web Ontology Language overview, http://www.w3.org/TR/owl-features/
27. Pollock, J.: Defeasible reasoning. Cognitive Science 11(4), 481–518 (1987)
28. Prakken, H., Sartor, G.: A dialectical model of assessing conflicting argument in legal reasoning. Artificial Intelligence and Law 4(3-4), 331–368 (1996)
29. Prakken, H., Sartor, G.: Modelling reasoning with precedents in a formal dialogue game. Artificial Intelligence and Law 6(2-4), 231–287 (1998)
30. T. P. S. Project. PLT Scheme, http://www.plt-scheme.org/.
31. Rawls, J.: A Theory of Justice. Belknap Press of Harvard University Press (1971)
32. Rissland, E.L., Ashley, K.D., Loui, R.P.: AI and law: A fruitful synergy. Artificial Intelligence 150(1–2), 1–15 (2003)
33. Routen, T., Bench-Capon, T.: Hierarchical formalizations. International Journal of Man-Machine Studies 35, 69–93 (1991)
34. Sartor, G.: Reasoning with factors. Technical report, University of Bologna (2005)
35. TopQuadrant. Topbraid composer (October 2007), http://www.topbraidcomposer.org/
36. Verheij, B.: Dialectical argumentation with argumentation schemes: An approach to legal logic. Artificial Intelligence and Law 11(2-3), 167–195 (2003)
37. Walton, D.: Fundamentals of Critical Argumentation. Cambridge University Press, Cambridge (2006)
38. Wyner, A., Bench-Capon, T.: Argument schemes for legal case-based reasoning. In: JURIX 2007: The Twentieth Annual Conference on Legal Knowledge and Information Systems (2007)

Assumption-Based Argumentation for Epistemic and Practical Reasoning

Francesca Toni

Imperial College London, Department of Computing

Abstract. Assumption-based argumentation can serve as an effective computational tool for argumentation-based epistemic and practical reasoning, as required in a number of applications. In this paper we substantiate this claim by presenting formal mappings from frameworks for epistemic and practical reasoning onto assumption-based argumentation frameworks. We also correlate these mappings to formulations of epistemic and practical reasoning in abstract argumentation terms.

1 Introduction

Argumentation has proven to be a useful abstraction mechanism for understanding several problems in artificial intelligence. In particular, several mechanisms for non-monotonic reasoning have been proven to be instances of argumentation frameworks (e.g. see [6,2]), and defeasible logic can be understood in argumentation terms (e.g. see [13]). Moreover, argumentation has been extensively applied in legal settings (e.g. see [19,17,20]) requiring defeasible reasoning. Finally, argumentation has been studied as a powerful tool for understanding several forms of reasoning needed to be performed by rational agents (e.g. see [15]). In this paper, we consider two forms of reasoning that rational agents may need to perform, namely reasoning as to which beliefs they should hold (referred to here as *epistemic reasoning*) and reasoning as to which course of action/decision they should choose (referred to here as *practical reasoning*). Both forms of reasoning may be defeasible as the information available to the agent may be conflicting in general (e.g. if this information comes from different sources). Both forms of reasoning rely upon manipulating rules and preferences amongst rules. These preferences may themselves be defined in terms of defeasible rules, as often the case in the literature [3,19].

We show how a particular form of argumentation, known as *assumption-based argumentation* (ABA) [2,5,7,8,9], can be used to model and realise both epistemic and practical reasoning. Whereas in abstract argumentation [6] the notions of *argument* and *attack* are primitive, in ABA they are derived from notions of *deductive system* and corresponding *deductions*, *assumptions* and *contrary* of assumptions: intuitively, an argument is a deduction supported by a set of assumptions, and an argument attacks another if the first argument supports the contrary of an assumption in the second. Standard argumentation semantics [6] can then be ascribed to ABA frameworks. Finally, computational mechanisms

P. Casanovas et al. (Eds.): Computable Models of the Law, LNAI 4884, pp. 185–202, 2008.
© Springer-Verlag Berlin Heidelberg 2008

for computing these semantics exist [7,8,9,10,11], with several computational advantages deriving from singling out assumptions in arguments and avoding re-computation when these assumptions have already been encountered earlier in the computation [7,8,9].

In this paper, for each of epistemic and practical reasoning, we define a concrete rule-based representation framework and a mapping of this framework into ABA. By virtue of this mapping, epistemic and practical reasoning can be equipped with a number of argumentation semantics, inherited from ABA (as well as abstract argumentation) [2,6]. Moreover, epistemic and practical reasoning can be realised in practice by deploying the computational mechanisms that ABA is equipped with [7,8,9,10,11]. The mapping is defined by associating new assumptions to defeasible rules (for epistemic and practical reasoning) and decisions (for practical reasoning) and appropriately setting the contrary of these assumptions.

In the case of epistemic reasoning without preference rules, we also provide a mapping into abstract argumentation, and prove a formal correspondence between this mapping and the mapping into ABA.

The paper is organised as follows. In section 2 we give some background for abstract argumentation and ABA. In section 3 we provide a formulation of epistemic reasoning with defeasible rules and rule-based preferences amongst them, and provide a mapping into ABA frameworks first ignoring preferences, in section 3.1, and then considering preferences, in section 3.2. In both cases, for simplicity, we ignore strict rules. In section 4 we provide a formulation of practical reasoning with defeasible rules and decisions, and provide a mapping into ABA frameworks. In section 5 we exemplify our approach, by adapting an example from [1] to combine practical and epistemic reasoning. In section 6 we conclude.

2   Abstract and Assumption-Based Argumentation

Definition 1. *An* abstract argumentation framework *is a pair (Arg, attacks) where Arg is a finite set, whose elements are referred to as* arguments, *and* attacks ⊆ Arg × Arg *is a binary relation over Arg. Given sets* $X, Y \subseteq Arg$ *of arguments,* X attacks Y *iff there exists* $x \in X$ *and* $y \in Y$ *such that* $(x, y) \in$ attacks.

Given an abstract argumentation framework, several notions of *"acceptable" sets of arguments* can be defined [6].

Definition 2. *A set* X *of arguments is*

- conflict-free *iff it does not attack itself;*
- admissible *iff* X *is conflict-free and* X *attacks every set of arguments* Y *such that* Y *attacks* X;
- preferred *iff* X *is maximally admissible;*
- sceptically preferred *iff* X *is the intersection of all preferred sets of arguments;*
- complete *iff* X *is admissible and* X *contains all arguments* x *such that* X *attacks all attacks against* x;

- grounded *iff X is minimally complete;*
- ideal *iff X is admissible and it is contained in every preferred set of arguments.*

The last notion was not in the original [6], but has been proposed recently [8,9] as an alternative, less sceptical semantics than the grounded semantics.

The abstract view of argumentation does not deal with the problem of actually finding arguments and attacks amongst them. Typically, arguments are built by connecting rules in the belief set of the proponent of arguments, and attacks arise from conflicts amongst such arguments. In *assumption-based argumentation* (ABA), arguments are obtained by reasoning backwards with a given set of inference rules (the belief set), from conclusions to premises that are assumptions, and attacks are defined in terms of a notion of "contrary" of assumptions. Belief set and backward reasoning are defined in terms of a *deductive system*:

Definition 3. *A deductive system is a pair $(\mathcal{L}, \mathcal{R})$ where*

- $\mathcal{L}$ *is a formal language consisting of countably many sentences, and*
- $\mathcal{R}$ *is a countable set of inference rules of the form*

$$\frac{x_1, \ldots, x_n}{x}$$

where $x \in \mathcal{L}$ is called the conclusion *and $x_1, \ldots, x_n \in \mathcal{L}$ are called the* premises *of the inference rule, and $n \geq 0$.*

If $n = 0$, then the inference rule represents an axiom. Note that a deductive system does not distinguish between domain-independent axioms/rules, which belong to the specification of the logic, and domain-dependent axioms/rules, which represent a background theory. For notational convenience, throughout the paper we write $x \leftarrow x_1, \ldots, x_n$ instead of $\dfrac{x_1, \ldots, x_n}{x}$ and x instead of $x \leftarrow$.

Definition 4. *Given a deductive system $(\mathcal{L}, \mathcal{R})$ and a selection function [1] f, a (backward) deduction of a conclusion x based on (or supported by) a set of premises P is a sequence of multi-sets $S_1, \ldots, S_m$, where $S_1 = \{x\}$, $S_m = P$, and for every $1 \leq i < m$, where σ is the sentence occurrence in S_i selected by f:*

1. *If σ is not in P then $S_{i+1} = S_i - \{\sigma\} \cup S$ for some inference rule of the form $\sigma \leftarrow S \in \mathcal{R}$. [2]*
2. *If σ is in P then $S_{i+1} = S_i$.*

Each S_i is referred to as a step *in the deduction.*

---

[1] A selection function is any function from sets of elements to elements. The definition of backward deduction relies upon some chosen selection function. However, note that if a backward deduction for a conclusion exists for some selection function, then a backward deduction for that conclusion will exist for any other selection function. This result follows from the analogous result for SLD-resolution for Horn clauses.

[2] We use the same symbols for multi-set membership, union, intersection and subtraction as we use for ordinary sets.

In the remainder of this paper we will use the following notation: $P \vdash c$ will stand for "there exists a deduction of c supported by P". This notation is simplistic as it does not allow to distinguish different deductions to the same conclusions and supported by the same premises, but it is a useful shorthand when the steps in the deduction are not of interest (see the discussion in [9]).

Deductions are the basis for the construction of arguments in ABA, but to obtain an argument from a backward deduction the premises are restricted to *assumptions* only. Moreover, to specify when one argument attacks another, we need to specify *contraries* of assumptions.

Definition 5. *An ABA framework is a tuple* $\langle \mathcal{L}, \mathcal{R}, \mathcal{A}, {}^{\overline{}} \rangle$ *where*

- $(\mathcal{L}, \mathcal{R})$ *is a deductive system.*
- $\mathcal{A} \subseteq \mathcal{L}, \mathcal{A} \neq \{\}$. $\mathcal{A}$ *is the set of candidate* assumptions.
- *If* $x \in \mathcal{A}$, *then there is no inference rule of the form* $x \leftarrow x_1, \ldots, x_n \in \mathcal{R}$.
- ${}^{\overline{}}$ *is a (total) mapping from* $\mathcal{A}$ *into* $\mathcal{L}$. $\overline{x}$ *is the* contrary *of* x.

Note that ABA frameworks are still abstract, in the sense that in order to be deployed they need to be instantiated. Several instances have been studied already [2,17]. In this paper we study some additional instances, for epistemic and practical reasoning. Note that, by the third bullet, following [7] we restrict ourselves to *flat* frameworks [2], whose assumptions do not occur as conclusions of inference rules. Flat frameworks are restricted but still interesting and general, as, for example, they admit default logic and logic programming as concrete instances [2], as well as all the instances we will consider in this paper.

In the ABA approach to argumentation, arguments are deductions to conclusions, based solely upon assumptions, and the attack relationship between arguments depends solely on sets of assumptions and their contraries.

Definition 6. *A set of assumptions A attacks a set of assumptions B iff there exists an assumption $x \in B$ and a deduction $A' \vdash \overline{x}$ such that $A' \subseteq A$: if this is the case, we say that A attacks B on x.*

This notion of attack between sets of assumptions implicitly gives a notion of attack between arguments supported by sets of assumptions: the attacking argument needs to have as conclusion the contrary of an assumption in the support of the attacked argument.

Within ABA, implicitly, a set of assumptions stands for the set of all arguments whose premises are contained in the given set of assumptions (see [9]). Thus, the computation of "acceptable" sets of arguments amounts to computing "acceptable" sets of assumptions:

Definition 7. *A set of assumptions A is*

- conflict-free *iff A does not attack itself;*
- admissible *iff A conflict-free and A attacks every set of assumptions B that attacks A;*
- preferred *iff it is maximally admissible;*

- sceptically preferred *iff* X *is the intersection of all preferred sets of assumptions;*
- complete *iff it is admissible and contains all assumptions* x *such that* A *attacks all attacks against* $\{x\}$;
- grounded *iff it is minimally complete;*
- ideal *iff* A *is admissible and it is contained in every preferred set of assumptions.*

3   Epistemic Reasoning

Epistemic reasoning may be performed within a framework consisting of defeasible and strict rules and facts, some of which may express preferences over the application of rules and the use of facts (thus some of these preferences may be themselves defeasible). The use of rules to represent preferences rather than fixed partial orders is advocated by many, e.g. [3,19], driven by the requirements of applications, for example in a legal domain. Before defining our frameworks for epistemic reasoning, we give some preliminary notions.

Definition 8

- *A language* $\mathcal{L}$ *is a set of ground literals, which can be atoms* A *or negations of atoms* $\neg A$. *We will refer to these literals as* basic literals.
- *A naming* $\mathcal{N}$ *is a bijective function associating a distinguished name* $\mathcal{N}(x)$ *to any element* x *in any given domain* X. *For any given* X, *we will refer to the set of all such names as* $\mathcal{N}(X)$.
- *A preference literal (wrt* X *and* $\mathcal{N}$) *is of the form* $N_1 \succ N_2$ *where* N_1, N_2 *are (different) names in* $\mathcal{N}(X)$.
- *A literal is either a basic or a preference literal.*

Intuitively, $N_1 \succ N_2$ stands for "the element named N_1 is preferred to the element named N_2". In the remainder of this paper, given a basic literal L, with an abuse of notation, $\neg L$ will stand for the complement of L, namely $\neg L$ if L is an atom, and A if L is a negative literal $\neg A$. Moreover, given a preference literal L of the form $N_1 \succ N_2$, $\neg L$ will stand for $N_2 \succ N_1$.

Definition 9. *Given a language* $\mathcal{L}$, *a set* X *and a naming* $\mathcal{N}$:

- *A basic rule (wrt* $\mathcal{L}$) *is of the form* $B_1, \ldots, B_n \rightarrow B_0$ *where* $B_0, \ldots, B_n$ *are basic literals in* $\mathcal{L}$ *and* $n \geq 0$.
- *A preference rule (wrt* $\mathcal{L}, \mathcal{N}$ *and* X) *is of the form* $B_1, \ldots, B_n \rightarrow B_0$ *where* B_0 *is a preference literal (wrt* X *and* $\mathcal{N}$), $B_1, \ldots, B_n$ *are literals in* $\mathcal{L}$ *or preference literals (wrt* X *and* $\mathcal{N}$), *and* $n \geq 0$.
- *A rule is either a basic rule or a preference rule.*

Given a rule $B_1, \ldots, B_n \rightarrow B_0$, B_0 is referred to as the conclusion and $B_1, \ldots, B_n$ as the premises. When $n = 0$ the rule is sometimes referred to as a fact.

Definition 10. *Let $\mathcal{L}$ be a language and $\mathcal{N}$ be a naming. An* epistemic framework *is a pair $\langle D, S \rangle$, with D the* defeasible *and S the* strict *components, such that D can be partitioned into sets $D_1, \ldots, D_d, S_1, \ldots, S_s$ (respectively), $d, s \geq 0$, and*

- *D_1 and S_1 are sets of* basic rules *(wrt $\mathcal{L}$)*
- *for each $i \geq 1$, D_i (S_i) is a set of* preference rules *wrt $\mathcal{L}$, $X = \cup_{j=1,\ldots,i-1} D_j$ ($X = \cup_{j=1,\ldots,i-1} S_j$, respectively) and $\mathcal{N}$.*

Rules in D (S) are referred to as defeasible *(strict, respectively).*

Note that preference rules may be strict or defeasible.

Intuitively, defeasible rules may or may not be chosen by a rational reasoner, whereas strict rules will always need to be included in all "reasoning lines" of these reasoner. A rational reasoner needs to avoid conflicts in its chosen reasoning lines. Conflicts in epistemic frameworks arise from "deriving" complementary conclusions from sets of chosen strict and defeasible rules, either of the form A and $\neg A$ or of the form $N_1 \succ N_2$ and $N_2 \succ N_1$. As strict rules cannot be disregarded ever, it is reasonable to assume that conflicts cannot arise amongst them alone. Conflicts will however typically arise from defeasible (and strict) rules. The semantics of defeasible frameworks needs to resolve these conflicts. In the remainder of this section, we will show how to provide this semantics for epistemic frameworks without strict rules. We will refer to an epistemic framework simply as D. We ignore strict rules for simplicitly, as they require special attention to guarantee "closedness" and "consistency" of epistemic reasoning [4]. For a treatment of strict rules in ABA see [22].

3.1 Epistemic Frameworks without Preference Rules

In this section we show how to provide a family of semantics for any epistemic framework without strict and preference rules as an instance of ABA first and of abstract argumentation then, and show the correspondence between the two different frameworks.

Below, we will assume given an epistemic framework $\epsilon = D$ wrt a language $\mathcal{L}$.

An Assumption-Based Argumentation View

Definition 11. *The ABA framework corresponding to $\epsilon = D$ is $\langle \mathcal{L}_\epsilon, \mathcal{R}_\epsilon, \mathcal{A}_\epsilon, \overline{} \rangle$ whereby*

- *$\mathcal{A}_\epsilon$ is a set of literals not already in $\mathcal{L}$ such that there exists a bijective mapping α from rules in D into $\mathcal{A}_\epsilon$;*
- *$\mathcal{L}_\epsilon = \mathcal{L} \cup \mathcal{A}_\epsilon$;*
- *$\mathcal{R}_\epsilon = \{B_0 \leftarrow B_1, \ldots, B_n, \alpha(B_1, \ldots, B_n \rightarrow B_0) | B_1, \ldots, B_n \rightarrow B_0 \in D\}$*
- *$\alpha(B_1, \ldots, B_n \rightarrow B_0) = \neg B_0$.*

Intuitively, any assumption in $\mathcal{A}_\epsilon$ correspond to the applicability of the corresponding rule, which is opposed by the complement of the conclusion of that rule being "derivable": this is expressed by the definition of contrary.

Example 1. Given $D = \{q; q \rightarrow p; r; r \rightarrow \neg p\}$, $\mathcal{R}_\epsilon$ may be [3]

$$\{\, q \leftarrow a_1;\ p \leftarrow q, a_2;$$
$$r \leftarrow a_3;\ \neg p \leftarrow r, a_4\}$$

$\mathcal{A}_\epsilon = \{a_1, a_2, a_3, a_4\}$, and $\overline{a_1} = \neg q$, $\overline{a_2} = \neg p$, $\overline{a_3} = \neg r$, and $\overline{a_4} = p$.

By virtue of this formulation, any notion of acceptable set of assumptions may be adopted to provide a semantics to D. For instance, in example 1, $\{a_1, a_2\}$ is an admissible set of assumptions with $\{q, p\}$ the corresponding "output", and $\{a_1, a_3\}$ is the grounded set of assumptions with $\{q, r\}$ the corresponding "output".

Note that the translation proposed in definition 11 could be "optimised", by associating the same assumption to different rules with the same conclusion and by dropping assumptions associated to rules whose conclusion cannot be objected to (namely, such that there is no rule with complementary conclusion). For instance, the "optimised" version of the ABA framework in example 1 might have $\mathcal{R}'_\epsilon =$

$$\{\, q;\ p \leftarrow q, a_2;$$
$$r;\ \neg p \leftarrow r, a_4\}$$

and $\mathcal{A}'_\epsilon = \{a_2, a_4\}$.

An Abstract Argumentation View. We define the notion of abstract argumentation framework corresponding to an epistemic framework, by using the notion of reasoning line, formalised below, to express arguments, and the negation in $\mathcal{L}$ to express attacks.

Definition 12
- A reasoning line *wrt a set of rules $P \subseteq D$ is a sequence of literals $x_1, \ldots, x_m$ where x_1 is a fact in P and for each x_i, $1 < i \leq m$, there exists a rule $Y \rightarrow x_i$ in P such that $Y \subseteq \{x_1, \ldots, x_{i-1}\}$.*
- A reasoning line for y is a reasoning line $x_1, \ldots, x_m$ such that x_m is y.

Intuitively, arguments in favour of conclusions are given by sets of rules underlying reasoning lines for these conclusions, satisfying some restrictions. One possible such restriction, advocated by some approaches (e.g. [12]), is minimality of the support of reasoning lines.

Definition 13. *A minimal argument for/supporting a conclusion c is a pair (P, c) where $P \subseteq D$ is a set of (defeasible) rules such that*

1. *there exists a reasoning line for c wrt P, and*
2. *there exists no $P' \subset P$ satisfying 1.*

The support of a minimal argument (P, c) is P.

---

[3] Other choices of $\mathcal{R}_\epsilon$ are possible, by choosing a different α and thus $\mathcal{A}_\epsilon$.

For example, given an epistemic framework $D = \{a; a \rightarrow b; b, c \rightarrow d; c\}$, there is no minimal argument for b with support $\{c; a; a \rightarrow b\}$, whereas there is a minimal argument for d with support $\{c; a; a \rightarrow b; b, c \rightarrow d\}$.

Requiring minimality of arguments is computationally expensive, as checking minimality requires a global search over D when constructing arguments for some conclusion. We adopt here an alternative restriction on reasoning lines, less demanding computationally but serving a similar purpose to minimality, of imposing that all rules used in an argument are "relevant" to the conclusion being supported and serves a purpose in constructing the corresponding reasoning line.

Definition 14. *Given a selection function f, a* relevant argument *for conclusion c based on (or supported by) a set of rules $P \subseteq D$ is pair (P, c) such that there exists a sequence of pairs of multi-sets $(M_1, P_1) \ldots, (M_m, P_m)$, where $M_1 = \{c\}$, $P_1 = \{\}$, $M_m = \{\}$, $P_m = P$ and for every $1 \leq i < m$, if x is the sentence occurrence in M_i selected by $f: M_{i+1} = M_i - \{x\} \cup X$ for some rule of the form $X \rightarrow x \in D$ and $P_{1+1} = P_i \cup \{X \rightarrow x\}$.*

The support *of a relevant argument (P, c) is P.*

Existence of a minimal argument for a conclusion guarantees existence of a relevant argument for the same conclusion, and wrt the same support. However, there may be no minimal argument based on the same support as that of a relevant argument, as illustrated by the following example:

Example 2. Consider $D = \{q, r \rightarrow p; s \rightarrow q; t \rightarrow q; q \rightarrow r; s; t\}$. Then, each of D, $\{s; q, r \rightarrow p; s \rightarrow q; q \rightarrow r\}$ and $\{t; q, r \rightarrow p; t \rightarrow q; q \rightarrow r\}$ supports a relevant argument for p, but no minimal argument exists with support D.

We will see later on that relevant arguments supported by sets of rules in abstract argumentation for epistemic reasoning are in direct correspondence with sets of assumptions in the ABA framework for epistemic reasoning.

From now on, if confusion does not arise, we will often refer to relevant arguments simply as arguments. Also, we will often equate an argument to its support, when this will cause no confusion.

Definition 15. *A* sub-argument *of an argument (P, c) is an argument (P', c') supported by $P' \subseteq P$.*

Except for the trivial sub-argument amounting to the argument itself, a sub-argument typically supports a conclusion other than the argument. For example, given $D = \{s; s \rightarrow t; s, p \rightarrow t; p\}$, the set $\{s\}$, supporting s, is a sub-argument of $\{s; s \rightarrow t\}$, supporting t.

Definition 16. *The abstract argumentation framework corresponding to $\epsilon = D$, wrt a language $\mathcal{L}$, is $(Arg_\epsilon, attacks_\epsilon)$ such that*

- *Arg_ϵ is the set of all relevant epistemic arguments wrt D,*
- *given $A, B \in Arg_\epsilon$, A attacks$_\epsilon$ B iff there exists a literal $L \in \mathcal{L}$ such that*

1. *there exists a sub-argument of A supporting L, and*
2. *there exists a sub-argument of B supporting ¬L.*

Note that conditions 1 and 2 in the earlier definition capture both undercutting and rebuttal attack (as understood, e.g., in [18]): for the undercutting, the $¬L$ could be just sanctioned by a defeasible fact or could be an intermediate conclusion in the argument B.

Correspondence Between Assumption-Based and Abstract Argumentation Views. Below we will assume given an epistemic framework $\epsilon = D$ wrt $\mathcal{L}$, without strict or preference rules, and the abstract and ABA frameworks $(Arg_\epsilon, attacks_\epsilon)$ and $\langle \mathcal{L}_\epsilon, \mathcal{R}_\epsilon, \mathcal{A}_\epsilon, \overline{} \rangle$ (respectively) corresponding to ϵ.

Definition 17. *Let $c \in \mathcal{L}$.*

- *Given an argument (P, c) in $(Arg_\epsilon, attacks_\epsilon)$, the corresponding set of assumptions in $\langle \mathcal{L}_\epsilon, \mathcal{R}_\epsilon, \mathcal{A}_\epsilon, \overline{} \rangle$ is $\alpha(P) = \{\alpha(r) | r \in P\}$.*
- *Given a set of assumptions Δ such that $\Delta \vdash c$ in $\langle \mathcal{L}_\epsilon, \mathcal{R}_\epsilon, \mathcal{A}_\epsilon, \overline{} \rangle$, the corresponding argument in $(Arg_\epsilon, attacks_\epsilon)$ is $(\alpha^{-1}(\Delta), c)$, where $\alpha^{-1}(\Delta) = \{r \in D | \alpha(r) \in \Delta\}$.*

It is easy to see that there exists a one-to-one correspondence between relevant arguments wrt $(Arg_\epsilon, attacks_\epsilon)$ and backward deductions from assumptions in $\langle \mathcal{L}_\epsilon, \mathcal{R}_\epsilon, \mathcal{A}_\epsilon, \overline{} \rangle$, since assumptions in $\langle \mathcal{L}_\epsilon, \mathcal{R}_\epsilon, \mathcal{A}_\epsilon, \overline{} \rangle$ are in one-to-one correspondence with rules in D. More precisely:

Lemma 1

- *Given a backward deduction $\Delta \vdash c$ in $\langle \mathcal{L}_\epsilon, \mathcal{R}_\epsilon, \mathcal{A}_\epsilon, \overline{} \rangle$, there exists a relevant argument $(\alpha^{-1}(\Delta), c)$ in $(Arg_\epsilon, attacks_\epsilon)$.*
- *Given a relevant argument (P, c) in $(Arg_\epsilon, attacks_\epsilon)$, there exists a backward deduction $\alpha(P) \vdash c$ in $\langle \mathcal{L}_\epsilon, \mathcal{R}_\epsilon, \mathcal{A}_\epsilon, \overline{} \rangle$.*

As a consequence, attacks in the two frameworks are in correspondence, as follows:

Theorem 1

1. *Given arguments $A_1, A_2 \in Arg_\epsilon$, if $A_1 attacks_\epsilon A_2$ then $\alpha(A_1)$ attacks $\alpha(A_2)$ wrt $\langle \mathcal{L}_\epsilon, \mathcal{R}_\epsilon, \mathcal{A}_\epsilon, \overline{} \rangle$.*
2. *Given sets of assumptions $\Delta_1, \Delta_2 \subseteq \mathcal{A}_\epsilon$ if Δ_1 attacks Δ_2 on an assumption x wrt $\langle \mathcal{L}_\epsilon, \mathcal{R}_\epsilon, \mathcal{A}_\epsilon, \overline{} \rangle$ and $\alpha^{-1}(x) = Y \to y$ and $\Delta \vdash y$ for some $\Delta \subseteq \Delta_2$, then there exist arguments A_1 supported by $\alpha^{-1}(\Delta_1)$ and A_2 supported by $\alpha^{-1}(\Delta_2)$ such that $A_1 attacks_\epsilon A_2$.*

Note that the formulation of part 2 above guarantees that the rule corresponding to the assumption x that is being attacked is "triggered" by all the rules corresponding to the assumptions in the attacked set. This is needed as assumptions may exist in isolation, whereas arguments always come as a package (premises plus conclusion).

Proof (theorem 1)

1. If $A_1 attacks_\epsilon A_2$ then a sub-argument of A_1 derives L and a sub-argument of A_2 derives $\neg L$ for some $L \in \mathcal{L}$. Then, in $(Arg_\epsilon, attacks_\epsilon)$, there exist a relevant reasoning line/argument for L supported by a subset P_1 of the support of A_1 and a relevant reasoning line/argument for $\neg L$ supported by a subset P_2 of the support of A_2. Then, by lemma 1, in $\langle \mathcal{L}_\epsilon, \mathcal{R}_\epsilon, \mathcal{A}_\epsilon, \overline{} \rangle$ there exists a backward deduction $\alpha(P_1) \vdash L$. Moreover, let r be the last rule in P_2 used to derive $\neg L$. $\alpha(r) \in \alpha(P_2)$ and $\overline{\alpha(r)} = L$. Then, $\alpha(A_1)$ attacks $\alpha(A_2)$ by definition of attack in $\langle \mathcal{L}_\epsilon, \mathcal{R}_\epsilon, \mathcal{A}_\epsilon, \overline{} \rangle$.

2. If Δ_1 attacks Δ_2 in $\langle \mathcal{L}_\epsilon, \mathcal{R}_\epsilon, \mathcal{A}_\epsilon, \overline{} \rangle$ on x then there exists a backward deduction $\Delta' \vdash \overline{x}$ for some $\Delta' \subseteq \Delta_1$ (and $x \in \Delta_2$). Then, by lemma 1, in $(Arg_\epsilon, attacks_\epsilon)$ there exists a relevant reasoning line/argument $(\alpha^{-1}(\Delta'), \overline{x})$. Since $\alpha^{-1}(x) = Y \to y$, by definition of α, $\overline{x} = \neg y$. Then, there exists a sub-argument of the argument A_1 supported by $\alpha^{-1}(\Delta_1)$ for $\neg y$.

   Since we have assumed that there exists a backward deduction/argument $\Delta \vdash y$ for some $\Delta \subseteq \Delta_2$, then there exists a sub-argument of the argument A_2 supported by $\alpha^{-1}(\Delta_2)$ for y.

   Thus there exist A_1 and A_2 as requested such that $A_1 \ attacks_\epsilon \ A_2$.

From this result we can prove a correspondence between "acceptable" sets of assumptions in ABA and corresponding "acceptable" sets of arguments in abstract argumentation, corresponding to an epistemic framework. For example:

Corollary 1

- *Let Δ be a set of assumptions in $\langle \mathcal{L}_\epsilon, \mathcal{R}_\epsilon, \mathcal{A}_\epsilon, \overline{} \rangle$ and let*
  $X = \{(\alpha^{-1}(\Delta'), c)|\Delta' \subseteq \Delta \text{ and } \Delta' \vdash c\}$
  be the set of corresponding arguments in $(Arg_\epsilon, attacks_\epsilon)$. Then, if Δ is conflict-free wrt $\langle \mathcal{L}_\epsilon, \mathcal{R}_\epsilon, \mathcal{A}_\epsilon, \overline{} \rangle$ then X is conflict-free wrt $(Arg_\epsilon, attacks_\epsilon)$.
- *Let X be a set of arguments wrt $(Arg_\epsilon, attacks_\epsilon)$ and let*
  $\Delta = \bigcup_{(P,c) \in X} \alpha(P)$
  be the set of corresponding assumptions in $\langle \mathcal{L}_\epsilon, \mathcal{R}_\epsilon, \mathcal{A}_\epsilon, \overline{} \rangle$. Then, if X is conflict-free in $(Arg_\epsilon, attacks_\epsilon)$ then Δ is conflict-free wrt $\langle \mathcal{L}_\epsilon, \mathcal{R}_\epsilon, \mathcal{A}_\epsilon, \overline{} \rangle$.

Proof

- By contradiction, assume X is not conflict-free. Then, there exists $A, B \in X$ such that $A \ attacks_\epsilon \ B$. By theorem 1, $\alpha(A)$ attacks $\alpha(B)$ and, by construction of Δ, Δ attacks itself: contradiction.
- By contradiction, assume Δ is not conflict-free. Then, there exists $\Delta' \subseteq \Delta$ and $\Delta' \vdash \overline{x}$ such that $x \in \Delta$. Let $\alpha^{-1}(x) = Y \to y$. By construction of Δ, since all assumptions come from arguments built from rules, there exists $\Delta'' \subseteq \Delta$ such that $\Delta'' \vdash y$. By theorem 1, there exists A_1, A_2 both supported by $\alpha^{-1}(\Delta)$ such that $A_1 \ attacks_\epsilon \ A_2$. Thus, $X \ attacks_\epsilon$ itself (by definition of $attacks_\epsilon$).

3.2 Epistemic Frameworks with Preferences

In this section we show how to provide a family of semantics for any epistemic framework without strict but with preference rules as an instance of ABA. For lack of space, we omit to show the corresponding instance of abstract argumentation.

Below, we will assume given an epistemic framework D wrt a language $\mathcal{L}$. We will use the following notation: given a set of rule X, a literal L is *defined* in X if a rule in X has L or $\neg L$ as its conclusion.

Definition 18. *The ABA framework corresponding to an epistemic framework* $\epsilon = D$ *is* $\langle \mathcal{L}_\epsilon, \mathcal{R}_\epsilon, \mathcal{A}_\epsilon, \overline{} \rangle$ *whereby*

- $\mathcal{A}_\epsilon$ *is a set of literals not already in* $\mathcal{L}$ *such that*
  - *there exists a bijective mapping* α *from rules in* D *into* $\mathcal{A}_\epsilon$;
- $\mathcal{L}_\epsilon = \mathcal{L} \cup \mathcal{A}_\epsilon \cup \mathcal{B}_\epsilon \cup \mathcal{C}_\epsilon$ *where* $\mathcal{B}_\epsilon$ *and* $\mathcal{C}_\epsilon$ *are distinct sets of literals not already in* $\mathcal{L} \cup \mathcal{A}_\epsilon$ *such that*
  - *there exists a bijective mapping* β *from rules in* D *into* $\mathcal{B}_\epsilon$;
  - *there exists a bijective mapping* χ *from assumptions in* $\mathcal{A}_\epsilon$ *into* $\mathcal{C}_\epsilon$;
- $\mathcal{R}_\epsilon = \{x \leftarrow \beta(X \rightarrow x) | X \rightarrow x \in D\} \cup$
  $\{\beta(X \rightarrow x) \leftarrow X, \alpha(X \rightarrow x) | X \rightarrow x \in D\} \cup$
  $\{\chi(a) \leftarrow n' \succ n, \beta(Y \rightarrow \neg x) | \, a = \alpha(X \rightarrow x), X \rightarrow x, Y \rightarrow \neg x \in D,$
  $n = \mathcal{N}(X \rightarrow x), \; n' = \mathcal{N}(Y \rightarrow \neg x),$
  $n' \succ n \text{ is defined in } D\} \cup$
  $\{\chi(a) \leftarrow \beta(Y \rightarrow \neg x) | \, a = \alpha(X \rightarrow x), X \rightarrow x \in D, Y \rightarrow \neg x \in D,$
  $\mathcal{N}(X \rightarrow x) \succ \mathcal{N}(Y \rightarrow \neg x) \text{ is not defined in } D\};$
- $\overline{a} = \chi(a)$.

Intuitively, assumptions in $\mathcal{A}_\epsilon$, as in the case of no preference rules, correspond to the applicability of the corresponding rules, sentences in $\mathcal{B}_\epsilon$ correspond to the actual application of the corresponding rules, and sentences in $\mathcal{C}_\epsilon$ correspond to objecting to the application of a rule, by a rule with higher preference and conflicting conclusion being "derivable": this is expressed by the definition of contrary and χ.

Example 3. Given $D = \{q \rightarrow p; q; \neg p; \neg q; r \rightarrow n_1 \succ n_3; r; \neg r\}$, where $n_1 = \mathcal{N}(q \rightarrow p)$ and $n_3 = \mathcal{N}(\neg p)$, $\mathcal{R}_\epsilon$ may be

$$\{ \, p \leftarrow b_1; \quad b_1 \leftarrow q, a_1; \quad c_1 \leftarrow n_3 \succ n_1, b_3;$$
$$q \leftarrow b_2; \quad b_2 \leftarrow a_2; \quad c_2 \leftarrow b_4;$$
$$\neg p \leftarrow b_3; \quad b_3 \leftarrow a_3; \quad c_3 \leftarrow n_1 \succ n_3, b_1;$$
$$\neg q \leftarrow b_4; \quad b_4 \leftarrow a_4; \quad c_4 \leftarrow b_2;$$
$$n_1 \succ n_3 \leftarrow b_5; \quad b_5 \leftarrow r, a_5;$$
$$r \leftarrow b_6; \quad b_6 \leftarrow a_6; \quad c_6 \leftarrow b_7;$$
$$\neg r \leftarrow b_7; \quad b_7 \leftarrow a_7; \quad c_7 \leftarrow b_6\}$$

$\mathcal{A}_\epsilon = \{a_1, a_2, a_3, a_4, a_5, a_6, a_7\}$, and $\overline{a}_i = c_i$, for $i = 1, \ldots, 7$. Note that no inference rule with conclusion c_5 appears in $\mathcal{R}_\epsilon$ since no rule for $n_3 \succ n_1$ (which is $\neg n_1 \succ n_3$) exists in D.

By virtue of this formulation, any notion of "acceptable" set of assumptions may be adopted to provide a semantics to D. For instance, in example 3, $\{a_1, a_2\}$ is an admissible set of assumptions, with "output" $\{p, q\}$. Indeed, this set of assumptions is conflict-free (there is no backward deduction from any of its subsets supporting any of c_1, c_2). Moreover, it attacks (by means of a deduction supporting c_4) the set of assumptions $\{a_4\}$ that attacks it (by means of a deduction supporting c_2). Note that the assumption a_1 is not attacked by any set of assumptions, as there is no backward deduction supporting c_1. The set of assumptions $\{a_3, a_4\}$ is also admissible, as it is conflict-free and it attacks all attacks against it: it is attacked by

- $\{a_1, a_2, a_5, a_6\}$ (supporting a backward deduction for c_3) which is counter-attacked by $\{a_4\}$,
- $\{a_2\}$ (supporting a backward deduction for c_4) which is also counter-attacked by $\{a_4\}$.

Note that the translation proposed in definition 18 can be "optimised", to eliminate newly introduced literals that serve no purpose. For instance, the "optimised" version for example 3 might have $\mathcal{R}_\epsilon =$

$$\{\, p \leftarrow b_1; \quad b_1 \leftarrow q, a_1; \quad c_1 \leftarrow n_3 \succ n_1, b_3;$$

$$q \leftarrow a_2;$$

$$\neg p \leftarrow b_3; \quad b_3 \leftarrow a_3; \quad c_3 \leftarrow n_1 \succ n_3, b_1;$$

$$\neg q \leftarrow a_4;$$

$$n_1 \succ n_3 \leftarrow r, a_5;$$

$$r \leftarrow a_6;$$

$$\neg r \leftarrow a_7 \}$$

and $\overline{a_1} = c_1$, $\overline{a_2} = \neg q$, $\overline{a_3} = c_3$, $\overline{a_4} = q$, $\overline{a_5} = n_3 \succ n_1$, $\overline{a_6} = \neg r$, $\overline{a_7} = r$. Note that further "optimisation" would be possible, e.g. by eliminating the rule with conclusion c_1 (as there is no possible backward deduction for c_1, since there is no rule with conclusion $n_3 \succ n_1$).

4   Practical Reasoning

Definition 19. *Given a language $\mathcal{L}$, a* framework for practical reasoning *is a tuple $\langle \epsilon, \mathcal{D}, \mathcal{G} \rangle$ where*

- *ϵ is an epistemic framework wrt $\mathcal{L}$;*
- *$\mathcal{D}$ is a set of sets $\mathcal{D}_1, \ldots, \mathcal{D}_m$ such that $\mathcal{D}_i \subseteq \mathcal{L}$ for each $i = 1, \ldots, m$ and no element of $\bigcup_{i=1,\ldots,m} \mathcal{D}_i$ occurs in the conclusion in any rule in ϵ;*
- *$\mathcal{G}$ is a sequence $\mathcal{G}_1, \ldots, \mathcal{G}_n$, $n \geq 1$, such that $\bigcup_{i=1,\ldots,n} \mathcal{G}_i \subseteq \mathcal{L}$ and there exists no $L \in \mathcal{L}$ such that $L \in \mathcal{G}_i \cap \mathcal{G}_k$ for $i \neq k$.*

Intuitively, $\mathcal{D}$ is the set of potential *decisions* that the agent may choose amongst in order to achieve its *goals* in $\mathcal{G}$. The decisions in each individual $\mathcal{D}_i$ are intended to be mutually exclusive, but decisions across different $\mathcal{D}_i$s are compatible (for example, a decision to jail a person is compatible with fining that person: jail and fine could thus belong to different $\mathcal{D}_i$s). The goals are grouped in a manner reflecting their degree of importance to the agent: the goals in a $\mathcal{G}_i$ earlier in the sequence are more important than the goals in a $\mathcal{G}_k$ later in the sequence. All goals in the same group $\mathcal{G}_i$ have the same degree of importance.

Note that the ranking in $\mathcal{G}$ could be specified, more generally, in terms of a set of (strict and defeasible) rules, including preference rules, e.g. as in [16]. Moeover, the mutual exclusion amongst decisions could be specified, more generally, in terms of sets of (strict and defeasible) rules specifying the context in which the mutual exclusion might take place. Both extensions are left for future work.

For simplicity, from now on we will assume that $\mathcal{G}$ and $\mathcal{D}$ consist each of a single set. We will also assume given $\langle \epsilon, \mathcal{D}, \mathcal{G} \rangle$ wrt $\mathcal{L}$.

Frameworks for practical reasoning can be modelled within a generalised form of assumption-based frameworks, whereby contrary is a (total) mapping from assumptions into *sets* of literals in $\mathcal{L}$. Given such a framework, the notion of attack between sets of assumptions is modified as follows

- a set of assumptions A attacks a set of assumptions B iff there exists an assumption $x \in B$, a sentence $y \in \overline{x}$ and an argument $A' \vdash y$ such that $A' \subseteq A$.

Definition 20. *The (generalised) ABA framework corresponding to* $\pi = \langle \epsilon, \mathcal{D}, \mathcal{G} \rangle$ *is* $\langle \mathcal{L}_\pi, \mathcal{R}_\pi, \mathcal{A}_\pi, \overline{} \rangle$ *whereby, given that* $\langle \mathcal{L}_\epsilon, \mathcal{R}_\epsilon, \mathcal{A}_\epsilon, \overline{} \rangle$ *is the ABA framework corresponding to* ϵ,

- $\mathcal{A}_\pi = \mathcal{A}_\epsilon \cup \mathcal{D}$;
- $\mathcal{L}_\pi = \mathcal{L}_\epsilon$;
- $\mathcal{R}_\pi = \mathcal{R}_\epsilon$;
- *if* $x \in \mathcal{A}_\epsilon$, *then* $\overline{x} = \{y\}$ *where* y *is the contrary of* x *in* $\langle \mathcal{L}_\epsilon, \mathcal{R}_\epsilon, \mathcal{A}_\epsilon, \overline{} \rangle$; *if* $x \in \mathcal{D}$, *then* $\overline{x} = \mathcal{D} - \{x\}$.

Below, whenever $\overline{x} = \{y\}$, for $x, y \in \mathcal{L}$, we will write simply $\overline{x} = y$.

Example 4. Consider $\pi = \langle \epsilon, \mathcal{D}, \mathcal{G} \rangle$ where $\epsilon = \mathcal{D}$ with

$$p; \quad d_1, p \rightarrow q;$$
$$d_2 \rightarrow s; \quad s \rightarrow t; \quad t \rightarrow \neg p$$

and $\mathcal{D} = \{d_1, d_2\}$ ($\mathcal{G}$ is left unspecified for the time being, see below for possible choices). Then, in $\langle \mathcal{L}_\pi, \mathcal{R}_\pi, \mathcal{A}_\pi, \overline{} \rangle$, $\mathcal{R}_\pi$ is [4]:

$$p \leftarrow a_1; \quad q \leftarrow p, d_1, a_2;$$
$$s \leftarrow d_2, a_3; \quad t \leftarrow s, a_4; \quad \neg p \leftarrow t, a_5$$

where $\mathcal{A}_\pi = \{a_1, \ldots, a_5, d_1, d_2\}$ and $\overline{a_1} = \neg p, \overline{a_2} = \neg q, \overline{a_3} = \neg s, \overline{a_4} = \neg t, \overline{a_5} = p$, $\overline{d_1} = d_2, \overline{d_2} = d_1$.

---

[4] We adopt here the simpler translation given in section 3.1, as there are no preference rules in ϵ.

By virtue of this formulation, any notion of "acceptable" set of assumptions may be adopted to provide a semantics to $\langle \epsilon, \mathcal{D}, \mathcal{G} \rangle$. For instance, in example 4, $\{a_1, a_2, d_1\}$ and $\{a_3, a_4, a_5, d_2\}$ are both admissible sets of assumptions, with corresponding "outputs" $\{p, q\}$ and $\{s, t, \neg p\}$.

Amongst all acceptable sets of arguments, we want to consider solely those having the goals in $\mathcal{G}$ in their "output".

Definition 21. *Given* $\langle \mathcal{L}_\pi, \mathcal{R}_\pi, \mathcal{A}_\pi, \overline{} \rangle$, *an "acceptable" set of assumptions* Δ *wrt* $\langle \mathcal{L}_\pi, \mathcal{R}_\pi, \mathcal{A}_\pi, \overline{} \rangle$ *is* desired *iff* $\mathcal{G} \subseteq O(\Delta)$, *where* $O(\Delta) = \{x \in \mathcal{L}_\pi | \Delta' \vdash x, \Delta' \subseteq \Delta\}$.

Thus, practical reasoning may be realised within ABA by identifying acceptable sets of assumptions that contain a support for the desired goals. For instance, given $\mathcal{G} = \{\neg p\}$ in example 4, $\{a_3, a_4, a_5, d_2\}$ is desired admissible, whereas $\{a_1, a_2, d_1\}$ is not.

Note that in some cases no desired "acceptable" set of assumptions may exist, if the goals are incompatible. This may happen for example 4, for instance, given $\mathcal{G} = \{\neg p, q\}$. The use of stratification in $\mathcal{G}$ (that we have ignored here for simplicity) will help in general with identifying desired "acceptable" sets. For example, if $\mathcal{G} = \langle \{\neg p\}, \{q\} \rangle$, then $\{a_3, a_4, a_5, d_2\}$ is desired admissible, if instead $\mathcal{G} = \langle \{q\}, \{\neg p\} \rangle$, then $\{a_1, a_2, d_1\}$ is desired admissible.

Note that the translation proposed in definition 20 can be "optimised", to eliminate assumptions that serve no purpose. For instance, the optimised version of the ABA framework in example 4 might have $\mathcal{R}'_\epsilon=$

$$\{ p \leftarrow a_1; \quad q \leftarrow p, d_1;$$
$$s \leftarrow d_2; \quad t \leftarrow s, a_4; \quad \neg p \leftarrow t, a_5\}$$

with $\mathcal{A}'_\pi = \{a_1, a_4, a_5, d_1, d_2\}$ and $\overline{a_1} = \neg p$, $\overline{a_4} = \neg t$, $\overline{a_5} = p$, $\overline{d_1} = \{d_2, \neg q\}$, $\overline{d_2} = \{d_1, \neg s\}$.

5 Example

We show here how to deal with a variant of the example given in [1], whereby a judge needs to decide how best to punish a criminal found guilty, while deterring the general public, rehabilitating the offender, and protecting society from further crime. The judge can choose amongst three forms of punishment: (i) imprisonment, (ii) a fine, or (iii) community service. The judge believes that: (i) promotes deterrence and protection to society, but it demotes rehabilitation; (ii) promotes deterrence but has no effect on rehabilitation and protection of society; (iii) promotes rehabilitation but demotes deterrence.

This problem can be expressed as a practical reasoning framework $\pi = \langle \epsilon, \mathcal{D}, \mathcal{G} \rangle$ where $\epsilon = D$ with

$$prison \rightarrow punish; \quad prison \rightarrow deter;$$
$$service \rightarrow rehabilitate; \quad fine \rightarrow punish;$$

$$fine \rightarrow deter; \quad prison \rightarrow \neg rehabilitate;$$
$$service \rightarrow punish; \quad service \rightarrow \neg deter;$$
$$prison \rightarrow protect$$

and $\mathcal{D} = \{prison, fine, service\}$. We can represent the problem as a generalised ABA framework $\langle \mathcal{L}_\pi, \mathcal{R}_\pi, \mathcal{A}_\pi, \overline{} \rangle$ where $\mathcal{R}_\pi$ is

$$\{ punish \leftarrow prison, a_1; \quad deter \leftarrow prison, a_2;$$
$$rehabilitate \leftarrow service, a_3; \quad punish \leftarrow fine, a_4;$$
$$deter \leftarrow fine, a_5; \quad \neg rehabilitate \leftarrow prison, a_6;$$
$$punish \leftarrow service, a_7; \quad \neg deter \leftarrow service, a_8;$$
$$protect \leftarrow prison, a_9\}$$

$\mathcal{A}_\pi = \{prison, fine, service, a_1, \ldots, a_9\}$, and

$$\overline{prison} = \{fine, service\}, \quad \overline{fine} = \{prison, service\},$$
$$\overline{service} = \{prison, fine\}, \quad \overline{a_1} = \overline{a_4} = \overline{a_7} = \neg punish,$$
$$\overline{a_2} = \overline{a_5} = \neg deter, \quad \overline{a_3} = \neg rehabilitate,$$
$$\overline{a_6} = rehabilitate, \quad \overline{a_7} = \neg punish,$$
$$\overline{a_8} = deter, \quad \overline{a_9} = \neg protect.$$

If $\mathcal{G} = \{punish\}$ then $\{prison, a_1\}$, $\{fine, a_4\}$ and $\{service, a_7\}$ are all desired admissible, giving decision $prison$, $fine$ and $service$, respectively. If $\mathcal{G} = \{punish, deter, rehabilitate\}$ then no desired admissible set of assumptions exists, and thus no decision.

Note that our analysis of this example focuses on illustrating the argumentation approach taken in this paper. However, from a legal perspective, one would need to weigh the general semantic, argumentation notions in the light of the suspect and the crime. This would amount, in the case of $\mathcal{G} = \{punish\}$ for example, to choose one amongst the (equally) desired admissible $\{prison, a_1\}$, $\{fine, a_4\}$ and $\{service, a_7\}$.

Assume now that the third rule $service \rightarrow rehabilitate$ is replaced by the rule $service, motivation \rightarrow rehabilitate$, and the following rules are added to $\mathcal{D}$:

$$motivation$$
$$\neg motivation$$
$$n_{11} \succ n_{10}$$

where $n_{10} = \mathcal{N}(motivation)$ and $n_{11} = \mathcal{N}(\neg motivation)$. Then, the corresponding generalised ABA framework is $\langle \mathcal{L}'_\pi, \mathcal{R}'_\pi, \mathcal{A}'_\pi, \overline{} \rangle$ where $\mathcal{R}'_\pi = \mathcal{R}_\pi -$

$\{rehabilitate \leftarrow service, a_3\} \cup New$ with

$$New = \{ \ rehabilitate \leftarrow service, motivation, a_3;$$
$$motivation \leftarrow b_{10}; \quad b_{10} \leftarrow a_{10}; \quad c_{10} \leftarrow n_{11} \succ n_{10}, b_{11};$$
$$\neg motivation \leftarrow b_{11}; \quad b_{11} \leftarrow a_{11}; \quad c_{11} \leftarrow n_{10} \succ n_{11}, b_{10};$$
$$n_{11} \succ n_{10} \leftarrow a_{12}\}$$

$\mathcal{A}'_\pi = \mathcal{A}_\pi \cup \{a_{10}, a_{11}, a_{12}\}$, and contraries are defined as before with, in addition: $\overline{a_{10}} = c_{10}$, $\overline{a_{11}} = c_{11}$, $\overline{a_{12}} = n_{10} \succ n_{11}$. Then, for example, if $\mathcal{G} = \{punish, rehabilitate\}$ then no set of assumptions is admissible (and thus no decision is possible).

6   Conclusions

We have proposed concrete instances of assumption-based argumentation for epistemic reasoning, with defeasible rules and preferences between them, also specified by means of defeasible rules, and for practical reasoning, with defeasible rules, preferences between them, and decisions affecting beliefs. With respect to other argumentation-based approaches to these kinds of reasoning (e.g. [21]), our approach benefits from the availability of an implemented system (to be used for experimentation and for realising applications) [10,11], and the possibility of using many diverse semantics for argumentation (rather than committing to a specific one from the onset).

For both kinds of reasoning, we have ignored strict rules and their interplay with defeasible rules. An extension of our approach to accommodate strict rules can be found in [22].

We have illustrated our approach by means of a legal example borrowed from [1]. Within the ARGUGRID project, our approach to (epistemic and) practical reasoning can be used to model decisions concerning the orchestration of services available over the grid, taking into account preferences by the users and/or the service providers.

Like [1], our approach adopts an abductive approach to practical reasoning, but this can be directly modelled within assumption-based argumentation that is, fundamentally, abductive by its very nature.

Most approaches dealing with dynamic preferences in epistemic reasoning (e.g. [19,3]), rely upon the framework of extended logic programming, whereby default negation may occur in the premises of rules, interpreted as clauses. Here, as in [16], we omit default negation. Indeed, the effects of this kind of negation may be obtained in our framework by means of negative defeasible facts with top preference, as indicated in [14].

In the case of practical reasoning, for simplicity, we have assumed that all given goals and decisions are equally preferred: future work includes addressing cases where preferences over goals and decisions can be specified, either in terms of fixed partial orders or via defeasible rules in turn, following [16].

Acknowledgements

The author has been supported by a UK Royal Academy of Engineering/ Leverhulme Trust Senior Research Fellowship and by the Sixth Framework IST programme of the EC, under the 035200 ARGUGRID project. Many thanks to Adil Hussain and anonymous referees for useful comments on an earlier version of this paper.

References

1. Bench-Capon, T., Prakken, H.: Justifying actions by accruing arguments. In: COMMA 2006 (2006)
2. Bondarenko, A., Dung, P., Kowalski, R., Toni, F.: An abstract, argumentation-theoretic framework for default reasoning. Artificial Intelligence 93(1-2), 63–101 (1997)
3. Brewka, G.: Well-founded semantics for extended logic programs with dynamic preferences. Journal of Artificial Intelligence Research 4, 19 (1996)
4. Caminada, M., Amgoud, L.: An axiomatic account of formal argumentation. In: Proc. AAAI (2005)
5. Dimopoulos, Y., Nebel, B., Toni, F.: On the computational complexity of assumption-based argumentation for default reasoning. Artificial Intelligence 141, 57–78 (2002)
6. Dung, P.: The acceptability of arguments and its fundamental role in non-monotonic reasoning and logic programming and n-person game. Artificial Intelligence 77, 321–357 (1995)
7. Dung, P., Kowalski, R., Toni, F.: Dialectic proof procedures for assumption-based, admissible argumentation. Artificial Intelligence 170, 114–159 (2006)
8. Dung, P., Mancarella, P., Toni, F.: A dialectic procedure for sceptical, assumption-based argumentation. In: 1st International Conference on Computational Models of Argument (COMMA 2006) (September 2006)
9. Dung, P.M., Mancarella, P., Toni, F.: Computing ideal sceptical argumentation. Artificial Intelligence, Special Issue on Argumentation in Artificial Intelligence 171(10-15), 642–674 (2007)
10. Gaertner, D., Toni, F.: A credulous and sceptical argumentation system. In: Proceedings of ArgNMR (2007), www.doc.ic.ac.uk/~dg00/casapi.html
11. Gaertner, D., Toni, F.: On computing arguments and attacks in assumption-based argumentation. IEEE Intelligent Systems, Special Issue on Argumentation Technology 22(6), 24–33 (2007)
12. Garcia, A., Simari, G.: Defeasible logic programming: An argumentative approach. Journal of Theory and Practice of Logic Prog. 4(1-2), 95–138 (2004)
13. Governatori, G., Maher, M.J., Billington, D., Antoniou, G.: Argumentation semantics for defeasible logics. Journal of Logic and Computation 14(5), 675–702 (2004)
14. Kakas, A., Mancarella, P., Dung, P.: The acceptability semantics for logic programs. In: Hentenryck, P.V. (ed.) Proc. ICLP, pp. 504–519. MIT Press, Cambridge (1994)
15. Kakas, A., Moraitis, P.: Argumentation based decision making for autonomous agents. In: Proceedings of AAMAS 2003, pp. 883–890 (2003)
16. Kakas, A.C., Moraitis, P.: Argumentation based decision making for autonomous agents. In: AAMAS, pp. 883–890 (2003)

17. Kowalski, R.A., Toni, F.: Abstract argumentation. Journal of Artificial Intelligence and Law, Special Issue on Logical Models of Argumentation 4(3-4), 275–296 (1996)
18. Prakken, H., Sartor, G.: On the relation between legal language and legal argument: assumptions, applicability and dynamic priorities. In: Proc. of the 5th ICAIL, pp. 1–10. ACM Press, New York (1995)
19. Prakken, H., Sartor, G.: Argument-based extended logic programming with defeasible priorities. Journal of Applied Non-Classical Logics 7(1), 25–75 (1997)
20. Prakken, H., Sartor, G.: The role of logic in computational models of legal argument: a critical survey. In: Kakas, A.C., Sadri, F. (eds.) Computational Logic: Logic Programming and Beyond. LNCS (LNAI), vol. 2408, pp. 342–381. Springer, Heidelberg (2002)
21. Rahwan, I., Amgoud, L.: An argumentation-based approach for practical reasoning. In: Proc. AAMAS 2006, pp. 347–354. ACM Press, New York (2006)
22. Toni, F.: Assumption-based argumentation for closed and consistent defeasible reasoning. In: Nitta, K., Tojo, S., Satoh, K. (eds.) Proceedings First International Workshop on Juris-informatics (JURISIN 2007), in association with The 21th Annual Conference of The Japanese Society for Artificial Intelligence (JSAI 2007), 19 June (2007)

Computing Argumentation for Decision Making in Legal Disputes

Maxime Morge

Dipartimento di Informatica, Università di Pisa
via F. Buonarroti, 2 I-56127 Pisa, Italy
morge@di.unipi.it
http://maxime.morge.org

Abstract. In this paper, we present a decision support system for lawyers. This system is built upon an argumentation framework for decision making. A logic language is used as a concrete data structure for holding the statements like knowledge, goals, and decisions. Different priorities are attached to these items corresponding to the uncertainty of the knowledge about the circumstances, the lawyer's preferences, and the expected utilities of sentences. These concrete data structures consist of information providing the backbone of arguments. Due to the abductive nature of practical reasoning, arguments are built by reasoning backwards, and possibly by making suppositions over missing information. Moreover, arguments are defined as tree-like structures. In this way, our computer system, implemented in Prolog, suggests some actions and provides an interactive and intelligible explanation of this solution.

1 Introduction

Since legal disputes are resolved by confronting and evaluating the justifications of parties' positions, argumentation is central to law. This is the reason why many works in the area of Artificial Intelligence & Law focus on the computational model of argumentation. In particular, nonmonotonic logic techniques have been used to model the vagueness, indeterminacy and adversarial nature of the law with hierarchies of possibly conflicting rules (see [1] for a survey). However, even if modern techniques are used, this logical approach is still limited to the epistemic reasoning and do not encompass practical reasoning. The point is that a legal dispute in criminal cases is not only limited to draw conclusions (e.g. the guilt or the innocence of an accused) but must determine a sentence, i.e. take a decision.

In this paper, we present a decision support system for lawyers. This system is built upon an Argumentation Framework (AF) for decision making. A logic language is used as a concrete data structure for holding the statements like knowledge, goals, and decisions. Different priorities are attached to these items corresponding to the uncertainty of the knowledge about the circumstances, the lawyer's preferences, and the expected utilities of sentences. These concrete data structures consist of information providing the backbone of arguments. Due

P. Casanovas et al. (Eds.): Computable Models of the Law, LNAI 4884, pp. 203–218, 2008.

to the abductive nature of practical reasoning, arguments are built by reasoning backwards, and possibly by making suppositions over missing information. Moreover, arguments are defined as tree-like structures. In this way, our computer system, implemented in Prolog, suggests some actions and provides an interactive and intelligible explanation of this solution.

Section 2 introduces the walk-through example. In order to present our AF, we will browse the following fundamental notions. Firstly, we define the *object language* (cf Section 3) and the associated priorities (cf Section 4). Secondly, we will focus on the internal structure of *arguments* (cf Section 5). We present in Section 6 the *interactions* between them. These relations allow us to give a declarative model-theoretic *semantics* to our AF (cf Section 7) and we adopt a *dialectical proof procedure* to implement it (cf Section 8). Section 9 discusses some related works. Section 10 draws some conclusions and directions for future work.

2   Walk-through Example

Inspired by [2], we consider here criminal sentencing. Such a decision making problem requires a proper understanding of all relevant aspects. The goals for the sentences such as the punishment, the deterrence, the rehabilitation, and so on, as well as the knowledge about the surrounding circumstances, such as the influence of alcohol or drugs is also of vital importance. The judge is responsible for sentencing, based on the explicit goals and on her knowledge.

We assume that the user provides, via the GUI, *influence diagrams* [3]. These are simple graphical representations of multi-attribute decision problems. Here, they are used by the judge to display the structure of the decision problem related to the criminal case. In addition, the GUI allows the justifications to communicate specific details, in particular facts and preferences.

In legal disputes about criminal cases, the main goal, that consists in an appropriate sentence (denoted jdg), is addressed by a set of decisions, i.e. a choice amongst some sentences. The accused can (or cannot) be put in prison (Prison(yes), Prison(no)), the accused can (or cannot) do a community service (Service(yes) or Service(no)). The accused can (or cannot) be fine without payment (Fine(yes) or Fine(no)). The main goal is split into independent subgoals. The judgement must punish the offender (pu), rehabilitating the offender (re), protecting the society from crime (pt), and deterring the general public (de). The knowledge about the crime is expressed with predicates such as: guilty (the accused is found guilty), alcohol/drug (the crime is influenced by alcohol/drugs), driving (the crime is driving related) or mobile (the criminal was using a mobile phone[1]).

Figure 1 provides a simple graphical representation of the decision problem called influence diagram. The elements of the decision problem, i.e. *values* (represented by rectangles with rounded corners), *decisions* (represented by rectangles)

---

[1] We consider here the particular french jurisdiction where the usage of mobile is forbidden during driving.

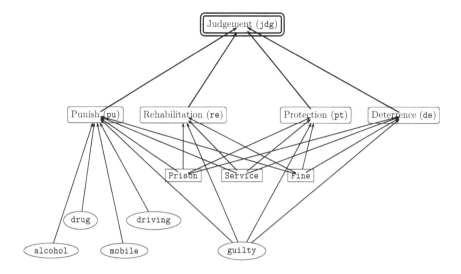

Fig. 1. Influence diagram for criminal sentencing

and *knowledge* (represented by ovals), are connected by arcs where predecessors affect successors. We consider here a multiattribute decision problem captured by a hierarchy of values where the abstract value (represented by rectangles with rounded corner and double line) aggregates the values in the lower level. When the structure of the decision is built, the alternatives must be identified, the preferences must be expressed and the knowledge gathered.

The judge also provides, through the GUI, the facts and her preferences. For example, the judge knows that the crime was influenced by alcohol and she does not know if the crime was made during the usage of a mobile. Due to conflicting sources of information, the judge has conflicting information about the influence of drugs and the fact that the crime is driving related. These sources of information are more or less reliable. The judge can (or cannot) have preferences over them.

In order to represent the structure of the decision and to express preferences and constraints, an object language is required.

3   The Object Language

Since we want to provide a computational model of argumentation for decision making and we want to instantiate it for our example, we need to specify a particular logic.

The object language expresses rules and facts in logic-programming style. In order to address a decision making problem, we distinguish:

- a set of *abstract goals*, i.e. some propositional symbols which represent the abstract features that the decisions must exhibit (in the example jdg is the only abstract goal);
- a set of *concrete goals*, i.e. some propositional symbols which represent the concrete features that the decisions must exhibit (in the example pu, re, pt and de);
- a set of *decisions*, i.e. some predicate symbols which represent the actions which must be performed or not (in the example Prison, Service, and Fine);
- a set of *alternatives*, i.e. some constants symbols which represent the mutually exclusive solutions for each decision (in the example yes, or no);
- a set of *beliefs*, i.e. some propositional symbols which represent epistemic statements (in the example guilty, alcohol, ...). In the language, we explicitly distinguish *assumable* beliefs (respectively *non-assumable*) beliefs, which can (respectively cannot) be taken for granted. Since we cannot make the supposition that the accused is guilty, guilty is non-assumable.

Since we want to consider conflicts in this object language, we need some forms of negation. For this purpose, we consider strong negation, also called explicit or classical negation, and weak negation, also called negation as failure. A strong literal is an atomic first-order formula, possible preceded by strong negation $\neg$. A weak literal is a literal of the form $\sim L$, where L is a strong literal. $\neg L$ says "L is definitely not the case", while $\sim L$ says "There is no evidence that L is the case". In order to express in a compact way the mutual exclusion between statements, such as the different alternatives for a decision, we define the incompatibility relation (denoted by $\mathcal{I}$) as a binary relation over atomic formulas which is asymmetric. Whatever the atom L is a belief or a goal, we have $L \mathcal{I} \neg L$ and $\neg L \mathcal{I} L$, while we have $L \mathcal{I} \sim L$ but we do not have $\sim L \mathcal{I} L$. Obviously, $D_1(a_1) \mathcal{I} D_1(a_2)$ and $D_1(a_2) \mathcal{I} D_1(a_1)$, D_1 being a decision predicate, a_1 and a_2 being different[2] alternatives for D. Moreover, some sentences can be incompatible. For instance, Prison(yes) $\mathcal{I}$ Fine(yes) and Fine(yes) $\mathcal{I}$ Prison(yes) but neither Prison(no) $\mathcal{I}$ Fine(no) and nor Fine(no) $\mathcal{I}$ Prison(no). Similarly, we say that two sets of sentences Φ_1 and Φ_2 are incompatible ($\Phi_1 \mathcal{I} \Phi_2$) iff there is a sentence ϕ_1 in Φ_1 and a sentence ϕ_2 in Φ_2 such as $\phi_1 \mathcal{I} \phi_2$. A theory gathers the statements about the decision making problem.

Definition 1 (Theory). *A theory $\mathcal{T}$ is an extended logic program, i.e a finite set of rules such as $R : L_0 \leftarrow L_1, \ldots, L_j, \sim L_{j+1}, \ldots, \sim L_n$ with $n \geq 0$, each L_i being a strong literal. The literal L_0, called the* head *of the rule, is denoted head(R). The finite set $\{L_1, \ldots, \sim L_n\}$, called the* body *of the rule, is denoted body(R). The body of a rule can be empty. In this case, the rule, called a fact, is an unconditional statement. R, called the name of the rule, is an atom in the language $\mathcal{L}$. All rules are ground.*

---

[2] Notice that in general a decision can be addressed by more than two alternatives.

Considering a decision making problem, we distinguish:

- *goal rules* of the form $R : G_0 \leftarrow G_1, \ldots, G_n$ with $n > 0$. Each G_i is a goal literal (or its negation). According to this rule, the goal is promoted (or demoted) by the combination of goal literals in the body;
- *epistemic rules* of the form $R : B_0 \leftarrow B_1, \ldots, B_n$ with $n \geq 0$. Each B_i is a belief literal. According to this rule, the belief B_0 is true if the conditions $B_1, \ldots, B_n$ are satisfied;
- *decision rules* of the form $R : G \leftarrow D(a), B_1, \ldots, B_n$ with $n \geq 0$. The head of the rule is a concrete goal (or its strong negation). The body includes a decision literal $(D(a))$ and a possible empty set of belief literals. According to this rule, the goal is promoted (or demoted) by the decision $D(a)$, provided that conditions $B_1, \ldots, B_n$ are satisfied.

Considering statements in the theory is not sufficient to take a decision.

4   Priority

In order to evaluate the previous statements, all relevant pieces of information should be taken into account, such as the likelihood of beliefs, the preferences between goals, or the expected utilities of the decisions.

In Mathematics, order relations are binary relations on a set. Since these relations classify the elements from the 'best' to the 'worst', with or without *ex æquo*, they are qualitative. For this purpose, we can consider either a preorder, i.e. a reflexive and transitive relation considering possible *ex æquo*, or an order, i.e. an antisymmetric preorder relation. The preorder (respectively the order) is total iff all elements are comparable. In this way, we consider that the *priority* $\mathcal{P}$ is a (partial or total) preorder on the rules in $\mathcal{T}$. $R_1 \mathcal{P} R_2$ can be read "R_1 has priority over R_2". $R_1 \not{\mathcal{P}} R_2$ can be read "R_1 has no priority over R_2", either because R_1 and R_2 are *ex æquo* (denoted $R_1 \sim R_2$), i.e. $R_1 \mathcal{P} R_2$ and $R_2 \mathcal{P} R_1$, or because R_1 and R_2 are not comparable, i.e. $\neg(R_1 \mathcal{P} R_2)$ and $\neg(R_2 \mathcal{P} R_1)$.

In this work, we consider that all rules are potentially defeasible and that the priorities are extra-logical and domain-specific features. The priority over concurrent rules depends of the nature of rules. Rules are *concurrent* if their heads are identical or incompatible. We define three priority relations:

- the priority over *goal rules* comes from the *preferences* overs goals. The priority of such rules corresponds to the relative importance of the combination of (sub)goals in the body as far as reaching the goal in the head is concerned;
- the priority over *epistemic rules* comes from the *uncertainty* of knowledge. The prior the rule is, the more likely the rule holds;
- the priority over *decision rules* comes from the *expected utility* of decisions. The priority of such rules corresponds to the expectation of the conditional decision in promoting the goal literal.

In order to illustrate the notions introduced previously, let us consider the example. The goal theory, the epistemic theory, and the decision theory are

Table 1. The goal theory (at upper left),the epistemic theory (at lower left), and the decision theory (at right)

r_{11} : pu ← Prison(yes), drug, driving, guilty
r'_{11} : pu ← Prison(yes), alcohol, driving, guilty
r''_{11} : pu ← Prison(yes), mobile, driving, guilty
r_{12} : pu ← Service(yes), guilty
r_{13} : pu ← Fine(yes), ∼ guilty
r_{212} : re ← Prison(yes), Service(yes), guilty
r'_{212} : ¬re ← Prison(yes), Service(no), guilty
r_{422} : de ← Prison(yes), Service(yes), guilty
r'_{422} : de ← Prison(no), Service(yes), guilty

Goal theory (upper left):

r_{01} : jdg ← pu, re, pt, ¬de
r_{02} : jdg ← pu, re, ¬de
r_{03} : jdg ← pu, ¬re, pt, de
r_{04} : jdg ← pu, re, de

Epistemic theory (lower left):

f_0 : driving ←
f_1 : guilty ←
f_2 : drug ←
f'_2 : ¬drug ←
f_3 : alcohol ←
f'_0 : ¬driving ←

Decision theory (right, continued):

r_{21} : ¬re ← Prison(yes), guilty
r'_{21} : ¬re ← Prison(yes), ∼ guilty
r_{22} : re ← Service(yes), guilty
r'_{22} : ¬re ← Service(yes), ∼ guilty
r'_{23} : re ← Fine(yes), ∼ guilty
r_{31} : pt ← Prison(yes), guilty
r_{33} : ¬pt ← Fine(yes), guilty
r'_{33} : pt ← Fine(yes), ∼ guilty
r_{41} : de ← Prison(yes), guilty
r'_{41} : de ← Prison(yes), ∼ guilty
r_{42} : ¬de ← Service(yes), guilty
r'_{42} : de ← Service(yes), ∼ guilty
r_{43} : ¬de ← Fine(yes), guilty
r'_{43} : de ← Fine(yes), ∼ guilty

represented in Table 1. A rule above another one has priority over it. To simplify the graphical representation of the theories, they are stratified in non-overlapping subsets, i.e. different levels. The *ex æquo* rules are grouped in the same level. Non-comparable rules are arbitrarily assigned to a level.

According to the decision theory, the community service is relevant to punish a criminal (r_{12}), being fined is relevant for the (non)punishment of an innocent (r_{13}), and the prison is relevant to punish a driving related crime influenced by drugs (r_{11}) or alcohol (r'_{11}), or made during the usage of a mobile phone (r''_{11}). Actually, the utilities of these alternatives with respect to pu depends on the surrounding circumstances. "Do a community service" is stronger in promoting re than "go to prison" in demoting re (r_{212} $\mathcal{P}$ r_{21}). Similarly, "go to prison" is stronger in promoting de than "do a community service" in demoting deterrence (r_{422} $\mathcal{P}$ r_{42}). Our formalism allows to capture the mutual influence of decisions over the independent goals.

According to the goal theory, achieving the goals pu, re, and pt and avoiding de is required to reach jdg (cf r_{01}). However, these constraints can be relaxed. We make pu an essential goal by requiring it also in r_{02}, in r_{03}, and in r_{04}. The achievement of pt can be relaxed (r_{01} $\mathcal{P}$ r_{02}). Moreover, the achievement of re is more important than de and pt put together (r_{01} $\mathcal{P}$ r_{03} and r_{02} $\mathcal{P}$ r_{03}) and promoting re while demoting de is preferable to promoting de (r_{02} $\mathcal{P}$ r_{04}).

Our formalism allows to capture complex and incomplete information about the preferences amongst goals.

According to the epistemic theory, the judge finds the accused guilty (f_1), knows that the crime was influenced by alcohol (f_3), and does not know if the crime was made during the usage of a mobile phone. Due to conflicting sources of information, the judge has conflicting information about the influence of drugs $(f_2$ and $f_2')$ and the fact that the crime is driving related $(f_0$ and $f_0')$. The sources of information can be more or less reliable. For instance, we have $f_0 \mathcal{P} f_0'$ but there is no strict priority between f_2 and f_2'. Our formalism allows to capture complex (and incomplete) information about the likelihood of the surrounding circumstances. We will build now arguments upon these (incomplete) statements in order to compare the alternatives.

5   Arguments

Due to the abductive nature of the practical reasoning, we define and construct arguments by reasoning backwards, and possibly by making suppositions over missing information. Since we adopt a tree-like structure of arguments, our framework not only suggests some actions but also provides an intelligible explanation of them.

The simplest way to define an argument is by a pair ⟨ premises, conclusion ⟩ as in [4]. This definition leaves implicit that the underlying logic validates a proof of the conclusion from the premises. When the argumentation framework is built upon an extended logic program, an argument is often defined as a sequence of rules [5]. These definitions ignore the recursive nature of arguments: arguments are composed of subarguments, subarguments for these subarguments, and so on. For this purpose, we adopt the tree-like structure for arguments proposed in [6] and we extend it with suppositions on the missing information.

Definition 2 (Argument). *An* argument *is composed by a conclusion, a top rule, some premises, some suppositions, and some sentences. These elements are abbreviated by the corresponding prefixes. An argument A is:*

1. *a hypothetical argument built upon an unconditional ground statement. If L is a assumable belief, then the argument built upon this assumable belief is defined as follows[3] $conc(A) = L$, $top(A) = \theta$, $premise(A) = \emptyset$, $supp(A) = \{L\}$, $sent(A) = \{L\}$.*
   or
2. *a built argument built upon a rule such that all the literals in the body are the conclusion of subarguments.*
   1) If f is a fact in $\mathcal{T}$ (i.e. $body(f) = \emptyset$), then the trivial *argument A built upon this fact is defined as follows: $conc(A) = head(f)$, $top(A) = f$, $premise(A) = \emptyset$, $supp(A) = \emptyset$, $sent(A) = \{head(f)\}$.*

---

[3] θ denotes that no literal is required.

2) If r is a rule in $\mathcal{T}$, we define the tree *argument A built upon this rule as follows. Let $body(r) = \{L_1, \ldots, L_j, \sim L_{j+1}, \sim L_n\}$ and $sbarg(A) = \{A_1, \ldots, A_n\}$ be the collection of arguments such that, for each strong literal $L_i \in body(r)$, $conc(A_i) = L_i$ with $i \leq j$ or $conc(A_i) = \sim L_i$ with $i > j$ (each A_i is called a subargument of A). Then: $conc(A) = head(r)$, $top(A) = r$, $premise(A) = body(r)$, $supp(A) = \cup_{A' \in sbarg(A)} supp(A')$, $sent(A) = \cup_{A' \in sbarg(A)} sent(A') \cup body(r) \cup \{head(r)\}$.*

As in [6], we consider *composite* arguments (3) and *atomic* arguments (2) where the top rule is a fact. Contrary to the other definitions of arguments (pair of premises - conclusion, sequence of rules), our definition considers that the different premises can be challenged and can be supported by subarguments. In this way, arguments are intelligible explanations. Moreover, we distinguish *hypothetical* arguments (1) and *built* arguments (2/3). While built arguments are built upon a top rule which is a rule or a fact of the theory, hypothetical arguments are built upon missing information. In this way, our framework allows to reason further by making suppositions related to the unknow beliefs and over possible decisions under which arguments can be built. Due to the abductive nature of practical reasoning, we define and construct arguments by reasoning backwards. Therefore, arguments do not include irrelevant information such as sentences not used to derive the conclusion.

Let us consider the previous example. Some of the arguments concluding pu are depicted in Figure 2. According to the argument B, the accused will be punish if he is guilty and we suppose that he does a community service. According to the argument A_1 (respectively A_2), the accused will be punish if he is guilty of a driving-related crime, if we suppose he goes in Prison, and if he is influence by drugs (respectively we suppose he was using a mobile phone). An argument can be represented as tree where the root is the conclusion (represented by a triangle) directly connected to the premises (represented by losanges) if they exist, and where the leefs are either some suppositions (represented by circles) or emptyset. Each plain arrow corresponds to a rule (or a fact) where the head node corresponds to the head of the rule and the tall nodes are in the body of the rule. The tree argument A^1 is composed of four subarguments: one hypothetical argument and three trivial arguments. The tree argument A^2 is composed of four subarguments: two hypothetical arguments and two trivial arguments. Neither trivial arguments nor hypothetical arguments contain subarguments. Due to their structures and their natures, arguments interact with one another.

6   Interactions between Arguments

The interactions between arguments may come from the incompatibility of their sentences, from their nature (hypothetical or built) and from the priority over rules. We examine in turn these different sources of interaction.

Since their sentences are conflicting, arguments interact with one another. For this purpose, we define the attack relation.

Definition 3 (Attack relation). *Let A and B be two arguments. A attacks B (denoted by attacks (A, B)) iff $sent(A) \mathcal{I} sent(B)$.*

This relation encompasses both the direct (often called *rebuttal*) attack due to the incompatibility of the conclusions, and the indirect (often called *undermining*) attack, i.e. directed to a "subconclusion". According to this definition, if an argument attacks a subargument, the whole argument is attacked. The attack relation is useful to build arguments which are homogeneous explanations.

Due to the nature of argument, arguments are more or less hypothetical. This is the reason why we define the size of their suppositions.

Definition 4 (Supposition size). *Let A be an arguments. The size of suppositions for A, denoted $\mathbf{suppsize}(A)$, is defined such that:*

1. *if A is a hypothetical argument, then $\mathbf{suppsize}(A) = 1$;*
2. *if A is a trivial argument, then $\mathbf{suppsize}(A) = 0$;*
3. *if A is a tree argument and $sbarg(A) = \{A_1, \ldots, A_n\}$ is the collection of subarguments of A, then $\mathbf{suppsize}(A) = \Sigma_{A' \in sbarg(A)} \mathbf{suppsize}(A')$.*

The size of suppositions for an argument does not only count the number of hypothetical subarguments which compose the argument but also counts the number of hypothetical subarguments of these subarguments, and so on.

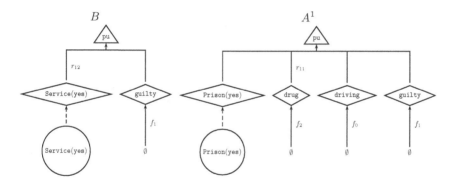

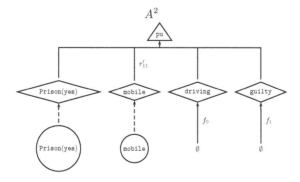

Fig. 2. Some arguments concluding **pu**

Since arguments have different natures (hypothetical or built) and the top rules of built arguments are more or less strong, they interact with one another. For this purpose, we define the strength relation.

Definition 5 (Strength relation). *Let A_1 be a hypothetical argument, and A_2, A_3 be two built arguments.*

1. *A_2 is stronger than A_1 (denoted $A_2 \; \mathcal{P}^{\mathcal{A}} \; A_1$);*
2. *If $(top(A_2) \; \mathcal{P} \; top(A_3)) \wedge \neg(top(A_3) \; \mathcal{P} \; top(A_2))$, then $A_2 \; \mathcal{P}^{\mathcal{A}} \; A_3$;*
3. *If $(top(A_2) \mathcal{P} top(A_3)) \wedge (suppsize(A_2) < suppsize(A_3))$, then $A_2 \; \mathcal{P}^{\mathcal{A}} \; A_3$;*

Since $\mathcal{P}$ is a preorder on $\mathcal{T}$, $\mathcal{P}^{\mathcal{A}}$ is a preorder on $\mathcal{A}(\mathcal{T})$. Built arguments are preferred to hypothetical arguments. An argument is stronger than another argument if the top rule of the first argument has a proper higher priority that the top rule of the second argument or if the top rule of the first argument does not have a proper higher priority but the number of suppositions made in the first argument is properly smaller than the number of suppositions made in the second argument. The strength relation is useful to choose (when it is possible) between homogeneous concurrent explanations, i.e. non conflicting arguments with the same conclusions.

The two previous relations can be combined to choose (if possible) between non-homogeneous concurrent explanations, i.e. conflicting arguments with the same conclusions.

Definition 6 (Defeats). *Let A and B be two arguments. A defeats B (written defeats (A, B)) iff:*

1. *attacks (A, B);*
2. *$\neg(B \; \mathcal{P}^{\mathcal{A}} \; A)$.*

Similarly, we say that a set S of arguments defeats an argument A if A is defeated by one argument in S.

Let us consider our previous example. The arguments in favor of prison (A^1 and A^2) and the argument in favor of community service (B) attack each other. Since the top rule of A^1 (i.e. R_{11}), the top rule of A^2 (i.e. R''_{11}), and the top rule of B (i.e. R_{12}) are not stronger than each other, A^1/A^2 defeat B and B defeats A^1/A^2. If we only consider these three arguments, the judge cannot decide what the best alternatives are, and the best arguments to explain the choices. However, A^1, which is composed of one hypothetical argument and three trivial arguments, is "better" than A^2, which is composed of two hypothetical arguments and two trivial arguments. Determining whether a suggestion and an explanation are ultimately suggested requires a complete analysis of all arguments and subarguments. In this section, we have defined the interactions between arguments in order to give them a status.

7 Semantics

We can consider our AF abstracting away from the logical structures of arguments. This abstract AF consists of a set of arguments associated with a binary defeat relation.

Given an AF, [7] and [8] define the following notions of "acceptable" sets of arguments:

Definition 7 (Semantics). *An AF is a pair $\langle \mathcal{A}, \text{defeats} \rangle$ where $\mathcal{A}$ is a set of arguments and defeats $\subseteq \mathcal{A} \times \mathcal{A}$ is the defeat relationship[4] for AF. For $A \in \mathcal{A}$ an argument and $S \subseteq \mathcal{A}$ a set of arguments, we say that:*

- *A is* acceptable *with respect to S (denoted $A \in S_{\mathcal{A}}^{S}$) iff $\forall B \in \mathcal{A}$, defeats (B, A) $\exists C \in S$ such that defeats (C, B);*
- *S is* conflict-free *iff $\forall A, B \in S \neg$ defeats (A, B);*
- *S is* admissible *iff S is conflict-free and $\forall A \in S, A \in S_{\mathcal{A}}^{S}$;*
- *S is* preferred *iff S is maximally admissible;*
- *S is* complete *iff S is admissible and S contains all arguments A such that S defeats all defeaters against A;*
- *S is* grounded *iff S is minimally complete;*
- *S is* ideal *iff S is admissible and it is contained in every preferred sets.*

The semantics of an admissible (or preferred) set of arguments is credulous, in that it sanctions a set of arguments as acceptable if it can successfully dispute every arguments against it, without disputing itself. However, there might be several conflicting admissible sets. Various sceptical semantics have been proposed for AF, notably the grounded semantics, the ideal semantics, and the sceptically preferred semantics, whereby an argument is accepted if it is a member of all maximally admissible sets of arguments.

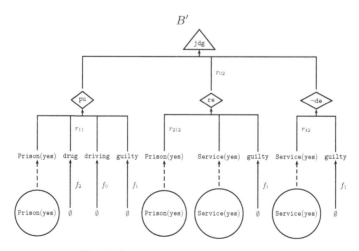

Fig. 3. Some arguments concluding `jdg`

---

[4] Actually, the defeat relation is called attack in [7] and in [8].

Since some ultimate choices amongst various admissible sets of alternatives are not always possible, we consider in this paper only the credulous semantics. Let us focus on the goal pu in the previous example. Since $\{A^1\}$, $\{B\}$ are admissible and $\{A^2\}$ is not admissible, different alternatives and explanations for different decisions can be suggested to reach pu. If we consider now the whole problem, the argument depicted in Figure 3 is the only one reaching jdg which is admissible.

In our example, there is only one admissible argument deriving the main goal. However, in the general case, a decision $D_1(a_1)$ is *suggested* iff $D_1(a_1)$ is a supposition of one argument in an admissible set deriving the main goal. Therefore, our AF involves some ultimate choices of the lawyer between various admissible sets of alternatives. In this section, we have given a status to the arguments.

8 Procedure

A dialectical proof procedure is required to compute the model-theoretic semantics of our argumentation framework. The procedures proposed in [8,9] compute the credulous semantics. Since our practical application requires to specify the internal structure of arguments, we adopt the procedure proposed in [8].

In order to compute admissible arguments in our AF, we have translated our AF in an Assumption-based AF (ABF for short). CaSAPI[5] [10] computes the admissible semantics in the ABF by implementing the procedure originally proposed in [11]. Suppose we wish to investigate whether an argument is preferred, i.e. it belongs to a preferred set. We know that it suffices to check that this argument is in an admissible set, since, by definition, a preferred set is a maximal admissible set and obviously all admissible sets are contained in a maximal admissible set. If the procedure succeeds, we know that the argument is contained in a preferred set. Moreover, we have developed a CaSAPI meta-interpreter to relax the goals achievements in the priority order and to make suppositions in order to compute the admissible semantics in our concrete AF[6]. We can easily extend it to compute the competing semantics which have been proposed in [8]. The implementation of our framework, called MARGO (Multiattribute ARGumentation framework for Opinion explanation), is written in Prolog and available in GPL (GNU General Public License) at http://margo.sourceforge.net/. In this section, we have shown how to compute admissible arguments in our AF.

9 Related Works

Argumentation has been put forward as a promising approach to support decision making [12]. While influence diagrams and belief networks [13] require that all the factors relevant for a decision are identified *a priori*, arguments are defeasible or reinstantiated in the light of new information not previously available.

---

[5] http://www.doc.ic.ac.uk/~dg00/casapi.html

[6] For brevity, we do not describe this mechanism in the paper.

Contrary to the theoretical reasoning, practical reasoning is not only about whether some beliefs are true, but also about whether some actions should or should not be performed. The practical reasoning [14] follows three main steps: i) *deliberation*, i.e. the generation of goals; ii) *means-end reasoning*, i.e. the generation of plans; iii) *decision-making*, i.e. the selection of plans that will be performed to reach the selected goals. For instance, [15] proposes an AF focusing on the deliberation (closed to the principle of [16] where argumentation is implicit) and [17,18] have provided formal models for deliberation and means-end reasoning. While some frameworks are based upon defeasible logic programming (e.g. [19,20]), most of them instantiate the abstract argumentation framework of Dung [7]. Since the latter abstracts away from the internal structure of arguments in order to focus on the manner in which arguments interact, [21] instantiates an argument scheme in the context of practical reasoning in order to capture the interaction in terms of internal structure.

In this work, we have proposed an AF for decision-making. In this perspective, [22] proposes a critical survey of some computational models of argumentation over actions. For this purpose, [23,24] have considered several principles according to the different types of arguments which are considered are aggregated. However, contrary to our approach, the potential interaction amongst arguments, as studied in the seminal work of Dung [7] is not considered. In this paper we have extended the legal example borrowed from [2] and we have adopted like [2] an abductive approach to the practical reasoning which is directly modelled within in our framework.

Finally, to the best of our knowledge, few implementation of argumentation over actions exist. CaSAPI and DeLP [7] are restricted to the theoretical reasoning. PARMENIDES[8] is a software to structure the debate over actions by adopting a particular argumentation scheme. GORGIAS [9] implements an argumentation based framework to support the decision making of an agent within a modular architecture. Like the latter, MARGO incorporate abduction on missing information. Moreover, we can easily extend it to compute the competing semantics which have been proposed in [8] since we have instantiated the abstract argumentation framework of Dung.

10   Conclusions

In this paper we have presented a concrete and implemented AF for practical reasoning in legal disputes which suggests different alternative courses of actions and provides an interactive and intelligible explanation of the choices. A logic language is used as a concrete data structure for holding the statements like knowledge, goals, and decisions. Different priorities are attached to these items

---

[7] http://lidia.cs.uns.edu.ar/DeLP
[8] http://cgi.csc.liv.ac.uk/~katie/Parmenides.html
[9] http://www.cs.ucy.ac.cy/~nkd/gorgias/

corresponding to the uncertainty of the knowledge about the circumstances, the lawyer's preferences between goals, and the expected utilities of sentences. These concrete data structures consist of information providing the backbone of arguments. Due to the abductive nature of practical reasoning, arguments are built by reasoning backwards, and possibly by making suppositions over missing information. To be intelligible, arguments are defined as tree-like structures. The interactions between arguments may come from the incompatibility of their sentences, from their nature (hypothetical or built) and from the priority over rules. Since an ultimate choice amongst various admissible sets of alternatives is not always possible, we have adopted a credulous semantics. In order to compute it, we have implemented our AF in Prolog.

In future works, we want to incorporate decision-theoretic techniques within the model. Standard decision theory weighs the cost and benefits of possible outcomes with their probabilities to produce a preference on the expected utilities of the alternatives. However in many practical applications, it is not natural to give a quantitative representation of many objectives, or it could not deal with the cases of decision makers that only have partial information. Further standard decision theory provides little support in giving intelligible explanation of the choices. For this purpose, it would be best to have a hybrid approach combining both quantitative and qualitative decision theory. Argumentation provides a natural framework for these hybrid systems by providing a link between qualitative objectives and its quantitative representation. In addition, sentencing is usually governed by some explicit statutory or regulatory rules. We want to take into account them in our framework.

Acknowledgements

We would like to thank the anonymous reviewers for their detailed comments on this paper. The author would like to thank Paolo Mancarella, Phan Minh Dung and Francesca Toni for their comments on a previous version of this paper. This work is supported by the Sixth Framework IST programme of the EC, under the 035200 ARGUGRID project.

References

1. Prakken, H., Sartor, G.: The role of logic in computational models of legal argument: a critical survey. In: Kakas, A.C., Sadri, F. (eds.) Computational Logic: Logic Programming and Beyond. LNCS (LNAI), vol. 2408, pp. 343–380. Springer, Heidelberg (2002)
2. Bench-Capon, T., Prakken, H.: Justifying actions by accruing arguments. In: Proc. of the 1st International Conference on Computational Models of Argument, pp. 247–258. IOS Press, Amsterdam (2006)
3. Clemen, R.T.: Making Hard Decisions. Duxbury Press (1996)
4. Amgoud, L., Cayrol, C.: A reasoning model based on the production of acceptable arguments. Annals of Maths and AI 34(1-3), 197–215 (2002)

5. Schweimeier, R., Schroeder, M.: Notions of attack and justified arguments for extended logic programs. In: van Harmelen, F. (ed.) Proc. of the 15th European Conference on Artificial Intelligence (ECAI), pp. 536–540. IOS Press, Amsterdam (2002)
6. Vreeswijk, G.: Abstract argumentation systems. Artificial Intelligence 90(1-2), 225–279 (1997)
7. Dung, P.M.: On the acceptability of arguments and its fundamental role in nonmonotonic reasoning, logic programming and n-person games. Artificial Intelligence 77(2), 321–357 (1995)
8. Dung, P.M., Mancarella, P., Toni, F.: Computing ideal sceptical argumentation. Artificial Intelligence, Special Issue on Argumentation in Artificial Intelligence 171(10-15), 642–674 (2007)
9. Vreeswijk, G., Prakken, H.: Credulous and sceptical argument games for preferred semantics. In: Brewka, G., Moniz Pereira, L., Ojeda-Aciego, M., de Guzmán, I.P. (eds.) JELIA 2000. LNCS (LNAI), vol. 1919, pp. 239–253. Springer, Heidelberg (2000)
10. Gartner, D., Toni, F.: CaSAPI: a system for credulous and sceptical argumentation. In: Simari, G., Torroni, P. (eds.) Proc. of Workshop on Argumentation for Non-monotonic Reasoning, pp. 80–95 (2007)
11. Dung, P.M., Kowalski, R.A., Toni, F.: Dialectic proof procedures for assumption-based, admissible argumentation. Artificial Intelligence 170(2), 114–159 (2006)
12. Fox, J., Parsons, S.: On using arguments for reasoning about actions and values. In: Doyle, J., Thomason, R.H. (eds.) Proceedings of the Working Papers of the AAAI Spring Symposium on Qualitative Preferences in Deliberation and Practical Reasoning, pp. 55–63. Standford, Menlo Park (1997)
13. Oliver, R.M., Smith, J.Q.: Influence Diagrams, Belief Nets and Decision Analysis. John Wiley and Sons, Chichester (1988)
14. Raz, J. (ed.): Practical Reasoning. Oxford University Press, Oxford (1978)
15. Leila Amgoud, S.K.: On the generation of bipolar goals in argumentation-based negotiation. In: Rahwan, I., Moraïtis, P., Reed, C. (eds.) ArgMAS 2004. LNCS (LNAI), vol. 3366, pp. 192–207. Springer, Heidelberg (2005)
16. Thomason, R.H.: Desires and defaults: A framework for planning with inferred goals. In: Proc. of the seventh International Confenrence on Principle of Knowledge Representation and Reasoning (KR), pp. 702–713 (2000)
17. Hulstijn, J., van der Torre, L.W.N.: Combining goal generation and planning in an argumentation framework. In: Proc. of the 9h International Workshop on Non-Monotonic Reasoning (NMR), pp. 212–218 (2004)
18. Rahwan, I., Amgoud, L.: An argumentation-based approach for practical reasoning. In: Proc. of the 5th International Joint Conference on Autonomous Agents and Multiagent Systems (AAMAS), Hakodate, Japan, pp. 347–354. ACM Press, New York (2006)
19. Guillermo, R., Simari, A.J., García, M.C.: Actions, planning and defeasible reasoning. In: Proc. of the 10th International Workshop on Non-Monotonic Reasoning (NMR), Whistler BC, Canada, pp. 377–384 (2004)
20. Kakas, A., Moraitis, P.: Argumentative-based decision-making for autonomous agents. In: Proceedings of the 2nd International Joint Conference on Autonomous Agents and Multi-Agent Systems (AAMAS), pp. 883–890. ACM Press, New York (2003)
21. Atkinson, K., Bench-Capon, T., McBurney, P.: Computational representation of practical argument. Synthese, special issue on Knowledge, Rationality and Action 152(2), 157–206 (2006)

22. Ouerdane, W., Maudet, N., Tsoukias, A.: Arguing over actions that involve multiple criteria: A critical review. In: Mellouli, K. (ed.) ECSQARU 2007. LNCS (LNAI), vol. 4724, pp. 308–319. Springer, Heidelberg (2007)

23. Amgoud, L., Prade, H.: Comparing decisions in an argumentation-based setting. In: Proc. of the 11th International Workshop on Non-Monotonic Reasoning (NMR), Session on Argumentation, Dialogue, and Decision Making, Lake District, UK, pp. 426–432 (2006)

24. Amgoud, L., Prade, H.: Explaining qualitative decision under uncertainty by argumentation. In: Proc. of the 21st National Conference on Artificial Intelligence (AAAI), Boston, pp. 16–20 (2006)

Deterrence and Defeasibility in Argumentation Process for ALIS Project

Michel Rudnianski and Hélène Bestougeff

ORT France
10 Villa d'Eylau, Paris 75116, France
michel.rudnianski@wanadoo.fr,
helene.bestougeff@wanadoo.fr

Abstract. Argumentation issues, which are of core importance to ALIS, are addressed through a particular category of qualitative games called Games of Deterrence. The graphs associated with those games are interpreted as sets of inferences sequences between statements in the framework of non-monotonic logic. Thus an argumentation process is interpreted as a game of deterrence, which resolution determines the truth or falsity of statements, and the possible argumentation strategies of the parties.

Keywords: acceptability, ALIS, argumentation, attack, consistency, defeasibility, game of deterrence, graph of deterrence, playability, rebutting, relevance, strategy.

1 Introduction

ALIS (acronym for Automated Legal Intelligent System) is an EU FP6 program which aims at developing a legal reasoning and decision making system in the field of Intellectual Property Rights[1]. The project is based on the development of a synergy between three different tools:

- Legal Reasoning
- Computational Logic
- Game Theory

Bilateral links between these three fields have already been established at different levels. In particular, with respect to Logic and Game Theory, bridges have been set up through representing quantitative games under their extensive form and analyzing the resulting graph as a logical system (in particular by Johan van Benthem [1, 2] and others [3, 4, 5]).

The aim of the present paper is to focus on a relation between the theoretical framework of argumentation [6, 7, 8] and a specific type of qualitative games called games of deterrence. These games were initially developed to model issues pertaining to nuclear strategy [9], before being applied to a variety of fields, spanning from control congestion in point-to-point communication networks, to

---

[1] www.alisproject.com

P. Casanovas et al. (Eds.): Computable Models of the Law, LNAI 4884, pp. 219–238, 2008.
© Springer-Verlag Berlin Heidelberg 2008

business strategy, negotiations [10], and inference systems [11]. It has been established that a game of deterrence and a particular structure of logical propositions could be linked through a one-to-one mapping [12].

This connection is a two stage one:

- between matrix games of deterrence and a particular kind of graph called graph of deterrence representing the relations between the players strategies playablilities, on the one hand
- between graphs of deterrence and a set of inferences sequences on the other

This qualitative approach, based not on optimization but on the analysis of strategies playability, is an alternative to the standard quantitative one, in line with some classical game theoretic typologies [13]. It presents the advantage of paving the way to a direct representation of sets of inferences sequences as games.

Thus leaning on the results obtained so far, we shall show that there is a one-to-one mapping between a game of deterrence and a particular structure of arguments. This mapping will be used to assess the relevance of a set of statements made by parties involved in a dispute. The methodology will be illustrated on the example of an issue regarding intellectual property rights. After summarizing the core aspects of ALIS project concerning the synergy between legal reasoning and decision making, we shall recall the core concepts of games of deterrence, focusing on the case of matrix games. In particular, we shall introduce graphs of deterrence, associated in a one-to-one mapping with matrix games of deterrence, and provide a typology of games, each type being associated with a particular solution set of the game. This mapping will then make possible to link argumentation and deterrence, through interpreting an argumentation process as a dynamic of statements, which will in turn be analyzed as a game of deterrence. Solutions of this game of deterrence will provide the sets of statements that should be selected by parties involved in an argumentation. In other words, the methodology proposed will enable to introduce strategic considerations in the argumentation analysis.

2 ALIS Project Core Features with Regard to Argumentation

The core features of ALIS regarding argumentation are to be found mainly in the two layer structure of the system, enabling to complement a system of legal reasoning by a system of decision making, through integrating game theoretic tools. Thus the vision conveyed by ALIS of a regulatory or legal issue, is a two stage one. The first stage aims at analyzing compliance of behaviours or arguments with laws and regulations. Based on the knowledge of what is compliant and what is not, as well as on the knowledge of the risks incurred by the infringer in case of no compliance, the second stage aims at selecting strategies for the parties involved and especially for ALIS end users. It follows that rational selection of stakeholders strategies should heavily depend on the results of compliance analysis. It is therefore of most interest to explore possible connections between the strategic stage and the logic stage, that is the stage at which a logical analysis of argumentation is performed. In turn, one can think of the nature of such connection at two levels at least.

The first level is heuristic: the findings of compliance analysis are interpreted in terms of decisions and/or strategies (in the game theoretic sense of the term). At this level the connection between the logical and the strategic stage has to be re-explored for each new case under consideration, even if lessons can be drawn from previous cases and lead through some learning process, to more general conclusions.

The second level is theoretical: what is looked for here is the existence of a theoretical coupling between the two stages (i.e. compliance on the one hand, strategy on the other). This is a matter of importance for the development of ALIS. Indeed, should the existence of such coupling be proved and its properties unravelled, there would be a robust basis for automating the passage from one problematic to the other, within of course the limit drawn by the uncertainties inherent to the case under consideration:

- thus, assessment of behaviours or arguments with respect to compliance would tell which strategies are playable for a rational decision maker
- conversely, on the sole basis of knowing the defeating relations between behaviours or arguments at a bilateral level, it will be possible to determine through a strategic analysis the sequence of behaviours or arguments a stakeholder could or should deploy.

3   Games of Deterrence: An Introduction

3.1   Bounded Rationality

Games of deterrence are qualitative games in which players can only distinguish between acceptable outcomes (noted 1) and unacceptable outcomes (noted 0). The relevance of such games comes from the fact that in real life a decision maker facing numerous alternatives will not develop the standard optimizing approach prescribed by quantitative Game Theory, but rather think in terms of what is acceptable and what is not. It follows that in a game of deterrence, the objective of a rational player is to get an acceptable outcome.

3.2   Playability and Deterrence: Non Formal Definitions

The player strategies properties stem from this assumption. Thus, a strategy e of a player E is termed:

- *safe* if it guarantees player E an acceptable outcome, whatever the other players strategies
- *dangerous* if it is not safe

Similarly, a strategy e of a player E is termed:

- *positively playable* if it guarantees E an acceptable outcome, provided the other players are rational (any safe strategy is positively playable)
- *playable by default* if E has no positively playable strategy
- *playable* if it is either positively playable or playable by default

It clearly stems from the above definitions that a safe strategy is positively playable.

Given a binary matrix game, strategy r of player R is said to be *deterrent* vis-à-vis strategy e of player E, iff the three following conditions apply:

1. r is playable
2. implementation of strategic pair (r,e) implies an unacceptable outcome for E
3. E has another strategy e* positively playable

The main idea conveyed by the third condition is that, in order for R to deter E from doing something that the former doesn't like, it is necessary to let E with an alternative that is acceptable to him/her. This is no more than common sense: you have everything to fear from the one who has nothing to lose.

Example 1

$$Player R$$

	r_1	r_2
e_1	$(1,0)$	$(1,1)$
e_2	$(0,1)$	$(0,1)$

$Player E$

- e_1 is safe, hence playable
- (e_1, r_1) leads to an unacceptable outcome for player R
- Player R has an alternative positively playable strategy r_2

Consequently, e_1 is deterrent vis-à-vis r_1. Similarly one can show that r_2 is deterrent vis-à-vis e_2

Example 2

$$Player R$$

	r_1	r_2
e_1	$(1,0)$	$(1,1)$
e_2	$(0,1)$	$(1,0)$

$Player E$

- e_1 is safe, hence positively playable
- (e_1, r_1) leads to an unacceptable outcome for player R, hence r_1 is not positively playable

- It follows from the definition of positive playability that r_1 is either not playable or playable by default
- If r_1 is not playable, then although (r_1, e_2) leads to an unacceptable outcome for player E, e_2 is positively playable.
- Consequently, since (e_2, r_2) leads to an unacceptable outcome for player R, r_2 is not positively playable.
- So under the above assumption, player R has no positively playable strategy.
- But then it stems from the definitions of playability, that both strategies of player R are playable by default, which implies in turn that strategy e_2 of player E is not playable

On the whole, e_1 is not deterrent vis-à-vis r_1 or r_2, which are playable by default. Moreover r_1 is deterrent vis-à-vis e_2.

The two above examples indicate a strong relation between the concepts of playability and deterrence.

Indeed, it has been shown [9] that a strategy e of player E is playable if and only if there is no strategy r of player R which is deterrent vis-à-vis e.

3.3   Playability System

The concepts introduced above can be restated more formally.

Consider a finite binary matrix game where players E and R have respective strategic sets S_E and S_R, and outcome functions A and B.

For any pair (e,r) $\in S_E \times S_R$:

- $A(e,r) \in \{0, 1\}$ denotes the outcome for E
- $B(e,r) \in \{0, 1\}$ denotes the outcome for R

A strategy e of E is said to be safe iff for all r $\in S_R$, $A(e,r) = 1$

Let J(e) be an index, called index of positive playability of e ,such that :

(i) If e is safe then J(e)=1

(ii) $J(e) = \prod_{r \in S_R}[1 - (1 - A(e,r))J(r)](1 - j_E)(1 - j_R)$

with

$$j_E = \prod_{e \in S_E}(1 - J(e)) \ ; \ j_R = \prod_{r \in S_E}(1 - J(r))$$

j_E and j_R are termed respectively index of playability by default of E and R Strategy e is said to be:

- positively playable if J(e) =1
- playable by default if $j_E = 1$

It should be noticed that according to the above definitions, if one strategy of player is playable by default, then all the player's strategies are.

The system S of equations enabling to compute all J(e), e $\in S_E$ and J(r), r $\in S_R$, is called the *playability system* of the game.

A solution of S is a set of binary values $J(e_1), J(e_2), , J(e_n), J(r_1), J(r_2), , J(r_p)$ satisfying S. In general there is no uniqueness of the solution.

A strategic pair (e,r) $\in S_E \times S_R$ is said to be an *equilibrium* if both strategies are playable for some solution of the playability system of the game.

Then a game of deterrence can be defined by the structure $< S_E, S_R, S >$

It follows straightforwardly that given:

- two games of deterrence G $= < S_E, S_R, S >$ and G' $=< S_E, S_R, S' >$
- a strategic pair (e,r) $\in S_E \times S_R$,

strategy e might be deterrent vis-à-vis strategy r with respect to S, and not deterrent vis-à-vis r with respect to S'.

Likewise, given a game G $= < S_E, S_R, S >$, let Sol(S) be the set of solutions of S, sol^1 and sol^2 two elements of Sol(S). Strategy e can be deterrent vis-à-vis r in sol^1 and not deterrent vis-à-vis r in sol^2.

Example 3

$$PlayerR$$

		r_1	r_2
	e_1	$(0,0)$	$(1,1)$
$PlayerE$			
	e_2	$(1,1)$	$(0,0)$

Let u $= 1 - j_E$ and v $= 1 - j_R$

The playability system writes:

$J(e_1) = [1 - J(r_1)]uv; J(e_2) = [1 - J(r_2)]uv; 1 - u = [1 - J(e_1)][1 - J(e_2)]$
$J(r_1) = [1 - J(e_1)]uv; J(r_2) = [1 - J(e_2)]uv; 1 - v = [1 - J(r_1)][1 - J(r_2)]$

It can be easily shown that the game displays three solutions:
sol^1: $J(e_1) = 0, J(e_2) = 0, J(r_1) = 0, J(r_2) = 0$
sol^2: $J(e_1) = 1, J(e_2) = 0, J(r_1) = 0, J(r_2) = 1$
sol^3: $J(e_1) = 0, J(e_2) = 1, J(r_1) = 1, J(r_1) = 0$

sol^1 includes 4 equilibria, all playable by default. There is no deterrence relation associated with that solution

sol^2 displays a unique equilibrium, (e_1, r_2) which is positively playable. Hence, in this solution e_1 is deterrent vis-à-vis r_1, and r_2 is deterrent vis-à-vis e_2
sol^3 displays a unique equilibrium (e_2, r_1) which is positively playable. Hence, in this solution e_2 is deterrent vis-à-vis r_2, and r_1 is deterrent vis-à-vis e_1.

3.4 Graphs of Deterrence

Given a binary matrix game, a *graph of deterrence* is a bipartite graph such that:

- its vertices are the players strategies
- given the strategic pair (e,r) $\in S_E \times S_R$, there exists an arc of origin e (resp.r) and extremity r (resp.e) iff the outcome of R (resp. E) is 0.

Let us call :

- *E-path* (resp. *R-path*) a path which has a root (i.e. a vertex without predecessor) belonging to S_E (resp. S_R)
- *C-graph*, a graph including neither an E-path nor an R-path

It has been shown that (Classification Theorem [9]):

- the graph of deterrence can be broken down into connected parts, each one
being an E-path, an R-path, or a C-graph
- depending on the presence of these components in the graph, one can distinguish between 7 types of games, E, R, C, E-R, E-C, R-C, E-R-C, such that
each one is associated with a solution set displaying specific properties.

Determining the type of game through the graph of deterrence approach enables therefore to specify the game solution.

The graph of deterrence approach may thus shorten the way toward determining the game solution set, especially when the dimensions of the game matrix
are important.

For instance:

- if the game is an E-type (resp. R-type) one, then:
  - player E (resp. player R) has only one playable strategy per path: the
  root
  - all strategies of player R(resp. player E) are playable by default
- if the game is an E-R type one, then:
  - all strategies of player R (resp. player E) located on an R-path (resp. an
  E-path) are positively playable
  - all strategies of player R (resp. player E) located on an E-path (resp. an
  R-path) are not playable
- if the game is a C-type one, it then displays at least one solution for which
all the players strategies are playable by default.

Examples

According to the above definition, the graph of deterrence associated with the
game presented in example 1 is:

The game is of type E-R. Applying the results recalled here above leads to
the conclusions already found in section 3.2

Similarly, the graph of deterrence associated with the game of example 2 is:

The game is of type E, and again the results recalled here above lead to the conclusions already found in section 3.2.

Last, the graph of deterrence associated with the game of example 3 is:

The game is a C-type one, and the results recalled here above point out the solution 1 found in section 3.3 [2].

4   Games of Deterrence for Argumentation

4.1   The Classical Approach of Argumentation

An argumentation process aims at determining in a given context the truth or falsity of statements [14].

This context can be characterized by several dimensions:

- the information available
- the theory organizing the relations between the pieces of information
- the nature of the argumentation process, etc.

As far as this last point is concerned, we shall focus on two party argumentation processes for which a statement made by one party called the proponent is put in question by the other called the opponent. This type of argumentation has been already analyzed in a vast literature, including models of argumentation processes based on formal concepts of argument.

Given an information set I, an argument can be modelled as a pair (x, Ψ) where x is the conclusion written under the form of a statement, and $\Psi \subseteq I$ is a set of inferences sequences, which constitutes the ground or support for x.

In other words given (x, Ψ), $\Psi \vdash x$.

Leaning on this definition, and given two arguments (x_1, Ψ_1) and (x_2, Ψ_2) , the relations of attack, rebut, undercut and defeasibility between these two arguments are defined as follows:

1. (x_1, Ψ_1) *attacks* (x_2, Ψ_2) if $x_1 \equiv \neg x_2$
2. (x_1, Ψ_1) *rebuts* (x_2, Ψ_2) if x_1 attacks x_2
3. (x_1, Ψ_1) *undercuts* (x_2, Ψ_2) if x_1 attacks some $\varphi \in \Psi_2$
4. (x_1, Ψ_1)) *defeats* (x_2, Ψ_2)) if it either rebuts or undercuts it

The first two definitions resort only to the statement dimension of the argument, while the last two may also be considering using statements, as soon as an inference can be considered as a statement.

It follows that one might envisage to analyze argumentation processes at the level of statements rather than at the one of arguments, given of course that inferences need nevertheless to be taken into account.

[2] Unlike the previous example, exhaustive determination of the solution set - which in this case includes solutions 2 and 3, cannot be derived straightforwardly from the results of section 3.1, but requires a deeper breakdown of the game type into several sub-types.

4.2   From Deterrence to Statement Defeasibility

One can infer from the developments made about example 1 that the approach for establishing the truth or falsity of the statement "e_2 is positively playable", that is to say the game theoretic analysis, can be associated in a-one-to-one mapping with a particular argumentation process, in which the two players of the game of deterrence are considered as playing each one a specific role: player E is the proponent of the statement 'e_2 is positively playable' and player R, the opponent.

The reason for which e_2 is not positively playable lies in the fact that one vertex of the graph of deterrence - r_2 - which is an adjacent predecessor of e_2 is positively playable. In turn this means that two conditions are required for drawing this conclusion:

1. the statement 'r_2 is playable' is true
2. if the statement 'r_2 is playable' is true then the statement 'e_2 is positively playable' is false, which amounts to say that the statement 'e_2 is not positively playable' is true

These two conditions are equivalent to the argument {e_2 is not positively playable, {r_2 is playable, r_2 playable $\Rightarrow e_2$ not positively playable}}.

Now before going further, it must be noticed that the one-to-one mapping between the game of deterrence and the argumentation process doesn't mean that the two can simply be superposed. Indeed, in the game of deterrence, each player is supposed to select one strategy, while in the argumentation process nothing prevents the parties from selecting a set of arguments or statements rather than a single one [15]. Nevertheless, this distinction raises no difficulty, as soon as one considers instead of the original game of deterrence, another one called game of multi-strategies in which the set of strategies of a player is the set of parts of the strategic set in the original game[3].

What seems straightforward when considering in the above example the passage from games of deterrence to argumentation processes, requires nevertheless that an assumption about on whom falls the burden of proof in the argumentation process, be made explicit in order for the bridging to be generalized [16].

In the previous example, everything was as if it was implicitly assumed that the burden of proof is on R and not on E. In other words, it is on the opponent and not on the proponent that falls the burden of proof. At the game of deterrence level, this stems directly from the definitions of playability, and the dialectic relation between playability and deterrence. In other words, given the statement 'strategy e is playable', one can consider the latter as true, as long as there is no statement of the type 'there exists a strategy r which is deterrent vis-à-vis e', which can be declared true.

---

[3] For such a switch from games of strategies to games of multi-strategies to be relevant, it is necessary to associate with the set of parts of each player strategic set, a rule that enables to compute the outcome associated with a given pair of parts.

In the example considered, this means that :

- (positively) playable strategies are associated with true statements
- non (positively) playable strategies are associated with false statements

But this means in turn that if the status of a statement - true or false - is the result of existing relations or implications between all statements, then given:

- a matrix game of deterrence $< S_E, S_R, S >$
- the set (X, Y) of statement pairs associated with (S_E, S_R),
- the set of relations Ψ between the elements of $P(X \cup Y)$ derived from S
- a pair of statements (x,y) associated with the strategic pair $(e,r) \in S_E \times S_R$, such that the truth or falsity of each statement is determined by the playability properties of the corresponding strategy, we shall say that:

1. x *attacks directly* y iff e is an adjacent predecessor of r on the graph of deterrence of the associated game
2. x *attacks* y iff e is a predecessor of r on the graph of deterrence of the associated game, such that the number of hops defining the distance between e and r is odd
3. x *defeats directly* y with respect to Ψ, and we shall write $< x, \Psi > $ D y, iff e is deterrent vis-à-vis r.

In terms of statements, one can easily establish that the first two definitions here above mean that if x is true then y is not true. It is therefore the simple translation at the level of statements of the definition of attack at the level of arguments.

Now as far as the third definition is concerned, it stems from the three conditions for e to be deterrent vis-à-vis r, that if statement x defeats directly statement y with respect to Ψ, then:

$\forall \Psi' \subseteq \Psi$, the argument (x,$\Psi$) defeats the argument (y,Ψ').

But the reverse is not true, since for instance the argument {y is not true,{ x is true, x true $\Rightarrow$ y not true}} doesn't imply that e is an adjacent predecessor of r on the graph of deterrence, which is the second condition for e to be deterrent vis-à-vis r.

Hence, the concept of direct defeasibility between statements as introduced here above and the classical concept of defeasibility between arguments as recalled in the previous sub-section are not equivalent. For such an equivalence to be established we need to consider also the case when e is not adjacent to r on the graph of deterrence, while:

- e is playable
- e is playable $\Rightarrow$ r is not playable

We can consider these two conditions as defining some *indirect* relation of deterrence between e and r, and therefore we can define a general concept of defeasibility D_G between statements as follows - x *defeats y with respect to Ψ* - and we shall write :

$< x, \Psi > D_G$ y, iff e is playable, and e is playable $\Rightarrow$ r is not playable.

Then, there is clearly an equivalence between the concepts of defeasibility between statements, and defeasibility between arguments.

For instance, let us come back to example 1.

The playability system S writes:
$$J(e_1) = 1; J(e_2) = [1 - J(r_1)][1 - J(r_2)]$$
$$J(r_1) = [1 - J(e_1)]; J(r_2) = 1^4$$

If x_1, x_2, y_1 and y_2 are the statements associated respectively with positive playability of e_1, e_2,r_1,r_2, then :

x_2 is true iff y_1 and y_2 are false; y_1 is true iff x_1 is false

Furthermore the two statements x_1 and y_2 are true.

It follows for instance that, $\forall \Psi' \subseteq \Psi$:

- $< y_2, \Psi > D\ x_2$
- the argument(y_2, Ψ) defeats the argument (x_2, Ψ')

Now this definition of defeasibility extends straightforwardly to the case where defeasibility of a statement y is not obtained by a single statement x, but by a set of statements $\{x_1, x_2, x_n\}$.

What is then required, is to extend the definition of deterrence, by resorting for instance to games of multi-strategies.

It can be noticed that the definition of defeasibility proposed here above takes into consideration - albeit in a different way - the aspects of defeasibility analyzed by Prakken and Sartor [17] who consider that in the field of law, the concept of defeasibility covers three aspects:

- *theory based defeasibility*, which assesses and compares for a given set of available information, the various theories according to which an argument is true or not.
- *inference based defeasibility* which states that an argument which truth is supported by a given set of information, may not be considered true if the set under consideration is enriched.
- *process based defeasibility* which refers to the dynamic aspects and the framework in which argumentation takes place, for instance on whom falls the burden of proof

First, as far as theory based defeasibility is concerned, for the same pair of statements sets (X,Y), given two different sets of inferences sequences, Ψ_1 and Ψ_2 , it stems from the definition of defeasibility given here above that one may have simultaneously: $< X, \Psi_1 > D_G Y\ and\ < X, \Psi_2 > \neg D_G Y$.

Secondly, let:

- (X_1, X_2) be a pair of statements sets such that $X_1 \subseteq X_2$
- (Ψ_1, Ψ_2) be a pair of inferences sequences sets such that:
  - $\Psi_1 \subset (X_1 \cup Y)^2$ and $\Psi_2 \subset (X_2 \cup Y)^2$

---

[4] As each player has a safe strategy, j_E and j_R equal 0, and hence, for the sake of simplicity, are not represented in S.

- given $X \subseteq X_1$, any relation between subsets of $(X \cup Y)^2$ belonging to Ψ_1, is a relation between the same subsets belonging to Ψ_2, and vice-versa (Ψ_2 will be called *an extension* of Ψ to X_2),

then X_2 can be considered as an enrichment of X_1 in terms of information.

We see that bringing extra information implies modifying the initial set Ψ_1, since even if relations between existing statements are unchanged, Ψ_2 must include the new statements.

Last, the above definition doesn't address explicitly the issue of the defeasibility process, and therefore one can consider that this definition is consistent with multiple types of processes that can be envisaged, for instance with on whom falls the burden of proof.

4.3   Deterrence, Rebutting, Undercutting

As already noticed, besides defeasibility, literature on argumentation contains several categories of relations between arguments. We shall here focus on two of them.

The first one is "rebutting".

Given two arguments (x_1, Ψ_1) and (x_2, Ψ_2), (x_1, Ψ_1) is said to rebut (x_2, Ψ_2) if x_1 attacks x_2. It follows that at the level of statements, there is no relevant distinction between rebutting and attack.

For instance, let us come back to example 1. We easily deduce from the graph of deterrence that the statement $x_2 = $ "strategy e_2 is positively playable" will not be true if the statement $y_2 = $ "strategy r_2 is playable" is true, which is undoubtedly the case since r_2 is a root. Therefore we can conclude that the argument (y_2, Ψ) rebuts the argument (x_2, Ψ)

Let us then consider "undercutting". We shall say that *a statement x undercuts a statement y* with respect to Ψ if Ψ includes a statement φ, such that x attacks φ . We see that this is equivalent to say that argument (x, φ) undercuts argument (y, Ψ).

Let us for instance consider the following game of deterrence, which is a slight variation of example 1:

PlayerR

	r_1	r_2
e_1	$(1, 0)$	$(1, 1)$
e_2	$(0, 1)$	$(1, 1)$

PlayerE

And the corresponding graph of deterrence:

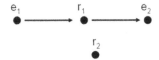

Assume that R makes the statement "e_2 is not playable" on the ground that r_1 is an adjacent predecessor of e_2 on the graph. Obviously, this ground is not a proper one, since the adjacent predecessor of r_1 on the graph of deterrence is the safe strategy e_1 which, as it can be established straightforwardly, is deterrent vis-à-vis r_1. This amounts to say that r_1 is not playable, and hence doesn't satisfy a necessary condition for being deterrent vis-à-vis e_2 which is in turn positively playable.

4.4 Decidability

As the game considered in example 1 is of type E-R, issues of playability just reduce in this case to those of positive playability. It has been seen on examples 2 and 3 that this is not the case in general.

Take for instance example 3, in which the game is of type C, and hence no player has a safe strategy. We know that this game displays three solutions, including one for which both players strategies are playable by default.

As far as the bridging between deterrence and defeasibility is concerned, this seems to raise two different problems: the first one dealing with the multiplicity of solutions, and the second one with the absence of deterrence in the case of the solution where the players strategies are playable by default.

As far as the first problem is concerned, if we consider the two solutions for which no player has playable by default strategies, we see that in terms of playability they contradict each other:

1. in solution 2, e_1 is positively playable and deterrent vis-à-vis r_1 while r_2 is positively playable and deterrent vis-à-vis e_2
2. in solution 3, on the opposite, r_1 is positively playable and deterrent vis-à-vis e_1 while e_2 is positively playable and deterrent vis-à-vis r_2

Applying then the equivalence pointed out above between deterrence and defeasibility leads to the same contradiction, as far as the truth of the associated statements is concerned. And there is of course no way to decide on the basis of the information available which of the two solutions should be selected.

But even, should this difficulty be overcome, we would be left with the problem of solution 1, in which all strategies are playable by default, and hence there is no deterrence relation.

So, on the whole, the conclusions to be drawn from the analysis of example 3 are simply that the case is not decidable in the sense that:

– either there is no defeasibility relation (solution 1)
– or there are defeasibility relations but which contradict each other (solutions 2 and 3)

In other words interpretation of the game matrix leads simply to the following conclusions:

– if x_1 is true then y_1 is not and vice-versa
– if x_2 is true, then y_2 is not and vice-versa

where x_1,x_2,y_1,y_2 are the statements associated with positive playability of e_1,e_2,r_1 and r_2 respectively.

Assuming that the information available is correct, the only way to solve the problem of lack of decidability is to enrich the information set till the resulting extension of the set of inferences sequences between statements leads to a unique solution of the game, in which no player has playable by default strategies. This means transforming the C-type game into another type, through finding facts, that is statements associated with safe strategies.

4.5   Consistency Condition

If the problem raised by the multiplicity of solutions can be solved through looking for statements such that the graphs of deterrence of the corresponding games would not include C-graphs anymore, such modification of the game structure is in general not sufficient for avoiding problems stemming from the existence of strategies playable by default.

Let us consider for instance the game of example 2 analyzed in section 3. The graph of deterrence is an E-path, and we have seen that in this case all strategies of player R are playable by default. So how can we represent the set of statements and inferences:

$\{(x_1$ is true) and $(x_1$ true $\Rightarrow y_1$ not true) and $(y_1$ true $\Rightarrow x_2$ not true) and $(x_2$ true $\Rightarrow y_2$ not true)$\}$?

On the one hand, given the assumption that the burden of proof is on the opponent of an statement, a direct analysis of the set here above leads to the (obvious) following conclusions: x_1 is true, hence y_1 is not true, hence x_2 is true, hence y_2 is not true.

On the other hand translating the above argumentation problem in terms of games of deterrence leads, as has already been seen, to the following conclusions:

- the only positively playable strategy for E is e_1
- all strategies of R are playable by default

We then see that, unless some changes are introduced, we are led to a contradiction as far as the bridging between argumentation process and game of deterrence is concerned. Indeed e_2 is not playable which should mean that the corresponding statement x_2 is not true, while we came to the opposite conclusion after proceeding to a direct analysis of the argumentation process. The non playability of e_2 stems of course from the playability by default of the strategies of R.

It follows that in order to restore the playability of e_2 what is required is to get rid of the playability by default for the strategies of player R.

This can be achieved simply by adding to the strategic set of player R a third strategy r_3 which would be an isolated vertex on the graph of deterrence, so that its introduction would change nothing to the set of relations between the existing vertices of the graph.

This means that at the level of the argumentation process, the set of information should be enriched by a fact, thus that this fact introduces no modification in the existing chain of inferences between the other statements, and is therefore irrelevant as far as the argumentation process is concerned. In other words this new fact is nothing more than an artefact, to which we shall refer to in the sequel as the *consistency condition*.

How should the consistency condition be selected ?

In the example under consideration, the consistency condition displays two properties:

- the corresponding strategy doesn't belong to any of the pre-existing parts of the graph of deterrence
- it is an isolated and hence a safe strategy.

Any strategy or set of strategies satisfying these two conditions will do. For example, one could have chosen, instead of a single strategy, a set of strategies of player R such that each one of them would have been an isolated vertex.

Obviously, whenever it is possible, one will choose for the sake of simplicity a unique strategy.

Now it may be possible that one may not easily find directly such a strategy, and therefore that it might be necessary to build up the consistency condition from existing strategies, through going to the statement analysis side. Consider for instance a statement P. Then whatever this statement and its validity, the statement "P or not P" is always true. Given what precedes, one can associate with this last statement a strategy r_3 that will have no connection with any other vertex on the graph of deterrence.

4.6   Game of Deterrence Representation and Processing of an Argumentation Issue

Consider for instance a dispute between two parties E and R, about the use of a software:

- R considers that he is entitled to use the software since he has done it openly for a long time without raising objection (statement y_2)
- E denies it on the ground that she didn't transfer any right to R (statement x_2)
- in turn R declares that E had no legal ground which would allow her to accept or refuse to transfer the rights pertaining to the use of the software (statement y_1)
- E establishes that she had full rights since she has asserted the software (statement x_1)

In this dispute, R is the proponent and E the opponent.

The structure of argumentation can be represented by the set Ψ, such that:
$\Psi = \{x_1, x_1 \Rightarrow \neg y_1, y_1 \Rightarrow \neg x_2, x_2 \Rightarrow \neg y_2\}$ Let X and Y be the sets of statements of the two parties respectively.

With each pair of statements $(x,y) \in X \times Y$, let us associate a pair of binary numbers
$(A(x,y), B(x,y))$ such that:
$A(x,y) = 0$ iff y defeats directly x with respect to Ψ
$B(x,y) = 0$ iff x defeats y directly with respect to Ψ

It follows that there is a one to one mapping between the chain of argumentation and the following matrix:

$$Proponent\ R$$

	y_1	y_2
x_1	$(1,0)$	$(1,1)$
x_2	$(0,1)$	$(1,0)$

$Opponent\ E$

In turn, this matrix can be interpreted as the matrix of the game of deterrence of example 2, with the players strategic sets $S_E = \{e_1, e_2\}$ and $S_R = \{r_1, r_2\}$ being such that e_1, e_2, r_1 and r_2, are associated with x_1, x_2, y_1, and y_2 respectively.

To avoid playable by default strategies, let us add a strategy r_3 for player R such that r_3 is an isolated vertex. The new game matrix is then:

$$Player\ R$$

	r_1	r_2	r_3
x_1	$(1,0)$	$(1,1)$	$(1,1)$
x_2	$(0,1)$	$(1,0)$	$(1,1)$

$Player\ E$

The graph of deterrence is:

In other words the game is now of type E-R, and it displays a unique solution such that:

1. E has only one positively playable strategy, which is the root e_1
2. Strategies r_1 and r_2 of player R are not playable (while strategy r_3 is of course playable by construction)

The conclusion in terms of the dispute under consideration is then straightforward: R has no right to use the software.

4.7   Consequences on the Argumentation Process

The procedure followed for solving the case can be interpreted at different levels.

First, in the game of deterrence, one can consider that the safe strategies associated with the argumentation process correspond to legal rules or established facts.

The argumentation procedure as described in the above example consists in finding paths connecting the statement under assessment to legal provisions or established facts.

If there is a legal provision or an established fact which is a vertex of even rank on one path, as it is the case in the example under consideration, then the statement can be declared false.

Conversely, if :

- either there is no path connecting the statement to a legal provision or an established fact
- or on all paths connecting the statement to be assessed to a legal provision or an established fact, the later is a vertex of odd rank

then, on the basis of the assumption made about the burden of proof, the statement can be declared true.

Now at a more "strategic" level, the above procedure enables to determine the appropriate sequence of statements to be deployed in order to win the case. In that sense, the sequences of statements can be considered as strategies in a meta-game built on the game of deterrence associated with the set of available arguments.

Different ways can be followed to build this meta-game. The one which seems the most appropriate at first sight resorts to a method already introduced above, and which consists in selecting not an statement but a sequence of statements. This can be considered as a method of brute force, since if a given party has n statements available, the number of sequences of k arguments that can be considered, is $!kC_n^k$ which leads to a total number of strategies per player equal to: $\sum_{k=1}^{n} !kC_n^k$

The number becomes astronomical as soon as k is higher than few units, with as an obvious consequence that the method becomes intractable.

Therefore one may think of an alternative and classical method, developed by Nigel Howard [9]. It consists in selecting a strategy on the basis of the anticipations made by the player under consideration about the strategic choice of the other player.

For instance if R thinks that E will choose to put on the table the statement according to which she never transferred any rights to R (statement x_2), the latter can invoke the statement according to which E has no legal ground to transfer rights concerning the software (statement r_1).

This can be considered as a piecewise approach given the fact that that the players choose only one statement, and do not consider *the statements map*, that is the representation of the set of relations between statements in a one-to-one mapping with the graph of deterrence of the associated game (not taking the consistency condition into account). This might look at first glance as a weakness, unless the parties at stake adopt a sophisticated - but risky - behaviour, in that sense that each one or at least one of them assumes that the other one doesn't master perfectly the statements map.

This assumption may not hold in the simple case considered here where the statements map is associated with a graph of deterrence which is a path.

But it could be considered in more complex cases where the statements map is associated with a graph of deterrence including C-graphs. In particular the question can be raised about structures of the statements map such that the proponent wins the argumentation and other for which the winner is the opponent.

That such structures exist is unquestionable. Consider for instance a statements map which is a single path. If the number of vertices on the path is an odd number, than the winner is the proponent, while if it is an even number, the winner is the opponent. It stems from the discussion here above that the conclusion can be generalized to more complex cases where there are several paths, provided, that all converge toward a single statement.

If the statements map is associated with a C-graph, then the result of the argumentation is a stalemate, either because as already seen, the corresponding game of deterrence displays a unique solution for which the strategies of both players are playable by default, or because it displays multiple solutions, among which no choice is possible with the available information.

The issue becomes more complex when there are multiple not converging paths, which means that either the proponent throws several disconnected statements on the table, or each party is the proponent for a set of statements and the opponent for another set disconnected from the first one. At first sight one can of course consider that this is an issue of separated argumentation processes. But observation of real life shows that some time a party which is not sure to win a specific argumentation, may open another one, either because it wants to leave the first one aside, or because in a more sophisticated way the second argumentation introduced is viewed as a strategic tool in the global confrontation or for a negotiation to come.

To that aim it is necessary that the parties proceed to an assessment of the various possible states of the world associated with a given set of argumentation processes.

Let us consider for instance a set of three argumentations processes A_1, A_2, A_3, each one being associated with a single path (not taking the consistency condition into account). Suppose that E is the winner of A_1, while R is the winner of A_2, and A_3. Assume furthermore that the parties know the winner of each argumentation process, and associate an outcome with each one.

Thus, for E losing A_2 is acceptable, but losing A_3 is unacceptable. Likewise for R losing A_1 is unacceptable while losing A_2 or A_3 is acceptable.

Then, it is obvious that a rational party will not be the proponent of an argumentation, which it will lose. It follows that in strategic terms each party has the choice between either not being a proponent, or being the proponent of an statement or a set of statements, which cannot be defeated. It stems from these assumptions that the interaction between the parties can then be considered as the following matrix game:

		Player R		
	Nothing	A_2	A_3	$A_2 and A_3$
Nothing	$(1,1)$	$(1,1)$	$(0,1)$	$(0,1)$
Player E				
A_1	$(1,0)$	$(1,0)$	$(0,0)$	$(0,0)$

Observation of the matrix shows straightforwardly that all strategies of player R are equivalent, hence have the same playability, and hence are playable. It follows that none of the two strategies of player E is positively playable, and hence that both are playable by default, which implies in turn that the same goes for the strategies of player R.

On the whole, the game displays a unique solution for which all the players strategies are playable by default.

This means in terms of the relation between the two parties, that confrontation through "shooting" argumentation processes toward the opposing party doesn't lead to a solution acceptable for both parties, and even worse, that such confrontation might lead on the opposite to a situation which is unacceptable for both parties as would be the case, if the players would select the strategic pairs (A_1, A_3) or $(A_1, A_2$ and $A_3)$.

On the other hand, as there is a state of the world which is acceptable for both (the status quo) there is ground for a negotiated solution, which can be implicit or explicit (each party simply does nothing as long as the other party does the same).

5   Conclusions

Based on the analysis of some core issue of the European research project ALIS, the present paper has proposed a bridging between argumentation issues and a particular category of qualitative games called games of deterrence. More precisely, it has been established that through focusing on statements, one could analyze argumentation between two parties as a game of deterrence which established properties could be used to solve the argumentation issue. In particular, it has been shown how the issue of defeasibility can be approached by the analysis of deterrence.

In the framework of the European research project ALIS, the results obtained enable the parties involved in the argumentation process to connect the analysis of compliance, with strategic considerations, dealing with how a party could or should select an argumentation strategy in the frame of a dispute, depending of course on which party falls the burden of proof. This is the first step on the way to connect compliance and strategy. The next one will deal with how to develop a theoretical link between the results of the compliance analysis and the strategies to be selected by the parties involved in the IPR issue for which argumentation about compliance of these parties have been analyzed.

References

1. van Benthem, J.: Logic in Games. Lecture Notes and Book Preversion. ILLC (2001)
2. van Benthem, J.: Open Problems in Logic and Games. In: Artemov, S., Barringer, H., d'Avila Garcez, A., Lamb, L., Woods, L. (eds.) Essays in Honour of Dov Gabbay, pp. 229–264. King's College Publications, London (2005)

3. Harrenstein, P., van der Hoek, W., Meyer, J.J., Witteveen, C.: Boolean games. In: van Benthem, J. (ed.) Theoretical Aspects of Rationality and Knowledge. Proceedings of the 8th Conference (TARK 2001), pp. 287–298. Morgan Kaufmann, San Francisco (2001)

4. Harrenstein, P.: Logic in Conflict. PhD thesis, Utrecht University (2004)

5. Bonzon, E., Lagasquie-Schiex, M.C., Lang, J., Zanuttini, B.: Boolean Games Revisited. In: European Conference on Artificial Intelligence (ECAI 2006), pp. 265–269. Springer, Heidelberg (2006)

6. Aspic Project: Theoretical Framework for Argumentation, Deliverable 2.1 (2004)

7. Prakken, H.: AI and Law, Logic and Argument Schemes. In Argumentation 19. Special Issue on The Toulmin Model Today, 303–320 (2005)

8. Bench-Capon, T.J.M., Dunne, P.E.: Argumentation in Artificial Intelligence. Artificial Intelligence 171, 619–641 (2007)

9. Rudnianski, M.: Deterrence Typology and Nuclear Stability: A Game Theoretic Approach. In: Avenhaus, R., Karkar, H., Rudnianski, M. (eds.) Defense Decision Making, pp. 37–168. Springer, Heidelberg (1991)

10. Rudnianski, M., Bestougeff, H.: Bridging Games and Diplomacy. In: Avenhaus, R., Zartmann, I.W. (eds.) Diplomacy Games: Formal Models and International Negotiation. Springer, Heildelberg (2007)

11. Rudnianski, M.: Deterrence, Fuzzyness and Causality. In: Proceedings ISAS 1996, Orlando, pp. 473–479 (1998)

12. Rudnianski, M., Bestougeff, H.: Multi-Agent Systems Reliability, Fuzzyness and Deterrence. In: Hinchey, M.G., Rash, J.L., Truszkowski, W.F., Rouff, C.A. (eds.) FAABS 2004. LNCS (LNAI), vol. 3228, Springer, Heidelberg (2004)

13. Isaacs, R.: Differential Games. Wiley and Sons, Chichester (1965)

14. Dung, P.M.: On the Acceptability of Arguments and its Fundamental Role in Non-Monotonic Reasoning, Logic Programming and N-Person Games. Artificial Intelligence 77, 321–357 (1995)

15. Holbech, N., Parsons, S.: A generalization of Dung's Abstract Framework for Argumentation Arguing with Sets of Attacking Arguments. In: Proceedings of Argmas, pp. 7–21 (2006)

16. Gordon, T., Prakken, H., Walton, D.: The Carneades Model of Argument and Burden of Proof. Artificial Intelligence 171, 875–896 (2007)

17. Prakken, H., Sartor, G.: The three faces of Defeasibility in the Law. Ratio Juris 17, 118–139 (2003)

18. Howard, N.: Paradoxes of Rationality: Theory of Metagames and Political Behavior. MIT Press, Cambridge (1971)

Temporal Deontic Defeasible Logic:
An Analytical Approach

Régis Riveret and Antonino Rotolo

CIRSFID and Law Faculty, University of Bologna, Via Galliera 3, 40121, Bologna, Italy,
{rriveret,rotolo}@cirsfid.unibo.it

Abstract. In [1,2,3] basic Defeasible Logic was extended to capture some temporal aspects in legal reasoning. All these extensions can be criticized in two respects: first, a synthetical approach with which all temporal and substantial elements of the norm are represented within the same sentence was adopted instead of an analytical approach in which one sentence represents the substantive content of the norm, and other sentences specify its temporal features. Second, no semantics was provided. This paper aims to provide a Temporal Deontic Defeasible Logic adopting an analytical approach and an argumentation semantics.

1 Introduction

In [1,2,3] basic Defeasible Logic (see [4]) was extended to capture some temporal aspects in legal reasoning. [1] proved useful in modeling temporal aspects of normative reasoning such as temporalised normative positions. [2] allowed for a logical account of the notion of the temporal viewpoint (the temporal position from which things are considered) and norm modifications. [3] provided a formal characterization of deadlines. All these variants can be criticized in two respects that this paper aims to repair.

First, as pointed out by [5], legal language may adopt either an analytical approach or a synthetical approach for the expression of time. In the synthetical approach all temporal and substantial elements of the norm are represented within the same sentence. For example:

- during the period from 10/6/1997 to 16/6/1997, anyone who parks in front of the station is liable to a fine.

In the analytical approach one sentence represents the substantive content of the norm, and other sentences specify its temporal features. In the above example, instead of having just one sentence, we could have had two sentences:

- Anyone who parks in front of the station is liable to a fine.
- Norm 1 is in efficacy from 10/6/1997 to 16/6/1997.

Some drawbacks and advantages of the analytical representations of time are listed by Marín and Sartor [5]. Drawbacks are that (i) more than one sentence needs to be considered in order to determine both the substantial content of the norm and its temporal status, (ii) inferences from analytical representations must take into account the interaction between substantive norms and norms which regulate temporal status. Besides,

P. Casanovas et al. (Eds.): Computable Models of the Law, LNAI 4884, pp. 239–253, 2008.

the analytical approach has a number of advantages: (i) it is modular, so that temporal features can be specified separately (e.g. they can be made dependent upon future and possibly not yet known facts), (ii) it better models the temporal structures usually adopted in legal language. The extensions [1,2,3] of basic Defeasible Logic adopted a synthetical approach. A first objective of this paper is to provide a variant of Temporal Deontic Defeasible Logic adopting an analytical approach as in [5].

Second, these extensions were defined by means of proof theories, and thus an appropriate semantics was needed to remove ambiguities. It is well known that non-monotonic reasoning can also be analyzed in terms of argumentation: non-monotonicity arises when an argument for a conclusion is defeated by a counter-argument. So a non-monotonic logic can be interpreted in terms of interacting arguments, giving it an argumentation semantics. An argumentation semantics is useful in applications where arguments are a natural feature of the problem domain, such as in law. A recent development is represented by argumentation and mediation systems which assist the users in expressing and organizing their arguments, in assessing their impact on controversial legal issues or in building up effective interactions in dialectical contexts. A second objective of this paper is to present an argumentation semantics for a variant of Temporal Deontic Defeasible Logic. The argumentation semantics is inspired by the semantics provided in [6], which allows us to provide intuitive explanations of conclusions in terms of arguments.

This paper is organized as follows. In section 2, we introduce the general conceptual model behind the framework. Section 3 describes a variant of Temporal Deontic Defeasible Logic that formalizes the legal model. Section 4 defines what arguments are. Section 5 focuses on the interaction amongst arguments. Section 6 provides the argumentation semantics.

2   Temporal Model

Our analysis shall be centered upon the distinction of several temporal aspects in legal norms. A first distinction deals with so-called *external times* and *internal times* of norms.

- The internal times are times associated to a norm or provision which is specified within the norm or provision.
- The external times are times associated to a norm or provision which is specified in a different norm or provision.

A second distinction deals with so-called *static times* and *dynamic times* of norms.

- The static times are times associated to a norm or provision which cannot change on the basis of events,
- The dynamic times are times associated to a norm or provision which can change on the basis of future events.

On this setting, we concentrate on two kinds of external and dynamic times associated to norms, namely the period of force and its period of efficacy:

- The period of force is the period during which the legal norm is in force or valid (if by validity of a norm we mean its partaking to the normative system). The period of force for each fragment may change over time as a function of the modifications the document goes through. By default, the beginning of the period of force is determined by other rules, and the end of the period of force is established by derogation rules.
- The period of efficacy is the period during which the provision *causes* its legal effect. For conditioned provisions, it is the period during which one can instantiate the provision's condition, thereby producing instances of the provisions effect. For instance, if the provision "if one smokes, then one is subject to a 10-euro fine" is in efficacy during $[t_1, t_2]$, then, during that period, the fact that a person smokes causes that persons subjection to the sanction. For unconditioned provisions, it is the period during which the provision is operative. For instance, if "it is forbidden to smoke" is operative during $[t_1, t_2]$, then, during that period, as an effect of the provision, smoking is prohibited. .

In general, a provision's period of efficacy coincide with its period of force, but in some cases they are different as for example in case of retroactive or ultra-active norms.

A commonly cited static time is the *date of existence* of a document when the law-making body (such as a senate or a lower house) stabilizes the document in its final form. The date of existence precedes the *date of enactment* when the competent authorities finalise the document by affixing their signatures to it (e.g. promulgation by a president, signature by a king or queen). In general, this date is clearly indicated in the document. The date of existence and date of enactment should not be confused with the *date of publication* when the normative document was published in an official journal. The official journal is designated as the source for establishing all public and legal. Another often cited static time is the *date of entry into force* of a document that marks the beginning of its period of force. In general, the date of entry of force is established on function of the date of document's publication in an official journal. For example, in the Italian normative system, the date of entry of force is establish after a 15-day period (called *vacatio legis*) starting at the date of publication.

3 The Logic Layer

3.1 The Language

We consider a linear discrete bounded set $\mathcal{T}$ of points of time termed "instants" and over it the order relation $> \subseteq \mathcal{T} \times \mathcal{T}$. We usually denotes the variables ranging over the elements of $\mathcal{T}$ by t and its eventual subscripts, the minimal unit by u. The lower and higher boundaries of $\mathcal{T}$ are denoted respectively by *min* and *max*.

Temporal intervals are defined as sets of instants between two indicated instants. Formally, an interval is a member of the set Inter $= \{[t_1, t_2] \in \mathcal{T} \times \mathcal{T} | t_1 \leq t_2\}$. As can be noted, this definition allows "punctual intervals" , i.e., intervals of the form $[t, t]$. We shall usually denote intervals by T.

We extend the basic language of Defeasible Logic with the operators OBL and PERM to indicate obligations and permissions. Each operator is temporalised with an interval $[t_1, t_2]$. For example, the formula $\text{OBL}^{\lfloor jan06,dec06 \rfloor} Tax$ means that taxes are due between January 2006 and December 2006. An assertion can also be prefixed by the operator @ parametrised by an interval $[t_1, t_2]$ to indicate when such assertion holds. For example, the fact $@^{\lfloor jan06,dec067 \rfloor} High\_Income$ means that $High\_Income$ holds between January 2006 and December 2006. Any operator in $\{@, \text{OBL}, \text{PERM}\}$ shall be called a modality because its behaviour is specific to the proof conditions. Since we allow sequences of temporalised operators, temporal modal assertions can have several temporal dimensions. For example, $@^{\lfloor jan07,dec07 \rfloor} \neg \text{OBL}^{\lfloor jan06,dec06 \rfloor} Tax$ says that between January 2007 and December 2007, no obligation holds between January 2006 and December 2006 to pay taxes.

A rule is a relationship between temporal modalised literals. A *strict rule* is an expression of the form $(\phi_1, \ldots, \phi_n \rightarrow \psi)$ such that whenever the premises are indisputable so is the conclusion . A *defeasible rule* is an expression of the form $(\phi_1, \ldots, \phi_n \Rightarrow \psi)$ whose conclusion can be defeated by contrary evidence. A *defeater* is a rule of the form $(\phi_1, \ldots, \phi_n \rightsquigarrow \psi)$ which cannot be used to draw any conclusion, their only use is to prevent some conclusions by defeating some defeasible rules. Rules can be temporalised by prefixing them with $@^{[t_1, t_2]}$, to indicate the period $[t_1, t_2]$ during which they hold and hence, when they can be used to derive conclusions.

Conflicts are managed by *superiority relations* $>$ among rules saying when a rule may override the conclusion of another rule.

In the following we define the language, that is, the set of rules specifying valid formulas composing a theory in this variant of Temporal Deontic Defeasible Logic.

Definition 1 (Language). *Let $\mathcal{T}$ a linear discrete bounded ordered set of instants of time, in which the minimal unit is u and the lower and higher boundaries of $\mathcal{T}$ are denoted respectively by min and max. Let* Prop *be a set of propositional atoms, and let* Mod $= \{@, \text{OBL}, \text{PERM}\}$ *be a set of modal operators, and* Lab *be a set of labels. The sets below are defined as the smallest sets closed under the following rules:*

Literals Lit $= \text{Prop} \cup \{\neg p | p \in \text{Prop}\}$
Temporal Modal Literals

$$\text{TempModLit} = \{X^T l, \neg X^T l | X \in \text{Mod}, T \in \text{Inter}, l \in \text{Lit}\}$$
$$\text{MTempModLit} = \{@^T \phi | T \in \text{Inter}, \phi \in \text{TempModLit}\}$$

Temporal Modal Rules

$$\text{TempModRul}_s = \{@^T (r: \quad \phi_1, \ldots, \phi_n \rightarrow \psi)$$
$$| T \in \text{Inter}, r \in \text{Lab}, \phi_1, \ldots, \phi_n, \psi \in \text{TempModLit}\}$$
$$\text{TempModRul}_d = \{@^T (r: \quad \phi_1, \ldots, \phi_n \Rightarrow \psi)$$
$$| T \in \text{Inter}, r \in \text{Lab}, \phi_1, \ldots, \phi_n, \psi \in \text{TempModLit}\}$$
$$\text{TempModRul}_{dft} = \{@^T (r: \quad \phi_1, \ldots, \phi_n \rightsquigarrow \psi)$$
$$| T \in \text{Inter}, r \in \text{Lab}, \phi_1, \ldots, \phi_n, \psi \in \text{TempModLit}\}$$
$$\text{TempModRul} = \{\text{TempModRul}_s \cup \text{TempModRul}_d \cup \text{TempModRul}_{dft}\}$$

Superiority Relations Sup $= \{s \succ r | s, r \in \text{Lab}\}$

We use some abbreviations: $A(r)$ denotes the set $\{\phi_1, \ldots, \phi_n\}$ of *antecedents* of the rule r, and $C(r)$ to denote the *consequent* ψ of the rule r. We indicate by $lab(r)$ the label of a temporal modal rule $r \in$ TempModRul. If q is a literal, $\sim q$ denotes the complementary literal (if q is a positive literal p then $\sim q$ is $\neg p$; and if q is $\neg p$, then $\sim q$ is p). Punctual intervals of the form $[t,t]$ shall be abbreviated by the instant t.

A theory consists of a discrete totally ordered set of instants of time, a set of *facts* or indisputable statements, a set of *rules*, a set of *superiority relations* $>$ among rules.

Definition 2 (Defeasible Theory). A defeasible theory is a structure $D = (\mathcal{T}, F, R, >)$ where $\mathcal{T}$ a discrete totally ordered set of instants of time, $F \subseteq$ TempModLit is a finite set of facts, $R \subseteq$ TempModRul is a finite set of rules such that each rule has a unique label, $> \subseteq$ Sup is a set of acyclic superiority relations.

Example 1. *We illustrate below a defeasible theory, and hence the language, by means of an example showing the handling of retroactive provisions.*

$$\mathcal{T} = \{min, jan05, dec05, jan06, dec06, jan07, dec07, max\}$$

$$
F = \{ @^{[jan06,dec06]} High\_Income,
$$
$$
@^{[jan06,dec06]} Force(ef1),\ @^{[jan06,dec06]} Efficace(ef1),
$$
$$
@^{[jan07,dec07]} Force(ef2),\ @^{[jan07,dec07]} Efficace(ef2) \}
$$

$$
R = \{ @^{[jan05,max]}(r_1: \quad @^t High\_Income \Rightarrow \mathrm{OBL}^t Tax),
$$
$$
@^{[jan07,max]}(r_2: \quad @^t High\_Income \Rightarrow \neg \mathrm{OBL}^t Tax),
$$
$$
@^{[jan06,dec06]}(fo1: \qquad \Rightarrow @^{[jan06,dec06]} Force(rl)),
$$
$$
@^{[jan07,dec07]}(fo2: \qquad \Rightarrow @^{[jan07,dec07]} Force(r2)),
$$
$$
@^{[jan06,dec06]}(ef1: \qquad \Rightarrow @^{[jan06,dec06]} Efficace(rl)),
$$
$$
@^{[jan07,dec07]}(ef2: \qquad \Rightarrow @^{[jan06,dec06]} Efficace(r2))\}
$$

$$>= \{r2 \succ r1\}$$

The period of force and efficacy a provision represented by a rule r are expressed by the predicates $Force(r)$ and $Efficace(r)$. Remark that these periods are defined by means of rules and not as facts because they are dynamic (see section 2), i.e., they can change on the basis of future events. The period T in a temporal rule $@^T r$ (where r is any rule in Rule) can be interpreted as the period of existence of the represented provision. The lower bound of this period of existence is the date of existence. The time of enactment and publication of the provision can be expressed by means of further predicates as $enactment(r)$ and $publication(r)$. Doing so, we can represent, for example, the vacatio legis establishing the date of entry of force of a provision after a X-day period starting at the date of publication (in Italy $X = 15$):

$$@^{[min,max]}(vl: \quad @^t Publication(X) \Rightarrow @^{[t+15,max]} Force(X)),$$

The times of force and efficacy of the vacatio legis rule holds for any time and can be expressed by the facts $@^{[min,max]} Force(r)$ and $@^{[min,max]} Efficace(r)$.

3.2 Proof Theory

A conclusion of a theory D is a tagged temporal modal literal or rule having one of the following forms:

$+\Delta\gamma$ meaning that γ is definitely provable in D.
$-\Delta\gamma$ meaning that γ is not definitely provable in D.
$+\partial\gamma$ meaning that γ is defeasible provable in D.
$-\partial\gamma$ meaning that γ is not defeasible provable in D.

Provability is based on the concept of a derivation (or proof) in D. A derivation is a finite sequence $P = (P(1),..,P(n))$ of tagged temporal modal literals or rules. $P(1..n)$ denotes the initial part of the sequence P of length n. Each tagged temporal modal literal or rule satisfies some proof conditions, which correspond to inference rules for the four kinds of conclusions we have mentioned above.

Before moving to the conditions governing derivability of conclusions, we need to introduce some preliminary notions. First, we provide by means of the conversion relations which modal expressions can be converted into which modal expressions.

Definition 3 (Modality conversion). For any $l \in \text{Lit}$ and $T_1, T_2 \in \text{Inter}$, let the sets:

$\text{convert}(\gamma) = \{\gamma | \gamma \in \text{TempModLit} \cup \text{MTempModLit}\}$,
$\text{convert}(\text{OBL}^{T_2}l) = \{\text{PERM}^{T_2}l, \neg\text{OBL}^{T_2}{\sim}l\}$,
$\text{convert}(\neg\text{OBL}^{T_2}{\sim}l) = \{\text{PERM}^{T_2}l\}$,
$\text{convert}(\text{PERM}^{T_2}l) = \{\neg\text{OBL}^{T_2}{\sim}l\}$,
$\text{convert}(\neg\text{PERM}^{T_2}{\sim}l) = \{\text{OBL}^{T_2}l, \text{PERM}^{T_2}l\}$,
$\text{convert}(Y^{[t_{2i},t_{2f}]}l) = \{@^{[t_{1i},t_{1f}]}Y^{[t_{2i},t_{2f}]}l | Y \in \text{Mod}, t_{1i} \geq t_{2i}, t_{1f} \geq t_{2i}\}$,
$\text{convert}(\neg Y^{[t_{2i},t_{2f}]}l) = \{@^{[t_{1i},t_{1f}]}\neg Y^{[t_{2i},t_{2f}]}l | Y \in \text{Mod}, t_{1i} \geq t_{2i}, t_{1f} \geq t_{2i}\}$,
$\text{convert}(@^{T_1}\gamma) = \{@^{T_1}\beta | \beta \in \text{convert}(\gamma), \gamma \in \text{TempModLit}\}$.

The first conversion indicates that any temporal modal literal can be converted into itself: this conversion is a mere trick allowing us to simplify the proof condition.

Defeasible provability deals with conflicts. This requires to state when two temporal modal literals are in conflict with each other:

Definition 4 (Complementary modal literals). For any $l \in \text{Lit}$ and $T_1, T_2, T_1', T_2' \in \text{Inter}$, let the sets:

$\mathscr{C}(@^{T_2}l) = \{\neg@^{T_2'}l, @^{T_2'}{\sim}l\}$
$\mathscr{C}(\neg@^{T_2}l) = \{@^{T_2'}l, \neg@^{T_2'}{\sim}l\}$
$\mathscr{C}(\text{OBL}^{T_2}l) = \{\neg\text{OBL}^{T_2'}l, \text{OBL}^{T_2'}{\sim}l, \neg\text{PERM}^{T_2'}l, \text{PERM}^{T_2'}{\sim}l\}$
$\mathscr{C}(\neg\text{OBL}^{T_2}l) = \{\text{OBL}^{T_2'}l, \neg\text{PERM}^{T_2'}l\}$
$\mathscr{C}(\text{PERM}^{T_2}l) = \{\text{OBL}^{T_2'}{\sim}l, \neg\text{PERM}^{T_2'}l\}$
$\mathscr{C}(\neg\text{PERM}^{T_2}l) = \{\neg\text{OBL}^{T_2'}{\sim}l, \text{PERM}^{T_2'}l, \neg\text{PERM}^{T_2'}{\sim}l, \text{OBL}^{T_2'}l\}$
$\mathscr{C}(@^{T_1}\gamma) = \{@^{T_1'}\beta | \beta \in \mathscr{C}(\gamma), \gamma \in \text{TempModLit}\}$.

Note that if $\beta \in \mathscr{C}(\gamma)$ then $\gamma \in \mathscr{C}(\beta)$. Next we move to temporal relations.

Definition 5 (Intervals: basic notions). Let 'start()' and 'end()' be the functions that return the lower and upper bounds of an interval respectively. Let u be the temporal unit. For any $T_1, T_2, T_3 \in \text{Inter}$, we have:

- $T_1 \sqsubseteq T_2$ iff $\text{start}(T_2) \leq \text{start}(T_1)$ and $\text{end}(T_1) \leq \text{end}(T_2)$;
- $\text{over}(T_1, T_2)$ iff $\text{start}(T_1) \leq \text{end}(T_2)$ and $\text{start}(T_2) \leq \text{end}(T_1)$;
- $T_1 \sqcup T_2 = T_3$ iff $\text{end}(T_1) + u = \text{start}(T_2)$, $\text{start}(T_3) = \text{start}(T_1)$ and $\text{end}(T_3) = \text{end}(T_2)$.

These relations are not meant to be a proposal of an algebra of intervals (e.g. [7]), instead, they have been chosen in order to make the proof conditions as simple as possible. To lighten the paper, we use some abbreviations consisting in placing temporal modal expression as an argument of the previous relations for intervals.

Definition 6. Let $(X_i^{T_{1i}})_{1..n}\gamma, (X_i^{T_{2i}})_{1..n}\gamma, (X_i^{T_{3i}})_{1..n}\gamma \in \text{TempModLit} \cup \text{MTempModLit} \cup \text{TempModRul}$, we have:

- $(X_i^{T_{1i}})_{1..n}\gamma \sqsubseteq (X_i^{T_{2i}})_{1..n}\gamma$ iff $\forall i \in \{1..n\}$, $T_{1i} \sqsubseteq T_{2i}$;
- $\text{over}((X_i^{T_{1i}})_{1..n}\gamma, (X_i^{T_{2i}})_{1..n}\gamma)$ iff $\exists i \in \{1..n\}$, $\text{over}(T_{1i}, T_{2i})$;
- $(X_i^{T_{1i}})_{1..n}\gamma \sqcup (X_i^{T_{2i}})_{1..n}\gamma = (X_i^{T_{3i}})_{1..n}\gamma$ iff $\exists i \in \{1..n\}$, $T_{1i} \sqcup T_{2i} = T_{3i}$, $\forall j \neq i$, $\text{start}(T_{1j}) = \text{start}(T_{2j}) = \text{start}(T_{3j})$, $\text{end}(T_{1j}) = \text{end}(T_{2j}) = \text{end}(T_{3j})$.

For example, we can write $@^{[10,15]}\neg\text{OBL}^{[30,80]}Tax \sqsubseteq @^{[0,50]}\neg\text{OBL}^{[0,80]}Tax$. We present separately the condition for a rule to be applicable so that it supports a temporal modal literal of the form $@^T Y^{T_Y} \gamma$:

If $@^{T_R}r$ is $+\Delta[@^T Y^{T_Y} \gamma]$-applicable then
(1) $C(r) = X^{T_X}\gamma'$, $T_Y \sqsubseteq T_X$, $Y^{T_X} \gamma \in \text{convert}(X^{T_X}\gamma')$, and
(2) (2.1) $+\Delta @^T r \in P(1..i)$, and
     (2.2) $\forall Z^{T_Z}\alpha \in A(r)$,
          (2.2.1) $+\Delta @^T Z^{T_Z}\alpha \in P(1..i)$, and
          (2.2.2) $+\Delta @^T @^{T_Z}Efficace(r) \in P(1..i)$, and
     (2.3) $+\Delta @^T Force(r) \in P(1..i)$ or $+\Delta @^T @^T Force(r) \in P(1..i)$.

A rule, which is not applicable, is discarded following the conditions below:

If $@^{T_R}r$ is $+\Delta[@^T Y^{T_Y} \gamma]$-discarded then
(1) $C(r) = X^{T_X}\gamma'$, $T_Y \not\sqsubseteq T_X$, or $Y^{T_X} \gamma \notin \text{convert}(X^{T_X}\gamma')$, or
(2) (2.1) $-\Delta @^T r \in P(1..i)$, or
     (2.2) $\exists Z^{T_Z}\alpha \in A(r)$,
          (2.2.1) $-\Delta @^T Z^{T_Z}\alpha \in P(1..i)$, or
          (2.2.2) $-\Delta @^T @^{T_Z}Efficace(r) \in P(1..i)$, or
     (2.3) $-\Delta @^T Force(r) \in P(1..i)$ and $-\Delta @^T @^T Force(r) \in P(1..i)$.

The conditions for a rule to be ∂-applicable(resp. ∂-discarded) are the same as those for Δ-applicable(resp. Δ-discarded), but where we replace Δ with ∂.

We are now ready to define the proof theory that is, the inference conditions to derive tagged conclusions from a given theory D. Note that the formalism we have introduced allows us to temporalise rules, thus we have to admit the possibility that rules are not only given but can be proved to hold for certain span of time. Accordingly we have to give conditions that allow us to derive rules instead of literals. We begin with the proof conditions to determine whether a rule is a definite conclusion of a theory D. A modal temporal rule r is definitely provable $(+\Delta)$ if (1) there exists a rule

r' in the set of rule R such that $r \sqsubseteq r'$, or (2) r is definitely provable in some temporal sub-intervals. A temporal modal rule r is not definitely provable if (1) there is not such rule in the set of rules defined in some sup-interval, or (2) r is not defined in any sub-intervals. Let $r \in R$:

If $P(i+1) = +\Delta r$ then
(1) $\exists r' \in R, r \sqsubseteq r'$ or
(2) $\exists r_1, r_2 \in R, r_1 \sqcup r_2 = r$,
$+\Delta r_1 \in P(1..i)$ and $+\Delta r_2 \in P(1..i)$.

If $P(i+1) = -\Delta r$ then
(1) $\forall r' \in R, r \not\sqsubseteq r'$, and
(2) $\forall r_1, r_2 \in R \; r_1 \sqcup r_2 = r$,
$-\Delta r_1 \in P(1..i)$ or $-\Delta r_2 \in P(1..i)$.

We can now move to definite conclusion of temporal modal literals.

If $P(i+1) = +\Delta \gamma$ then
(1) $\exists \gamma', \gamma \in \text{convert}(\gamma'), \gamma' \in F, \gamma \sqsubseteq \gamma'$, or
(2) $\exists r \in R_s, r$ is $\Delta \gamma$-applicable, or
(3) $\exists \gamma_1, \gamma_2, \gamma_1 \sqcup \gamma_2 = \gamma, +\Delta \gamma_1 \in P(1..i)$ and $+\Delta \gamma_2 \in P(1..i)$.

To prove that a temporal modal literal is not definitely provable we have to show that any attempt to give a definite proof fails.

If $P(i+1) = -\Delta \gamma$ then
(1) $\forall \gamma', \gamma \in \text{convert}(\gamma'), \gamma' \notin F, \gamma \sqsubseteq \gamma'$, and
(2) $\forall r \in R_s, r$ is $\Delta \gamma$-discarded, and
(3) $\forall \gamma_1, \gamma_2, \gamma_1 \sqcup \gamma_2 = \gamma, -\Delta \gamma_1 \in P(1..i)$ or $-\Delta \gamma_2 \in P(1..i)$.

We now turn our attention to defeasible derivations, that is, derivations giving an assertion as a defeasible conclusion of a theory D. We begin with the proof conditions to determine whether a rule is a defeasible conclusion $(+\partial)$ or not $(-\partial)$.

If $P(i+1) = +\partial r$ then $+\Delta r \in P(1..i)$ If $P(i+1) = -\partial r$ then $-\Delta r \in P(1..i)$

Defeasible provability $(+\partial)$ for temporal modal literals consists of three phases. In the first phase, we put forward a supported reason for the assertion that we want to prove. Then in the second phase, we consider all possible attacks against the desired conclusion. Finally in the last phase, we have to counter-attack the attacks considered in the second phase.

If $P(i+1) = +\partial \gamma$ then
(1) $+\Delta \gamma \in P(1..i)$, or
(2) $\forall \beta \in \mathscr{C}(\gamma), -\Delta \beta \in P(1..i)$, over$(\beta, \gamma)$, and
    (2.1) $\exists r \in R_{sd}, r$ is $\partial \gamma$-applicable,
    (2.2) $\forall \beta, \beta \in \mathscr{C}(\gamma)$, over$(\beta, \gamma), \forall s \in R$,
        (2.2.1) s is $\partial \beta$-discarded, and
        (2.2.2) $\exists w \in R_{sd}$,
            (2.2.1.1) w is $\partial \gamma$-applicable, and
            (2.2.1.2) $lab(w) \succ lab(s)$, or
(3) $\exists \gamma_1, \gamma_2, \gamma_1 \sqcup \gamma_2 = \gamma, +\Delta \gamma_1 \in P(1..i)$ and $+\Delta \gamma_2 \in P(1..i)$.

Let us illustrate the proof condition of the defeasible provability of γ. We have two cases: 1) We show that γ is already definitely provable; or 2) we need to argue using the defeasible part of D. In this second case, to prove γ defeasibly we must show that no complement temporal modal literals β definitely provable (2). We require then there must be a strict or defeasible (applicable) rule r providing γ (2.1). But now we need to consider possible attacks, i.e., that is, any rule s supporting attacking complement temporal modal literals. Note that here we consider defeaters, too, whereas they could not be used to support the conclusion γ. These attacking rules s have to be discarded (2.2.1), or must be defeated by a stronger rule w supporting γ (2.2.2). Finally, we have to cater for the case where γ is defeasible provable on sub-intervals making up γ (3).

To prove that a temporal modal literal is not defeasibly provable we have to show that any attempt to give a proof fails.

If $P(i+1) = -\partial\gamma$ then

(1) $-\Delta\gamma \in P(1..i)$, or

(2) $\exists\beta \in \mathscr{C}(\gamma)$, $+\Delta\beta \in P(1..i)$, over(β, γ), or

    (2.1) $\forall r \in R_{sd}$, r is $\partial\gamma$-discarded,

    (2.2) $\exists\beta$, $\beta \in \mathscr{C}(\gamma)$, over$(\beta, \gamma)$, $\exists s \in R$,

        (2.2.1) s is $\partial\beta$-applicable, and

        (2.2.2) $\forall w \in R_{sd}$,

            (2.2.1.1) w is $\partial\gamma$-discarded, or

            (2.2.1.2) $lab(w) \not\succ lab(s)$, and

(3) $\forall\gamma_1, \gamma_2, \gamma_1 \sqcup \gamma_2 = \gamma$, $-\Delta\gamma_1 \in P(1..i)$ or $-\Delta\gamma_2 \in P(1..i)$.

Example 2. *Let us illustrate the proof theory by considering the theory as exposed in example 1. By applying the rules $fo1$ and $ef1$, we derive:*

$+\Delta @^{[jan06,dec06]} @^{[jan06,dec06]} Force(r1)$, *and*
$+\Delta @^{[jan06,dec06]} @^{[jan06,dec06]} Efficace(r1)$.

From the fact and the rule $r1$, we derive:

$+\Delta @^{[jan06,dec06]} @^{[jan06,dec06]} High\_Income$ *and*
$+\Delta @^{[jan06,dec06]} r1$.

From these results, we can derive $+\Delta @^{[jan06,dec06]} OBL^{[jan06,dec06]} Tax$, *i.e., from the points of view January 2006 to December 2006, Tax is due between January 2006 to December 2006. By applying the rules $fo2$ and $ef2$, we derive:*

$+\Delta @^{[jan07,dec07]} @^{[jan07,dec07]} Force(r2)$, *and*
$+\Delta @^{[jan07,dec07]} @^{[jan06,dec06]} Efficace(r1)$.

From the fact and the rule $r2$, we derive

$+\Delta @^{[jan07,dec07]} @^{[jan06,dec06]} High\_Income$, *and*
$+\Delta @^{[jan07,dec07]} r2$.

From these results, we can derive $+\Delta @^{[jan07,dec07]} \neg OBL^{[jan06,dec06]} Tax$. *i.e., from the points of view January 2007 to December 2007, tax is not due between January 2006 to December 2006.*

4 The Argument Layer

The argument layer defines what arguments are. An argument for a temporal modal assertion (i.e. a literal or a rule) is a proof tree (or monotonic derivation) of that assertion in Temporal Modal Defeasible Logic. Nodes are labeled by either temporal modal literals or temporal modal rules. Inside an argument Arg, temporal modal literals and rule are tagged by ∂^{Arg} or Δ^{arg}. Nodes are connected by arrows that correspond to grounded inferences rules (see below). In the following, a temporal modal literal or rule shall be called an assertion denoted by Greek capital letters (Φ, Ω ...).

Definition 7. *An argument arg for an assertion Ω is a tree such that:*

- *each node is labeled by an assertion tagged by ∂^{arg} or Δ^{arg},*
- *the root node is labeled by a tagged assertion $+\partial^{arg}\Omega$ or $+\Delta^{arg}\Omega$,*
- *each arrow corresponds to a grounded inference rule of the following types:*

$$\frac{+\Delta^{arg}\Omega', \Omega \sqsubseteq \Omega'}{+\Delta^{arg}\Omega}$$

$$\frac{+\Delta^{arg}\Omega_1, +\Delta^{arg}\Omega_2, \Omega_1 \sqcup \Omega_2 = \Omega}{+\Delta^{arg}\Omega}$$

$$\frac{r \text{ is } \Delta^{arg}\gamma - \text{applicable}}{+\Delta^{arg}\gamma}$$

$$\frac{+\Delta^{arg}\gamma', \gamma' \in F, \gamma \in \text{convert}(\gamma')}{+\Delta^{arg}\gamma}$$

$$\frac{+\Delta^{arg}\Omega}{+\partial^{arg}\Omega}$$

$$\frac{+\partial^{arg}\Omega_1, +\partial^{arg}\Omega_2, \Omega_1 \sqcup \Omega_2 = \Omega}{+\partial^{arg}\Omega}$$

$$\frac{r \text{ is } \partial^{arg}\gamma - \text{applicable}}{+\partial^{arg}\gamma}$$

where for $\# \in \{\Delta, \partial\}$ if $@^{T_R}r$ is $+\#^{arg}[@^T Y^{T_Y}\gamma] -$ applicable then
(1) $C(r) = X^{T_X}\gamma'$, $T_Y \sqsubseteq T_X$, $Y^{T_X}\gamma \in \text{convert}(X^{T_X}\gamma')$, and
(2) (2.1) $+\#^{arg}@^T r \in P(1..i)$, and
(2.2) $\forall Z^{T_Z}\alpha \in A(r)$,
(2.2.1) $+\#^{arg}@^T Z^{T_Z}\alpha \in P(1..i)$, and
(2.2.2) $+\#^{arg}@^T @^{T_Z}Efficace(r) \in P(1..i)$, and
(2.3) $+\#^{arg}@^T Force(r) \in P(1..i)$ or, $+\#^{arg}@^T @^T Force(r) \in P(1..i)$.

- *a node labeled by a defeater cannot have an ascendant node labeled by a defeater.*

The last condition specifies that a defeater rule may only be used at the top of an argument; in particular, no chaining of defeaters is allowed. To illustrate the definition of argument, one can build from the theory D_{toy} of the previous section, the arguments (among others) shown in Figure 1.

Definition 8. *A (proper) sub-argument of an argument A is a (proper) sub-tree of the tree associated to A.*

Definition 9. *A tagged assertion $+\partial^A\Omega$ is a conclusion of an argument A if $+\partial^A\Omega$ labels a node of the argument A.*

A more usual alternative would be to regard only the root of an argument as its unique conclusion, but this choice would make the other definitions more complex. Since conclusions can be differently qualified depending on the rules used, arguments are differentiated as follows:

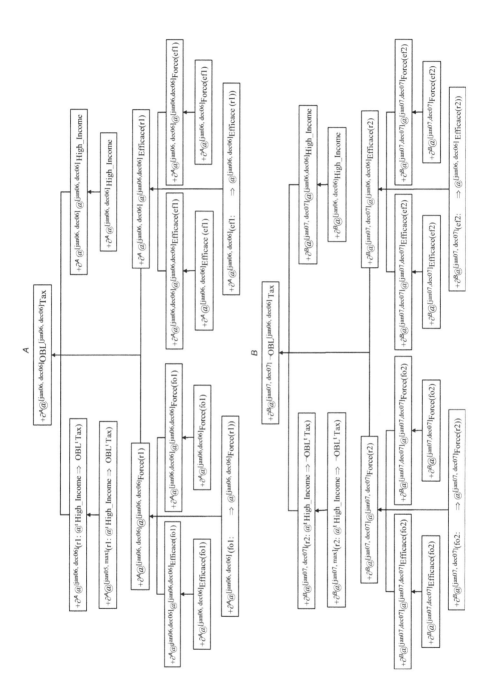

Fig. 1. Some arguments built from the theory exposed in the example 1

Definition 10. *A supportive argument is a finite argument in which no defeater is used. A strict argument is an argument of which all nodes are tagged by* $+\Delta$. *An argument that is not strict is defeasible.*

Example 3. *In the figure 1, the argument A is a supportive argument for* $+\partial^A @^{[2006,2007]}\text{OBL}^{[2006,2006]}Tax$. *It is not a strict argument and thus it is a defeasible argument.*

5 The Dialectical Layer

The precedent section defined the argument layer and isolated the concept of argument. This section presents the dialectical layer which is concerned with relations standing amongst arguments. It defines the notion of support and attack, and focuses on the interaction amongst arguments. Firstly, we introduce the notion of support:

Definition 11. *A set of arguments S supports a defeasible argument A if every proper sub-argument of A is in S.*

Note that, in our setting, the atomic arguments, constituted of a fact or a rule of the theory, are supported by the empty set.

The conditions that determine which argument can attack another argument are defined in the following. In the section of the proof theory, a defeasible conclusion is shown to have a proof condition consisting of three phases. In the first phase, a supporting rule r is provided.

In the second phase, all possible attacks against the desired conclusion are considered, that is, an attack consists of any rule s supporting a complemement temporal modal literal. In the third phase, counter-attacks are proposed by means of a defeating rule w.

So in the proof condition, the relation of attack between the first and second phase is somewhat different of the relation of attack between the second and third phase. To reflect this, we provide the notion of attack and defeat between arguments in the following.

Definition 12. *An argument S attacks a non-strict argument R iff* $+\partial^S\beta$ *and* $+\partial^R\gamma$ *are conclusions,* $\gamma \in \mathscr{C}(\beta)$ *and* over(γ,β).

Example 4. *In figure 1, the argument B attacks the argument A and vice versa.*

Definition 13. *An argument W defeats a non-strict argument S iff*

(1) $+\partial^W\gamma$ *and* $+\partial^S\beta$ *are conclusions respectively of a rule w and a rule s,* $\gamma \in \mathscr{C}(\beta)$, over($\gamma,\beta$), *and*
(2) $w \succ s$.

Example 5. *In the figure 1, the argument A attacks but does not defeat the argument B whereas the argument B attacks and defeats the argument A.*

Defeasible reasoning differentiates traditionally between rebuttal and undercutting. We stick to the tradition and define the notion of undercutting as follows:

Definition 14. *An argument A undercuts a defeasible argument B if A attacks a proper sub-argument of B.*

In this setting, an argument that is attacked but not undercut is said to be rebutted.

Definition 15. *A set of arguments S undercuts a defeasible argument B if there is an argument A supported by S that attacks a proper sub-argument of B.*

Definition 16. *A set of arguments S defeats a defeasible argument B if there is an argument A supported by S that defeats B.*

Comparing arguments by pairs is not enough since an attacking argument can in turn be attacked by other arguments. In the following, we will define justified arguments, i.e. arguments that have no viable attacking argument in the discourse, and rejected arguments that are attacked by justified argument. As in many argumentation systems, we base the status justified or rejected of arguments on the concept of acceptability of an argument w.r.t. to set of arguments S. That an argument A is acceptable w.r.t. to set of arguments S means that any attacker against A is defeated by an argument supported by S. In this line, we next present a slightly adapted version of [8]'s definition of acceptability.

Definition 17. *An argument A is acceptable w.r.t. a set of arguments S iff either*

(1) A is strict, or
(2) for any argument B attacking A
    (2.1) B is undercut by S, or
    (2.2) B is defeated by S.

The condition (2.1) aims to provide an ambiguity blocking semantics of the system, whereas the condition (2.2) aims to provide a team defeat feature of the system.

Based on the concept of acceptability we proceed to define justified arguments and justified assertion. That an argument A is justified means that it resists every refutation. Given a defeasible theory D, the set of arguments that can be generated from D is denoted by $Args_D$. The following definition is based on [9]'s definition of fixed point semantics.

Definition 18. *The set of justified arguments in a theory D is $JArgs_D = \bigcup_{i=0}^{\infty} J_{D,i}$ with*

- $J_{D,0} = \varnothing$,
- $J_{D,i+1} = \{arg \in Args_D | arg$ *is acceptable w.r.t.* $J_{D,i}\}$.

So, an argument A is acceptable w.r.t. $J_{D,i+1}$ if either A is strict, or any argument B attacking A is undercut by $J_{D,i}$ (i.e. there is an argument C supported by $J_{D,i}$ that attacks a proper sub-argument of B) or defeated by an argument supported by $J_{D,i}$.

Definition 19. *A tagged assertion $+\partial^{arg}\Omega$ is justified if it is the conclusion of a supportive argument in $JArgs_D$.*

A tagged assertion $+\partial^{arg}\Omega$ is justified means that it is provable $(+\partial)$. However, Defeasible Logic permits to express when a conclusion is not provable $(-\partial)$. Briefly, that a conclusion is not provable means that every possible argument for that conclusion has been refuted. In the following, this notion is captured by assigning the status rejected to arguments that are refuted. Roughly speaking, an argument is rejected if it has a rejected sub-argument or it cannot overcome an attack from a justified argument. Given an argument A, a set S of arguments (to be thought of as arguments that have already been rejected), and a set J of arguments (to be thought of as justified arguments that may be used to support attacks on A), we assume the following definition of the argument A being rejected by S and J:

Definition 20. *An argument A is rejected by the sets of arguments S and J when A is not strict and if (i) a proper sub-argument of A is in S or (ii) it is attacked by an argument supported by J.*

Definition 21. *The set of rejected arguments in a theory D w.r.t. J is $RArgs_D(J) = \bigcup_{i=0}^{\infty} R_{D,i}$ with*

- $R_{D,0}(J) = \varnothing$,
- $R_{D,i+1}(J) = \{arg \in Args_D | arg \text{ is rejected by } R_{D,i}(J) \text{ and } J\}$.

Definition 22. *A tagged assertion $+\partial^{arg}\Omega$ is rejected by J if there is no argument in $Args_D - RArgs_D(J)$ that ends with as a supportive rule for $+\partial^{arg}\Omega$.*

As shortcut, we say that an argument is rejected if it is rejected w.r.t. $JArgs_D$ and a literal is rejected if it is rejected by $JArgs_D$.

6   The Argumentation Semantics

An argumentation semantics with ambiguity blocking can now be provided by characterising conclusions in argumentation terms:

Definition 23. *Let D be a defeasible theory and Ω be an assertion,*

- $D \vdash +\Delta\Omega$ *iff there is a strict argument Arg supporting $+\partial^{Arg}\Omega$ in $Args_D$.*
- $D \vdash -\Delta\Omega$ *iff there is no strict argument Arg supporting $+\partial^{Arg}\Omega$ in $Args_D$.*
- $D \vdash +\partial\Omega$ *iff $+\partial^{arg}\Omega$ is justified.*
- $D \vdash -\partial\Omega$ *iff $+\partial^{arg}\Omega$ is rejected by $Jargs_D$.*

The proof theories are complete and consistent with the argumentation semantics, the proof not provided here for reason of space is similar to the one in [6]. It follows that for any defeasible theory, no argument is both justified and rejected, and thus no literal is both justified and rejected. Eventually, if the set $JArgs_D$ of justified arguments contains two arguments with conflicting conclusions then both arguments are strict. That is, inconsistent conclusions can be reached only when the strict part of the theory is inconsistent.

7   Conclusion

This paper is part of an ongoing effort in AI and Law to provide an appropriate representation of legal reasoning in a temporal setting. The proof theories and an argumentation semantics of a variant of Temporal Deontic Defeasible Logic with an analytical approach of temporal legal aspects has been provided. We addressed temporal non-monotonic reasoning and handling of legal temporal status: as a matter of future research, an interesting point is to accommodate the framework with other work on temporal reasoning in AI (c.f. [10]) to integrate, for example, temporal constraints and a richer temporal ontology allowing qualitative temporal reasoning.

Acknowledgements. The authors are indebted to G. Sartor, M. Palmirani and R. Rubino for discussion on the temporal model, and, G. Governatori for the logical part.

References

1. Governatori, G., Rotolo, A., Sartor, G.: Temporalised normative positions in defeasible logic. In: ICAIL 2005, Proceedings of 10th Int. Conf. on Artif. Intell. and Law, pp. 25–34. ACM, New York (2005)
2. Governatori, G., Rotolo, A., Riveret, R., Palmirani, M., Sartor, G.: Variants of temporal defeasible logics for modelling norm modifications. In: ICAIL 2007, Proceedings of the 11h Int. Conf. on Artif. Intell. and Law, pp. 155–159. ACM, New York (2007)
3. Governatori, G., Hulstijn, J., Riveret, R., Rotolo, A.: Characterising deadlines in temporal modal defeasible logic. In: AUSTAI 2007, 20th Australian Joint Conf. on Artif. Intell. Springer, Heidelberg (2007)
4. Antoniou, G., Billington, D., Governatori, G., Maher, M.J.: Representation results for defeasible logic. ACM Transactions on Computational Logic 2(2), 255–287 (2001)
5. Marín, R.H., Sartor, G.: Time and norms: a formalisation in the event-calculus. In: ICAIL 1999, Proceedings of the 7th Int. Conf. on Artif. Intell. and Law, pp. 90–99. ACM, New York (1999)
6. Governatori, G., Maher, M.J., Billington, D., Antoniou, G.: Argumentation semantics for defeasible logics. Journal of Logic and Computation 14(5), 675–702 (2004)
7. Allen, J.F.: Towards a general theory of action and time. Artif. Intell. 23(2), 123–154 (1984)
8. Dung, P.M.: On the acceptability of arguments and its fundamental role in nonmonotonic reasoning, logic programming and n-person games. Artif. Intell 77(2), 321–358 (1995)
9. Prakken, H., Sartor, G.: Argument-based extended logic programming with defeasible priorities. Journal of Applied Non-Classical Logics 7(1) (1997)
10. Vila, L., Yoshino, H.: Time in automated legal reasoning. Information and Communications Technology Law 7(3), 173–197 (1998)

Rulebase Technology and Legal Knowledge Representation

Giuseppe Contissa

CIRSFID – University of Bologna
via Galliera 3
40128 Bologna, Italy
giuseppe.contissa@unibo.it

Abstract. This paper reflects the results of a study conducted as a side work connected with the development of ALIS (Automated Legal Intelligent System), modeling a representation of legal knowledge in the area of intellectual property rights using the RuleBurst rule-based system technology. In this first stage, our work has been focused on Italian Copyright law, with the aim to develop a method that could be extended and applied, in a subsequent stage, to other IP legislations in Europe. The integration in the ALIS decision support system of the Ruleburst inferencing system with an advanced legal text retrieval engine and a game-theory strategy engine is facilitated by using a (quasi) natural language-knowledge representation, enhancing the benefits of isomorphism.

Keywords: ALIS, copyright law, Rule-based system, Knowledge Representation.

1 Introduction

This paper reflects a feasibility test carried out modeling a representation of legal knowledge in the area of intellectual property rights using the RuleBurst rule-based system technology. This test was conducted as a side work by CIRSFID team involved in the ALIS project[1].

ALIS consortium aims to combine the recent advances in game theory, artificial intelligence and law and regulation corpus structuring (semantics) in order to develop an innovative system providing European citizens and private companies with a transparent, fast, secure and reliable access to the European Legal Knowledge in the domain of IPR. Such a system will make the management of legal knowledge easier, in order to avoid conflicts through prevention, quicken judiciary decisions, facilitate compliance with laws and regulations and the drafting of new laws.

The representation in RuleBurst language has been developed in order not only to test the feasibility of integrating RuleBurst technologies in ALIS system, but also to help the development of functionalities of the latter, in particular those concerning the ALIS repository module, including a knowledge base representing the IPR legal domain in the form of production rules, and the correlated modules of the ALIS legal rules engine (the reasoning engine) and the user interface.

---

[1] This paper is partly based on research conducted for the EU Project ALIS (Contract no.: 027968) funded under the FP6 IST Programme.

P. Casanovas et al. (Eds.): Computable Models of the Law, LNAI 4884, pp. 254–262, 2008.

2 The Application Domain and Requirements

The choice of the IPR legal domain raised several issues to be faced and resolved before the development of the legal knowledge representation.

First of all, the large extent of the domain: Intellectual Property Laws are designed to protect different forms of intangible subject matter, and under this "umbrella-expression" are included the macro-categories of *strictu sensu* intellectual property rights (e.g. copyright), industrial property rights (e.g. patent and trademark protection) and commercial property rights (e.g. trade secrets normative). Each of them are peculiar and complex domains on their own, but at the same time highly linked to other legal domains (contract law, procedural law, etc).

Another strong cause of complexity consists in the fragmentary national legislations, even if they are part of a supranational (European in our case) legal order. In other words, any legal question that may arise on a given subject matter always needs to be faced and solved within its own national boundaries, even when it is possible to identify some common legal elements at a supranational level. This is so both because of the differences among domestic legislations (even on the same subject) that often occur when different countries and cultures are involved and because of the different procedural rules that may influence differently the approach to the subject matter itself from one country to another.

It is possible that this complexity is one of the reasons that discouraged the development of high number of legal knowledge representations in this wide and particular domain of the law, in favour of other more tractable ones (e.g. taxes or social security legislation).

To address the scope of the prototype, namely to test the potential of ALIS system, but with a limited effort in terms of time and work, the decision was therefore to focus on building a working prototype for a limited national subdomain of IPR law, The Italian copyright law (translated in English), with the aim to develop and validate a method to be extended and applied, in a subsequent stage, to other IP legislations in Europe.

From the technical perspective, some requirements have been raised for the application prototype, and the first issue has been to develop a representation that could fulfil the requirements of ALIS, a system that once realized will cover several areas of legal activities, some of which require a high degree of granularity in the representation of source norms.

Therefore, the knowledge representation should have been isomorphic to the maximum possible extent, as described by Bench-Capon[1] and Karpf [2], that is a representation where each legal source is represented separately, preserving its structure and the traditional mutual relations, references and connections with other legal sources, thanks to the fact that structural elements in the source texts correspond to specific elements in the representation; the representation of the legal sources and their mutual relations should also be separated from all other parts of the model, in particular the representation of queries and facts management.

Besides, considering the various typologies of potential users of the future ALIS system, some of which are not supposed to necessarily have any specific legal background (e.g. average citizens), the representation should have been easy to understand, develop, extend and refine, even for a legal domain expert with basic knowledge engineering competencies.

Finally, a prototype application based on the representation should also have been developed and tested with respect of its ability to interact with users through an easy interface, and to guide them through the integration of commentaries and step-by-step explanations. The application's user interface was also required to justify its conclusions in a human readable form.

3 Knowledge Representation

RuleBurst is a technology specifically devoted to develop legal rulebase systems, consisting in a suite of tools and a set of methodologies which support the creation and deployment of rule-based knowledge models[2], helping the rapid writing of rules through the use of the integrated rule editor and validation/mass testing tools, and an easy development and customization of user interfaces through other tools included in the suite. Moreover, the system allows a high degree of interoperability: rulebases are exportable in XML format, and the system can interact, through web services, with a wide range of other applications.

RuleBurst has not only been already extensively tested in "real-life" commercial applications worldwide, but has also raised a certain scientific interest in the past (e.g. the Hare application in Italy, developed with a previous version of the system [3] and several applications in UK and Australia [4]) and, in particular, is currently in use in the concurrent European funded project Estrella[3], which has shown several scientific points of contact with ALIS. For these reasons RuleBurst appeared suitable to address the requirements listed in the previous section.

The development of representation started with an analysis of the Italian Copyright Act aimed at identifying the core part of law to be represented: that was identified with the entire Title I (artt. 1 – 71-decies) named "Provisions on Copyright", covering protected works, holder of the right, content and duration of copyright, etc., and Title III, named "Common Provision" in particular artt. 107 - 114 (general provisions on transfer of exploitation rights) and artt. 156 – 174-quinquies (legal remedies and penalties).

Then, each norm from the core part of legal source text was represented in a correspondent rule (or set of rules). No deviations were made, as long it was possible, from the original structure of the text, even when it was redundant or confusing.

---

[2] The system is composed by the following components: RuleBurst Studio, an Integrated Development Environment (IDE) for managing multiple rule documents and developing rule-based applications, including support for screen development; RuleBurst Engine, RuleBurst's inference engine, coupled with its Application Programming Interface (API), for use with a number of existing RuleBurst tools and products, as well as providing an interface for applications developed on both Java and Microsoft .Net platforms; RuleBurst Interactive, providing the user interface for deploying rule sets using an interview-style approach.

[3] EU Project ESTRELLA(Contract no.: 027655) funded under the FP6 IST Programme, aimed to develop and validate an open, standards-based platform allowing public administrations to develop and deploy comprehensive legal knowledge management solutions, without becoming dependent on proprietary products of particular vendors.

In RuleBurst the rules in are maintained primarily in a customized Microsoft Word environment (even if the newest version of the system has been extended to include Microsoft Excel and Microsoft Visio as environments for managing rules), using the proprietary RuleBurst format: rules are written rules in (quasi) natural language using indentation and provided styles to format and enable them to be compiled.

A linguistic component (parser) takes into account the syntactic structure of phrases in order to identify their logical components and, lately, to automatically prepare questions and explanations for the user interface. Then, RuleBurst Studio compiles rule documents into an internal XML-based format, which is then used to build rulebase files for use with the RuleBurst Inference Engine.

Fig. 1. Ruleburst rule editing environment

The rules are the usual *if...then* production rules, where a conclusion is provided by a disjunction of alternative conditions. The conclusion of a rule is one literal. The body of a rule is composed of zero or more conditions, connected by 'AND' and 'OR' connectives. Negating all the literals, including those in the head, is supported. Well-formed sets of rules do not contain any cycles.

The formalism permits also the use of variables and entities. The latter data type allows for quantification, denoting a set of individuals which are presumed to share something in common. As a set of individuals, we may ascribe particular properties to elements of the set, or ascribe a property to the set as a whole, or indicate that individuals in the set may stand in specified relationships to other entities. Finally, as a set of individuals, we may quantify over the individuals as the set can be taken to restrict the domain of quantification.

A serious issue faced during the development of representation was the complexity and low-quality legislative drafting of the rules we set out to model: the Italian Copyright Act (Law 22 April 1941 n. 633) has been amended many times to include new provisions resulting not only from national sources, but also from the implementation of international treaties and EU directives[4]. This caused a high degree of complexity and ambiguity in the norms, partly worsened also by disputable choices in legal drafting by the Italian legislator.

This problem recurs several times in the Copyright Act; a typical example is provided by matching provisions set by articles 171, 171-bis, 172 and 174:

Art. 171. Without prejudice to the provisions of articles 171bis and 171ter, any person who, without having the right thereto, and for any purpose and in any form:

(a) reproduces, transcribes, recites in public [...] shall be liable to a fine of between 100,000 lire and 4,000,000 lire. The penalty shall be imprisonment of up to one year or a fine of not less than 1,000,000 lire if the acts referred to above are committed in relation to a work of another person [...]

Art. 171bis.-(1) Any person who unlawfully duplicates computer programs with remunerative intent, [...] shall be liable to imprisonment of between six months and three years and to a fine of between 5,000,000 and 30,000,000 lire. [...] The penalty shall be imprisonment of not less than two years and a fine of 30,000,000 lire if the offense is serious [...]

Art. 172. If the acts referred to in Article 171 are committed by negligence, the penalty shall be a fine of up to 1.032,00 euros.

Any person who:

(a) acts as an intermediary in violation of Articles 180 and 183,

(b) fails to carry out the obligations set out in Articles 153 and 154,

(c) violates the provisions of Articles 175 and 176,

shall be liable to the same penalty.

Art. 173. The penalties set out in the preceding Articles shall apply in all instances where the acts in question do not constitute a more serious offense under the Penal Code or other laws.

As you can note from the above example, the legislative provision in Art. 172 of Italian Copyright Law mixes some cases in which the penalty of a fine of up to 1.032,00 euros is a default penalty, with a case in which this penalty is to be considered an exception to the one provided in Art. 171, which at the same time statues its default penalty but provides another exception, but both applicable "without prejudice" of Art. 171-bis, which statues another standard penalty and another exceptional penalty, that can consequently both work as exceptions of the penalties described in

---

[4] e.g. the provisions of the Berne Convention for the Protection of Literary and Artistic Works, ratified and enforceable by Law No. 399, of June 20, 1978; WIPO Copyright Treaty (WCT). (adopted in Geneva on December 20, 1996); Directive 2001/29/EC of the European Parliament and of the Council of 22 May 2001 on the harmonisation of certain aspects of copyright and related rights in the information society; Directive 96/9/EC of the European Parliament and of the Council of 11 March 1996 on the legal protection of databases; etc.

art. 171. All this fines could, following Art. 173, be considered in some cases excluded by Art. 173[5].

Even in cases like these, we were quite successful in keeping a strict connection between the law text and the representation, and thus in maintaining a good level of isomorphism in the representation, thanks to the ability of RuleBurst to automatically generates structural attributes from source legislation (sections, paragraphs, articles, etc.). The following figure shows an excerpt from the knowledge base, consisting in the representation of one of previously cited articles, the Art. 172 of Italian Copyright Law.

172	**the penalty is a fine of up to 1.032,00 euros if**			
	either			
	(1)	both		
		the act is an act referred in article 171		
			the person is liable to a fine of between 100,000 lire and 4,000,000 lire	
			or	
			the penalty is imprisonment of up to one year	
			or	
			the penalty is a fine of not less than 1,000,000 lire	
		and		
		the act is committed by negligence		
	or			
	(2)	any		
		(a)	the person acts as an intermediary in violation of Articles 180 and 183	
		or		
		(b)	the person fails to carry out the obligations set out in Articles 153 and 154	
		or		
		(c)	the person violates the provisions of Articles 175 and 176	

Fig. 2. Representation of article 172 of Law 633/1941

In the rules showed in Fig. 2 the representation of structural elements from the legal text source (sections, paragraphs, articles, etc.) is obtained with the insertion of several attributes ("172", "(1)", "(2)", "(a)", etc. in the above example) next to the corresponding statement; compiling the rule will result in several structural attributes being automatically generated, of the type:

```
section 172 is satisfied;
section 172(2) is satisfied;
section 172(2)(c) is satisfied;
```

Structural elements are automatically added to the knowledge base as additional conclusions to the corresponding rules. This helps with construction, verification and maintenance of rules, making also possible to cross-reference them against the original material.

Another interesting feature in RuleBurst are the shortcut rules: they are a special kind of rules only used for forward chaining inference, not backwards chaining inference.

---

[5] The last consolidated text of the Copyright Act still mixes some fine expressed in liras, introduced with first version of the law, with some other fines expressed in euros, added by modification subsequent to the adoption of euro currency.

Furthermore, they do not participate in the question search[6]. Therefore, the user is never asked about their antecedent conditions during the backward chaining question search, but their conclusion is derived whenever these conditions are independently met. This, among other functions, is useful again to preserve isomorphism, not departing from the structure of the original text in those cases where in the law text different terms are used to express the same notion: in our representation this is the case of terms like "creator of work" and "author or work" or "creation" and "production".

Besides, as a general procedure also allowed also by RuleBurst suite, a separate maintenance of different types of rules was carried out:

- core legislative rules, which model core source copyright norms, being the closest in structure and wording to the underlying rules which are being represented;
- policy and interpretative rules, which, in the legislative context, are used to model policy implementations of the underlying legislation: this is for example the case of some concepts like "original work" which are recalled but not defined by the act itself, and therefore required the meaning to be elicitated from external jurisdictional sources;
- application rules, which provide an integration function. These rules were needed to support prototype's user interface development: this is the case of rules controlling the order of question to be presented to the user.

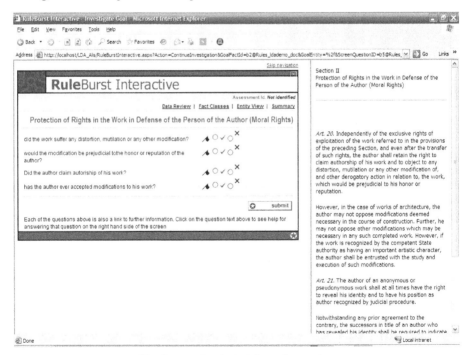

Fig. 3. Interview in RuleBurst Interactive

---

[6] The interview cycle is the process by which RuleBurst Engine determines the questions that require answers in order to reach a conclusion for the given goal attribute.

Finally, a prototype expert system was developed on the basis of the representation. Furthermore, the RuleBurst Interactive web-based tool has been customized to provide the user interface.

The user can investigate some of the main goals related to copyright legal issues (content and duration of copyright, compliance with special provisions in certain categories of works, applicability of civil remedies and penalties, etc.). Information is entered through a series of question screens: each screen (fig. 3) consists of one or more data entry fields and an additional pane of commentary, containing excerpts from the Italian Copyright Act or other helping material.

At the end of the interview, a report is presented to the user, where the reasoning applied in reaching the conclusion is justified also by including references to relevant source norms of the Italian Copyright Act.

The prototype has been validated using internal RuleBurst tools, and tested in a set of cases provided by domain experts of Italian IP legislation.

4 Conclusions

Even if the complexity and low-quality legislative drafting of source norms limited sometimes the isomorphism of representation, in general the result of the work was satisfactory in respect to the requirements of the prototype.

This research has been very useful, on the one hand, to help the developers further improve the tools used and, on the other hand, to rise interesting consideration from the legal drafting perspective.

A number of interesting issues should be nevertheless still investigated: the need to extend the dimension of the domain to be represented over the limited domain chosen for the prototype (to include IPR, contract law, procedural rules, general principles of law, etc.); some legal and logical issues relating to representation of time (both as time in the norm as described by Guastini [5] and time of existence/modification/repeal of the provision in the source text); the connection with a general ontology to represent concepts, action, relationships (this for the connection with the game theory strategy analyzer component) and structure of the law text (for the connection with the text retrieval component); the extension of the system to cover argument-based extended logic with defeasible priorities (see Prakken and Sartor [6] and Gordon [7][8] works).

In order to address most of these issues it would be interesting to study the interconnection between the studies carried out in the ALIS project and the results coming from Estrella project: the main technical objectives of the Estrella project are to develop a Legal Knowledge Interchange Format (LKIF) (see [9]and [10]), building upon XML-based standards of the Semantic Web, including RDF and OWL and translators between the LKIF format and several existing proprietary formats of LKBS vendors, including RuleBurst among them.

References

1. Bench-Capon, T.J.M., Coenen, F.P.: Isomorphism and legal knowledge based systems. Artificial Intelligence and Law 1(1), 65–86 (1992)
2. Karpf, J.: Quality assurance of Legal Expert Systems, Jurimatics No 8, Copenhagen Business School (1989)
3. Borsari, G., Cevenini, C., Contissa, G.S., Still, P.: Hare: An Italian Application of Soft-Law's STATUTE Expert Technology. In: Proceedings of the Tenth International Conference on Artificial Intelligence and Law, pp. 225–229. ACM Press, New York (2005)
4. Dayal, S., Harmer, M., Johnson, P., Mead, D.: Beyond Knowledge Representation: Commercial Uses For Legal Knowledge Bases. In: Proceedings of the Fourth International Conference on Artificial Intelligence and Law, ACM Press, New York (1993)
5. Guastini, R.: Teoria e dogmatica delle fonti. Giuffrè, Milan, Italy (1998)
6. Prakken, H., Sartor, G.: Argument-based extended logic programming with defeasible priorities. Journal of Applied Non-classical Logics 7, 25–75 (1997)
7. Gordon, T.f.: The Pleadings Game – an exercise in computational dialetics. Artificial Intelligence and Law 2(4), 239–292 (1994)
8. Gordon, T.f.: Constructing Arguments with a Computational Model of an Argumentation Scheme for Legal Rules. In: Proceedings of the Eleventh International Conference on Artificial Intelligence and Law, pp. 225–229. ACM Press, New York (2007)
9. Boer, A., Winkels, R., Vitali, F.: Proposed XML standards for law: Metalex and LKIF. In: Lodder, A.R., Mommers, L. (eds.) Legal Knowledge and Information Systems. Jurix 2007: The Twentieth Annual Conference, Frontiers in Artificial Intelligence and Applications, vol. 165, pp. 19–28. IOS Press, Amsterdam (2007)
10. Hoekstra, R., Breuker, J., Di Bello, M., Boer, A.: The LKIF Core Ontology of Basic Legal Concepts. In: Casanovas, P., Biasiotti, M.A., Francesconi, E., Sagri, M.T. (eds.) Proceedings of the Workshop on Legal Ontologies and Artificial Intelligence Techniques (2007), http://www.ittig.cnr.it/loait/LOAIT07-Proceedings.pdf

Source Norms and Self-regulated Institutions

Rossella Rubino[1] and Giovanni Sartor[1,2]

[1] CIRSFID Via Galliera, 3, 40121 Bologna - Italy
`rossella.rubino@unibo.it`
[2] European University Institute, Badia Fiesolana, Via dei Roccettini 9,
50016 San Domenico di Fiesole, Florence, Italy
`giovanni.sartor@eui.eu`

Abstract. In this paper we shall focus on an important class of constitutive norms, which we shall call *source-norms*, namely those norms establishing what norms, on basis of what properties, validly belong to a normative system. Institutions including their own source-norms – here called *Self-Regulated Institutions* – are able to incorporate dynamically and autonomously new norms in their normative systems. After describing these concepts, we shall present a formal model of source-norms built by exploiting the PRATOR system for defeasible argumentation and we shall try to apply it to electronic institutions.

Keywords: self-regulated institutions, source-norms, recognition rules, normative production.

1   Source-Norms

Source-norms establish what other norms, on basis of what properties, validly belong (or do not belong) to a normative system. This category of norms has a large scope, as we shall see, including norms empowering legislators to issue new statutes, norms enabling parties to regulate their relations through contracts, norms authorising judges to decide cases, and also norms determining the legal validity of customs, soft laws, doctrinal options, and so on. The general idea of a source-norm is related to, but not identical with, other concepts used in jurisprudence.

For instance, Hans Kelsen ([1]) introduces two ideas, namely, the idea of a *fundamental norm* (*Grundnorm*) and the idea of an *authorising* or empowering (*ermächtigend*, in German) norm. By a fundamental norm he means a single norm which is sufficient, together with all relevant facts (consisting in acts of enactment or customary practice), for identifying all legally valid norms. By an authorising norm, he means a norm providing for the validity of further norms, on the basis of the fact that such norms are produced by a certain (individual or collective) agent according to a certain procedure (in other words, the authorising norm gives that agent the power to create further norms). As he sees it, the fundamental norm is usually limited to provide for the validity of the constitution of a legal system, where a constitution, intended in a material sense, comprises the authorising norms enabling the production of valid general norms

P. Casanovas et al. (Eds.): Computable Models of the Law, LNAI 4884, pp. 263–274, 2008.
© Springer-Verlag Berlin Heidelberg 2008

(or rather the topmost of such authorising norms). By directly validating the constitution, however, the fundamental norms indirectly provides for the validity of all norms in the system: from the validity of a constitution (recognizing, for instance, legislation and customary practice as valid legal sources) we can infer the validity of rules issued by the legislator or customarily practised; from the validity of legislative rules (authorising administrative, judicial authorities, and contractual parties to issue further valid norms) we can infer the validity of administrative regulations, judicial decisions and private contracts.

Our notion of a source-norm is broader than Kelsen's idea of a fundamental norm, since it includes not only the (fundamental) norm which confers legal validity on a constitution, but also all authorising norms: constitutional norms conferring legal validity upon legislative norms, legislative norms conferring legal validity upon administrative regulations or upon contractual clauses, administrative regulations or contractual clauses conferring validity upon rules stated by other authorities or private organs, and so on. Our concept of a source-norm covers indeed all norms that enable the production of further norms, by different actors performing different kinds of acts: legislative bodies approving statutes, administrative authorities adopting general regulations or individual measures, judges issuing decisions, private parties making contracts, or citizens practising a shared custom. Note that source norms cover different kinds of norm-producing events: not only cases when certain agents intentionally state normative propositions in order to make such propositions binding (as for legislators and judges) but also cases when a social behaviour generates binding norms though the concerned individuals did not participate in the social behaviour in order to generate the corresponding norms (as in customs), or at least did not participate in the social behaviour with the intention of generating the norms as a result of that behaviour (as in legal doctrine).

Similarly, our concept of a source-norm can be related to Hart's ([2]) concept of *secondary rule*, by which he means a norm regulating the creation, modification or application of other norms, as distinct from the "primary" rule establishing what actions that individuals should or should not do. In fact, our source-norms include rules belonging to all the three categories in which Hart classifies his secondary rules: the *rules of recognition* specifying what features a rule should exhibit in order to be considered a source of the law; the *rules of change* empowering legislator to produce and change existing norms according to certain procedures, the *rules of adjudication* empowering certain individuals or bodies to settle disagreements about the primary rules or to punish the violation of primary norms. Our concept of source-norms, however, does not cover those rules addressed to legislator or judges, but which concern duties whose violation does not entail the invalidity of the concerned norms (for instance a legislator's duty to attend a certain number of session, or the judge's duty not to receive gifts from the parties). Thus, not all norms regulating a procedure for the production of legal norms qualify as source-norms, but only those which indicate (sufficient or necessary) conditions for such a procedure to be able to deliver legally valid norms.

2   A Taxonomy of Source-Norms

We can classify source-norms into different classes. With regard to the kind of event which is enabled to produce new norms, we can distinguish the following classes:

- *enactment-recognizing source-norms*, which empower the enactment of new norms through declarations aimed at creating such norms (as it happens in legislation, administrative regulations, contracts, and judicial decisions with regard to the parties of the case), and
- *practice-recognizing norms*, which give legal validity to the rules governing certain practices (on the basis of this very fact, as it is the case for customs and precedents).

Enactment-recognizing source-norms can be further distinguished into the following classes:

- *authority-based source-norms*, which enable the creation of norms unilaterally imposed on their addressees (as for statutes and regulations), and
- *agreement-based source-norms*, which provide for the creation of norms through the agreement of their addresses (as for contracts).

Note that, following Hans Kelsen ([3]), we assume that also statements regulating the behaviour of specific individuals (such as contracts or judicial decisions) produce legal norms, so that also the norms providing for the effects of such statements qualify as source-norms.

Practice-recognizing source-norms can be further distinguished into the following classes:

- *precedent-based source-norms*, which enable the creation of legal norms through precedents and
- *customary-based source-norms*, which provide for the creation of legal norms through customs.

With regard to the origin of the source-norm, we can distinguish the following classes:

- *fundamental* or *recognition source-norms* (the top level source-norms of an institution), whose validity does not depend on other source-norms of the institution, and
- *dependent source-norms*, which are qualified as valid by other source-norms of the institution (and in particular, by specifically enacted source-norms).

For instance, a law authorising a public agency to issue certain norms (for instance, the norm authorising the privacy authority to issue regulations concerning the security of personal data in the public administration) would be an enactment-recognizing source-norm and a dependent one (being valid on the basis of the higher level source-norm giving legislative power to the legislator). Note that the two qualifications (fundamental and dependent) are not really

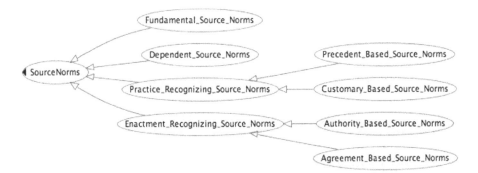

Fig. 1. A taxonomy of source norms

exclusive, since a source-norm having an independent validity can be reiterated as a rule which is valid on the basis of another source-norm: for instance the source-norm according to which private parties can make legally binding contracts can be stated by a legislator (and thus be valid on the basis of the fact that legislative statements produce valid norms), but such a source-norm would certainly be recognised as valid also in the absence of a legislative statement to this effect. In Fig. 1 you can see a graphical representation of the distinctions just introduced.

By taking into account the fact that a norm's validity depends on the validity of the norm that has enabled its creation (by conferring this effect upon a certain kind of event), we can establish a "genetic" hierarchy of norms. Such a hierarchy can be useful in case of conflict between norms, though such genetic relation of superiority (where the superior kind of norm is the one providing for the validity of the lower one, like when a legislative rule confers validity to administrative regulations) should not be immediately assimilated to the relation of superiority which is used to decide conflicts of laws issued by different sources (where the superior kind of norm is that which cannot be derogated by the lower one, while being able to derogate it).

2.1   A Formal Model of Source-Norms

To formally express source-norms (in a computable language) we refer to the PRATOR system for defeasible argumentation proposed in [4]. In such a system, the rules are expressions of the form

$$r : L_0 \wedge \cdots \wedge L_j \wedge \sim L_k \wedge \cdots \wedge \sim L_m \Rightarrow L_n$$

where r, a first-order term, is the name of the rule and each L_i $(0 \leqslant i \leqslant n)$ is a strong literal. The conjunction at the left of the arrow is the antecedent and the literal at the right of the arrow is the consequent of the rule.

Rules can be divided into two categories:

- defeasible rules, which express information that is intended by the user of the system to be subject to debate;

– strict rules, which represent the information that is intended to be beyond debate.

For example the norm "If the owner X sells object O to Y then Y becomes the owner of O" can be expressed as the defeasible rule:

$$r1 : owns(X, O) \wedge sells(X, O, Y) \rightarrow becomesOwner(Y, O)$$

This norm is correctly represented through a defeasible rule since only normally the purchase of an object determines the transferal of its property (there are cases where such transfer can be delayed or conditioned to future event).

An example of strict rule is "if X is a sale then X is a contract":

$$s1 : sale(X) \Rightarrow contract(X)$$

By using PRATOR's rule language, each source-norm can be modelled as a defeasible rule, having as a consequent $legal(X)$, namely, the legality (the legal validity) of the norms produced by the kind of source-event described in the antecedent of the source-norm. This source-event embeds the norm it creates (whose legality it produces). The kind of embedment depends upon the kind of source at issue. For instance, a legislative act (the act of approving a legislative text) embeds the norms it states, a judicial decision embeds its *rationes deci-dendi*, a custom embeds the norms of whose practice it consists, etc. In general a source-norm will have the form:

$$label : happens(X) \wedge \mathbf{sourceKind}(X) \wedge \mathbf{embeds}(X, Z) \rightarrow legal(Z)$$

where **sourceKind** is a kind of legal source, and **embeds** is a particular kind of embedment, depending on the kind of source at issue. For instance a source norm conferring legal validity to stated legislation can be expressed in a simplified way (omitting in particular temporal references) as

$$label : happens(X) \wedge legislativeAct(X) \wedge states(X, Z) \rightarrow legal(Z)$$

to be read as: if an event X happens, and X is a legislative Act (the approval of a law), and X states norm Z, then Z is a legal norm.

Correspondingly, a source norm conferring legal validity to *rationes decidendi* of precedents will be represented as:

$$label : happens(X) \wedge judicialDecision(X) \wedge basedOnRatio(X, Z) \rightarrow legal(Z)$$

to be read as: if an event X happens, and X is a judicial decision, and X is based upon ratio Z, then Z is a legal norm.

Only legal (legally valid) norms can be used in legal inferences. Legal norms include the fundamental norms whose legality (legal validity) is assumed, as well as the dependent norms whose legality depends on legally valid higher-level source-norms. Thus we are led to a chain of legality inferences where the legality of primary (in Hart's sense) norm n_1 depends on the legality of source-norm

n_2 (conferring legal validity to n_1, on the basis of its generating event), whose legality depends on the validity source-norm n_3 (conferring validity to n_2, on the basis of its generating event), and so on.

We can devise different ways to implement this procedure in an automatic reasoner. Consider the following knowledge base:[1]

$f1 : inPublicSpace(john)$
$f2a : happens(smokeActEnactment)$
$f2b : legislativeAct(smokeActEnactment)$
$f2c : states(smokeActEnactment, n1b)$
$f3a : happens(costitutionEnactment)$
$f3b : constitutionalAct(costitutionEnactment)$
$f3c : states(costitutionEnactment, n2)$
$n1a : inPublicSpace(X) \rightarrow forbidden(X, speakAgainstGovernment)$
$n1b : inPublicSpace(X) \rightarrow forbidden(X, smoke)$
$n2 : happens(X) \land legislativeAct(X) \land states(X, Y) \rightarrow legal(Y)$
$n3 : happens(X) \land constitutionalAct(X) \land states(X, Y) \rightarrow legal(Y)$
$a1 : fundamental(n3)$
$a2 : fundamental(X) \Rightarrow legal(X)$
$q1a : norm(n1a)$
$q1b : norm(n1b)$
$q2 : norm(n2)$
$q3 : norm(n3)$

A legally correct reasoning method should enable us to derive the substantive conclusion that John is forbidden to smoke (according to the law, namely, according to legally valid norms). He is forbidden to smoke on the basis of fact $f1$ combined with norm $n1b$, which is a legal norm. The legality of $n1b$ follows from norm $n2$ combined with facts $f2a$, $f2b$ and $f2c$, $n2$ being legal according to norm $n3$ combined with facts $f3a$, $f3b$ and $f3c$, $n3$ (the fundamental norm) being legal according to assumptions $a1$ and $a2$. On the other hand, a legally correct reasoning method should not deliver the conclusion that John is forbidden to speak against the government, since norm $n1a$, providing this conclusion, is not qualified as a legal one (this information is not entailed by the above knowledge base).

The most direct way to achieve this results consists in assuming that every legally valid inference embeds all steps required for establishing the legal validity

---

[1] This knowledge base corresponds to Kelsen's construction, where the top norm ($n3$) is a fundamental norm assumed to be legal (according to $a1$ and $a2$). We can however easily transform the Kelsenian knowledge base into a Hartian one: just substitute axioms $a1$ and $a2$ with the following:

$a1 : ruleOfRecognition(n3)$
$a2 : ruleOfRecognition(X) \land sociallyAccepted(X) \Rightarrow legal(X)$

where axiom $a2$ says that if a norm is a $ruleOfRecognition$ (is meant to identify the ultimate sources of a legal system), and it is socially accepted, then it is legal.

of the norms used, from the lowest norm at issue up to the fundamental rule at the top of its validation chain. However, this method overburdens substantive legal arguments with all inferences needed to establish the legal validity of the norms they use. This is not normally done in common legal reasoning, where inferences about substantive legal conclusions are usually distinguished from inferences about legal validity, though the failure to establish the legal validity of a norm undermines the legal acceptability of inferences using such norms. Moreover this approach would require us to modify the inference model of PRATOR, adding to it a check of legal validity (as in [5] and [6]).

We propose here a representation technique which avoids the just mentioned drawbacks: it separates substantive inference from legal-validity inferences, it make substantive inferences dependent on the legal validity of their normative premises, and it leaves PRATOR's inference model unchanged. We just add to the knowledge base a meta-norm stating that a norm is inapplicable if it is not provable that it is legal, that is

$$nl : norm(N) \land \sim legal(N) \Rightarrow \neg applicable(N)$$

Consequently, failure to establish the legal validity of a norm providing a legal conclusion will determine the inapplicability of that norm, and thus will strictly defeat the argument including such norm (since such an argument uses an inapplicable rule). Assume that we develop the following argument A_1 for the conclusion $forbidden(john, speakAgainstGovernment)$ (for the reader's ease, we include general rules in the arguments rather then the corresponding ground instance, when this helps readability):

$$A_1 = f1 : inPublicSpace(john)$$
$$n1a : inPublicSpace(X) \rightarrow forbidden(X, speakAgainstGovernment)$$

The argument supports the conclusion $forbidden(john, speakAgainstGovern-ment)$, which follows indeed from the premises in the argument: John is in a public space, and if one is in a public space one is forbidden to speak against the Government. However, this conclusion cannot be justified on the basis of the above knowledge base, integrated with rule nl: A_1 is strictly defeated (and overruled) by counterargument A_2, saying that rule $n1a$ is inapplicable, being a norm which is not legal

$$A_2 = q1a : norm(n1a)$$
$$nl : norm(n1a) \land \sim legal(n1a) \Rightarrow \neg applicable(n1a)$$

A_2 holds undefeated with regard to the knowledge base above, having no counterargument in it. Consider, on the other hand, the following argument for $forbidden(john, smoke)$:

$$A_3 = f1 : inPublicSpace(john)$$
$$n1b : inPublicSpace(X) \rightarrow forbidden(X, smoke)$$

Argument A_3 cannot be validly attacked by argument A_4 for $\neg applicable(n1b)$

$$A_4 = q1b : norm(n1b)$$
$$nl : norm(n1b) \wedge \sim legal(n1b) \Rightarrow \neg applicable(n1b)$$

since A_4 is strictly defeated by the argument A_5 for $legal(n1b)$, which concludes that $n1b$ is legal (contrary to what was assumed by nl in A_4):

$$A_5 = f2a : happens(smokeActEnactment)$$
$$f2b : legislativeAct(smokeActEnactment)$$
$$f2c : states(smokeActEnactment, n1b)$$
$$n2 : happens(X) \wedge legislativeAct(X) \wedge states(X, Y) \rightarrow legal(Y)$$

In its turn, A_5 cannot be successfully attacked by A_6 for $\neg applicable(n2)$

$$A_6 = q2 : norm(n2)$$
$$nl : norm(n2) \wedge \sim legal(n2) \Rightarrow \neg applicable(n2)$$

since A_6 is strictly defeated by A_7 for $legal(n2)$:

$$A_7 = f3a : happens(costitutionEnactment)$$
$$f3b : constitutionalAct(costitutionEnactment)$$
$$f3c : states(saleActEnactment, n2)$$
$$n3 : happens(X) \wedge constitutionalAct(X) \wedge states(X, Y) \rightarrow legal(Y)$$

Finally, argument A_7 cannot be successfully attacked by A_8

$$A_8 = q3 : norm(n3)$$
$$nl : norm(n3) \wedge \sim legal(n3) \Rightarrow \neg applicable(n3)$$

since A_8 is strictly defeated by A_9, which uses postulate $a2$, namely, the postulate specifying that fundamental norms are legally valid, combined with the assumption $a1$ that $n3$ is a fundamental norm (with regard to the considered legal system):[2]

$$A_9 = a2 : fundamental(X) \Rightarrow legal(Y)$$
$$a1 : fundamental(n3)$$

In conclusion, no attack can be successfully brought against A_3 by attacking the legal validity of its norm $n1b$, either directly or indirectly (namely, by attacking the legal validity of a norm whose legal validity is a precondition of $n1b$'s validity). Thus A_3's conclusion (John's obligation not to smoke) appears to be legally justified.

---

[2] We shall not consider here how the postulate that fundamental rules are valid and the assumption that a rule is indeed fundamental are to be viewed for a legal theoretical perspective. These assumption can indeed be viewed as neutral with regard to different legal theories (which according to different evaluative or theoretical assumptions may consider different norms to be fundamental). For a discussion of the concept of legal validity, see [7].

3   Norms in Agent Societies

Source-norms provide a mechanism for enabling autonomous agents participating in an institution to include autonomously new norms in their institution (without an external agent to intervene). Thus they provide an interesting device for enhancing the autonomy of electronic agents and of their institutions.

Many scholars are currently investigating how to make electronic institutions more autonomous, and such an autonomy is viewed as consisting in a chance being given to software agents, namely, the chance:

- to select the coordination mechanism during the run-time;
- to choose whether to be compliant with norms or not;
- to choose how to achieve individual and social goals within normative constraints.

For instance, in [8] a decision-making framework has been defined that enables agents to dynamically select the coordination mechanism in order to fit their prevailing circumstances and their current coordination needs, while in [9] the authors propose a set of strategies to be used by agents and analyse the effects of autonomous norm-compliance through simulation experiments.

In particular, the autonomy in the selection of coordination mechanism has been analysed from two points of view. The former concerns the organisational structure and the latter concerns the emergence of laws. In [10] a general diagnosis engine is defined to drive the adaptation of organisational structures, while in [11,12] two new reorganisation primitives have been introduced, composition and decomposition, to extend the possible architectures for Organization Self-Design (OSD). In [13] the MOISE+ organisation model is proposed as the cooperative framework of MAS reorganisation. On the other hand, the emergence of laws has been studied by many scholars among which [14,15] to name a few, who have focused on how norms emerge from behaviour. In [16] agents can select among alternative social laws by exploiting the notions of minimal and simple social laws.

We think that our model complements the second area of research, since it provides agents with a flexible model for identifying and creating new norms (in our framework, indeed emerging social law can be considered a special kind of customary law, which can be binding according to, and under the conditions specified by, an appropriate source-norm). Since agents can automatically detect source norms, norms need not constitute built-in constraints for agents: agents themselves can check the existence of norm-generating events, namely, normative sources, and can derive appropriate conclusion, concerning what norms exist and whether they are binding for them. Moreover, agents can produce such events (e.g. enact a regulatory act, make a contract, issue a decision based on a certain *ratio*, etc.) in order to create new norms.

4   Self-regulated Institutions

We can characterise different kinds of (electronic) institutions, having different kinds and degrees of normative autonomy, according to the kinds of source norms they have. First of all we can distinguish institutions according to the foundation of their source norms:

Self-regulated institutions only contain norms which are qualified as valid by source-norms belonging to the institution.

Other-regulated institutions also contains norms that are qualified as valid by source-norms not belonging to the institution.

Self-regulated institution can be further distinguished according to the origin of the source-events they take into consideration:

Non-delegating institutions only contains source-norms referring to source-facts taking place within the institution.

Delegating institutions also contains source-norms referring to source-facts taking place within other institutions.

In order to design the normative infrastructure of a self-regulated institution we need to include not only norms but also source-norms. This will enable agents autonomously and dynamically to know which norms are binding. Moreover they may be able to produce new norms by realising legal source-events, namely, by making so that the facts happen (e.g. a contract) that, according to the source norms, are able to produce further norms.

For agents to be able to appreciate the implications of the source norms, and thus to identify the legally valid norms so far produced, it is necessary to enable agents to reason with a knowledge base including both rules and facts. For experimenting with this feature we have used the ASPIC Argumentation Engine [17] an implementation – in Java – of the algorithms for defining the status of arguments defined in the European Project ASPIC[3]. The ASPIC engine determines the acceptability of arguments and constructs proofs using an argument-game approach. It allows a user to determine the yes (undefeated) or no (defeated) status of an argument, and presents a graph visualisation (an argument network) of the proof associated with such a determination. It also provides a machine readable version of the proof and results via AIFXML[4]. Since the ASPIC engine implements a subset of the logic of PRATOR, it sufficient for our purpose.

In ASPIC predicates are represented in a Prolog-like syntax and can be associated with a real number in the range (0,1] known as "degree of belief". Rules are also associated with a degree of belief, which allows us to separate strict knowledge from defeasible knowledge (strict knowledge has a degree of belief 1.0 while defeasible knowledge a degree of belief less than 1.0). Software agents may use the ASPIC Argumentation Engine in two different ways: if software agents are developed in Java then they can embed the engine in their Java application; otherwise, they can parse and interpret the AIFXML rulebase.

We have also experimented with Carneades [18], the Inference Engine currently being developed within the European Project ESTRELLA. Also Carneades distinguishes strict and defeasible rules but with Carneades it is not necessary to

---

[3] http://www.argumentation.org/
[4] http://aspic.acl.icnet.uk/

specify the degree of belief for each defeasible rule since undercutters are assumed to prevail over the undercut rules (namely, the rules declared to be inapplicable). Also in Carneades as in ASPIC, rules are reified and can be referred to by means of an identifier (more generally, rules have a set of meta-data properties). We cannot here present in detail Carneades's syntax. Let us just show how norm nl norm would be represented in Carneades:

$$(rule\ nl$$
$$(if\ (and\ (norm\ ?n)$$
$$(\sim\ legal\ ?n))$$
$$(not\ applicable\ ?n)))$$

5   Conclusions and Future Work

We have presented the source-norms, namely, norms establishing what other norms, on basis of what properties, validly belong (or do not belong) to a normative system. This idea is partially connected to Kelsen's fundamental and authorising norms and to Hart's recognition and secondary norms.

We have provided a taxonomy of source-norms, by distinguishing between enactment-recognizing source-norms and practice-recognizing norms and between fundamental and dependent source-norms. We have also shown how source norms can support self-regulated institutions, namely institutions composed by agents that not only obey rules, but also determine what rules are part of the institution's normative system and that create new rules. We have represented source-norms by using the logic of the PRATOR system and have tested their application through the ASPIC Argumentation Engine and the ESTRELLA inference engine.

In future work we intend to refine the model here presented and to use it to study the evolution of agent-societies.

Acknowledgements

The work reported in this paper was partially supported by the European Projects ESTRELLA (IST-027655) and ASPIC (IST-002307).

References

1. Kelsen, H.: Reine Rechtslehre. Einleitung in die rechtswissenschaftliche Problematik. Franz Deuticke (1934)
2. Hart, H.L.A.: The Concept of Law. Oxford University Press, Oxford (1997)
3. Kelsen, H.: The Pure Theory of Law. University of California Press, Berkeley (1967)
4. Prakken, H., Sartor, G.: Argument-based extended logic programming with defeasible priorities. Journal of Applied Non-Classical Logics 7, 25–75 (1997)
5. Yoshino, H.: The Systematization of Legal Metainference. In: Proceedings of the Fifth International Conference of Artificial Intelligence and Law (ICAIL). ACM, New York (1995)

6. Hernandez Marín, R., Sartor, G.: Time and norms: A formalisation in the event-calculus. In: Proceedings of the Seventh International Conference on Artificial Intelligence and Law (ICAIL), pp. 90–100. ACM, New York (1999)
7. Sartor, G.: Legal validity: An inferential analysis. Ratio Juris (forthcoming, 2008)
8. Excelente-Toledo, C.B., Jennings, N.R.: The dynamic selection of coordination mechanisms. Autonomous Agents and Multi-Agents Systems 9, 55–85 (2004)
9. Lopez y Lopez, F., Luck, M., d'Inverno, M.: Constraining autonomy through norms. In: Proceedings of the First International Joint Conference on Autonomous Agents and Multi-agent Systems (AAMAS 2002) - Session 6B: social order, pp. 674–681. ACM Press, New York (2002)
10. Horling, B., Benyo, B., Lesser, V.: Using self-diagnosis to adapt organizational structures. In: Proceedings of the Fifth International Conference on Autonomous Agents, pp. 529–536. ACM Press, New York (2001)
11. Gasser, L., Ishida, T.: A dynamic organizational architecture for adaptive problem solving. In: Proceedings of the Ninth National Conference on Artificial Intelligence (AAAI 1991), American Association for Artificial Intelligence, pp. 185–190. AAAI Press/MIT Press (1991)
12. Ishida, T., Gassend, L., Yokoo, M.: Organization self-design of distributed production systems. IEEE Transactions on Knowledge and Data Engineering 4, 123–134 (1992)
13. Hubner, F.J., Simao Sichman, J., Boissier, O.: Using the MOISE+ for a Cooperative Framework of MAS reorganisation. In: Bazzan, A.L.C., Labidi, S. (eds.) SBIA 2004. LNCS (LNAI), vol. 3171, pp. 506–515. Springer, Heidelberg (2004)
14. Conte, R.: Emergent (info)institutions. Journal of Cognitive System Research 2, 97–110 (2001)
15. Walker, A., Wooldridge, M.: Understanding the emergence of conventions in multi-agent systems. In: Lesser, V., Gasser, L. (eds.) Proceedings of the First International Conference on Multi-agent Systems (ICMAS 1995), pp. 384–389. AAAI Press/MIT Press (1995)
16. Fitoussi, D., Tennenholtz, M.: Choosing social laws for multi-agent systems: Minimality and simplicity. Artificial Intelligence 119, 61–101 (2000)
17. Amgoud, L., Bodenstaff, L., Caminada, M., McBurney, P., Parsons, S., Prakken, H., van Veenen, J., Vreeswijk, G.: Deliverable d2.6 - final review and report on formal argumentation system. Technical report, ASPIC - Argumentation Service Platform with Integrated Components (2006)
18. Gordon, T.F.: Constructing arguments with a computational model of an argumentation scheme for legal rules. In: Proceedings of the Eleventh International Conference on Artificial Intelligence and Law (ICAIL-2007), pp. 117–121. ACM, New York (2007)

Distributed Norm Enforcement: Ostracism in Open Multi-Agent Systems

Adrian Perreau de Pinninck, Carles Sierra, and Marco Schorlemmer

IIIA – Artificial Intelligence Research Institute
CSIC – Spanish National Research Council
Bellaterra (Barcelona), Catalonia, Spain
adrianp,sierra,marco@iiia.csic.es

Abstract. Enforcement in normative agent societies is a complex issue, which becomes more problematic as these societies become more decentralized and open. A new distributed mechanism is presented to enforce norms by ostracizing agents that do not abide by them in their interactions with other agents in the society. Simulations are run to check the mechanism's impact in different types of societies. The simulations have shown that complete ostracism is not always possible, but the mechanism substantially reduces the number of norm violations.

1 Introduction

In an open multi-agent system (MAS), there is no easy control over who is allowed to join the system. Open MAS are composed of autonomous agents of all types, without a pre-defined structure. In a normative MAS a set of norms are added to restrict the set of available actions in order to improve the coordination between agents. An autonomous agent has the choice whether or not to support a norm. It is up to the agent to decide if it benefits itself to abide by the norm. A utility maximizer agent will follow a norm if it is profitable for it, it is in the agent's own interest to act as the norm establishes. But some norms make it worthwhile for an agent to not abide by it if all other agents abide by them. For this kind of norms, an agent that does not adhere (*i.e.*, a violator) will profit at the expense of the agents that adhere.

Gnutella[1] is a suitable real life application to show how a multi-agent system may behave when norms are added. The scenario we will use in this paper is based on a simplification of this application. Gnutella is a pure peer-to-peer (P2P) file sharing application without centralized servers. Peers can share files on their hard drives so that others can download them. Each peer knows other peers (*i.e.*, friends or neighbours) with which it can interact. A peer can carry out two actions: search for peers that have a file it wants, and download the file from any of them. Peers are found by asking neighbouring peers if they or any of their neighbours have the file. This process is recursive. Once a list of peers that share the file are returned to the querying peer, it can choose one of them from

---

[1] http://www.gnutella.org

P. Casanovas et al. (Eds.): Computable Models of the Law, LNAI 4884, pp. 275–290, 2008.

which to download the file. Anyone with a Gnutella compliant software can join the society as long as it has a list of other peers with which to interact.

This system works even though users can use the network to download files without sharing any files themselves. Therefore, the burden of sharing is carried only by some of them, while the rest benefit from the service. In order to solve this problem the following norm could be added: "Users must share files in order to be allowed to download files themselves". But since the norm is not built into the Gnutella protocol, Gnutella peers can choose whether or not to adhere to it. A mechanism is needed so that norms can be enforced in any P2P network such as Gnutella.

This paper presents a new distributed mechanism that ostracises norm violating agents in an open MAS, thus attaining norm compliance. The test scenario in this paper allows agents in the MAS to interact with each other. The agents are structured in a network, in which agents can interact with the agents they are linked to directly or indirectly through a path of links (*i.e.*, agents can interact with direct neighbours, with neighbours of neighbours, and with their neighbours and so on...). The interaction initiator will search for a path in the society structure that leads to an interaction partner. All the agents in the path that are not the initiator or the partner agent are called mediator agents (*i.e.*, agents mediating the interaction).

A game-theoretic approach to interactions has been implemented. Interactions are modelled as a two-player game with two possible strategies; abide and violate. The utility function will be that of a prisoner's dilemma (see Figure 1), since the total utility gained by both players is maximized if both players abide by the norm, and the maximum utility to be gained by a single agent is maximized if it violates the norm while the other abides by it.

PD	Abide	Violate
Abide	3,3	0,5
Violate	5,0	1,1

Fig. 1. Prisoner's Dilemma Payoff Matrix

Violators are better off if they interact with norm-abiding agents, since they gain more utility. In order to attain norm enforcement the violators are not allowed to interact. Some agents in the society can enforce the norm through the ability to stop interacting with violators, and to stop them from interacting with the enforcer's own neighbours. When all the neighbours of a violator are enforcers, and they use this ability against it, it is ostracised.

The motivation for this technique comes from the study of enforcement in primitive societies [12]. A member of a community that repeatedly ignores its customs is forced to leave upon general consent. No one in the community will interact with the ostracized member from then on. Therefore, the harsh natural conditions surrounding those communities mean death for the ostracised member. Ostracism is achieved in human societies through force and physical constraint. If the ostracised member tried to return he may be killed. Achieving ostracism of electronic entities is a bit trickier, since they don't suffer pain. Inspiration has been sought from the network security area, where the most commonly used component is a firewall. Firewalls block those communications which appear to be harmful. The problem with firewalls is that they are usually set up by humans through complex rules, which must be updated manually. The enforcer agents in this paper will use gossip as a way to inform each other about the maliciousness of other agents. Thus building a distributed reputation measure.

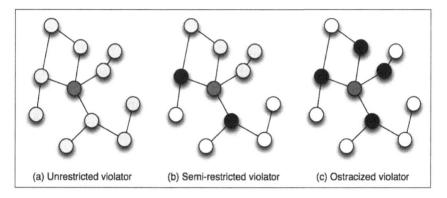

(a) Unrestricted violator   (b) Semi-restricted violator   (c) Ostracized violator

Fig. 2. Ostracizing a violator

The ostracism process is shown in Figure 2. Before a violator interacts, it is undetected (the dark gray node), and can interact with all the other agents (light gray nodes are liable to interact with the violator). When the violator interacts and violates the norm, if its partner is an enforcer, it will start blocking its interactions (black nodes are blocking agents, and white nodes are agents that the violator cannot interact with). When a violator is partially blocked, it is still able to reach part of the network. But when all the violator's neighbours block it, it is ostracised.

In order to find out information about other agents in a distributed environment, gossip between them can be used. The enforcement technique uses gossip as part of the enforcement strategy prior to ostracising agents. Since gossip should not take up too many resources, the outcome of interactions is only gossipped to the interaction mediators. If the violator agent interacts with an enforcer agent, the enforcer agent will spread this information to all mediator agents so they may block the violator in the future.

To study under which conditions the mechanism works, and give measures of its success (such as the violations received or the utility gained), a set of simulations have been run. The data extracted from them is used to support the following hypotheses:

- **H1** - Norm violations are reduced by applying a simple local blocking rule.
- **H2** - Network structure influences the enforcement capabilities.

- **H3** - The enforcement strategy used by enforcer agents can reduce the number of violations received by meek agents (*i.e.*, norm abiding agents which do not enforce the norm).
- **H4** - Enforcement makes abiding by the norm a rational strategy.

This paper is divided into five sections. Section 2 describes related work in the area of norm enforcement. Section 3 presents a detailed description of the scenario employed in the simulations. Section 4 describes the simulations and analyses the resulting data. Finally, Section 5 presents future work that will follow from this research.

2   Related Work

The problem of norm enforcement is not new. It has been dealt with in human society (also an open MAS) through the study of law, philosophy, and the social sciences. Recently studies in computer science deal with it, specially as a coordination mechanism for multi-agent systems. The application of norms from an evolutionary perspective was first studied by Axelrod in [1]. Where enforcement is seen as a meta-norm dictating that agents which do not punish violators should be punished themselves. The norm game is modelled as an N-Player Iterated Prisoner's Dilemma [1,8]. Since the norm is specified to maximise the society's utility, agents are expected to cooperate. Enforcement techniques are sought to ensure that agents prefer cooperation. In [4,7,13,16] norms are seen as a way to avoid aggression or theft. In these cases agents gain utility by collecting items that they find while they move around or by receiving them as gifts. But agents also have the ability to steal items from agents they find through aggression. A norm is added that dictates when a good is possessed by an agent. In which case it cannot be stolen by another. Therefore, a norm-abiding agent will not steal food possessed by another agent.

Two main lines of research in norm enforcement exist: sanctions and rewards to change the utilities of the game [2,3,8,15], and the spread of gossip in order to avoid interaction with violators [4,6,7,13,16]. Both approaches are based on making norm adopters better off than norm violators. But there is a downside to this [4,7], since all agents benefit from the norm while only normative agents bear the cost of enforcing it. Therefore, some agents are tempted to abide by the norm, but not to enforce it. Which makes it a recursive problem.

Norm enforcement models have been suggested in [2,6]. They show that norm-violation becomes an irrational strategy when non-normative behaviour is punished. Nonetheless these models assume the following: (i) agents are able to monitor other agents' activities; and (ii) agents have the ability to influence the resulting utility of interactions. Assumption (i) can be brought about by having a central agent mediate all interactions as done in [2]. Another way in which agents have information of other agents is trough direct interaction or gossip [4]. The first solution does not scale, since the mediator agent is the bottleneck in a large system. The second scales, but it is less efficient since detection of all violations is not always possible, furthermore gossip is an extra cost. Assumption

(ii) can be carried out through third-party enforcement [2], or self-enforcement [6]. Using a third party does not scale because the third party can easily be overwhelmed by enforcement petitions of agents. Also the third party must have access to a resource that all other agents need, and this is not always the case in real systems. Self-enforcement means that each agent takes care of those violators that affect it. Thus, all agents must have the ability to affect the outcome utility of interactions, by applying sanctions or spreading gossip.

Axelrod's mechanism for norm enforcement is based on self-enforcement and sanctions. He terms it the "shadow of the future" [1]. Defection by an agent is unlikely if it will interact often with the other agent. In which case the other agent will retaliate in future interactions. Nonetheless, this mechanism affects the utility of both agents because in the future they will both defect, and the utility will be less than if they had both cooperated. Futhermore, if the norm is to cooperate, then the enforcer is forced to violate the norm in order to retaliate, thereby becoming a violator.

Another mechanism for norm enforcement is the threat of ostracism. By avoiding interaction with violators, an agent can use the time to interact with a normative agent and achieve a higher payoff. Furthermore, violators eventually have no one with which to interact and may starve. Younger has studied [16] the possibility of avoiding interaction with norm-violators, but this is just one part of ostracism. An ostracised agent cannot interact with anyone in the society, which implies preventing it from interacting with anyone else. Human societies have used ostracism as a means to deal with norm violators [12]. In primitive societies the violator was expelled from the village and had to wander in no-man's land, or try to find another village that would take him. In modern societies, all land is owned by some state, thus violators are placed in a special area so that they cannot interact with the rest of society (e.g., prisons), but this measure has the associated cost of maintaining these areas. The electronic network in this article resembles a primitive society, an agent that has been ostracised wanders a sort of virtual no-man's land.

Emergence of norms in a structured multi-agent system has been studied in [9]. The first approach was to study regular graphs, hierarchies, and trees. This work was followed by another [5] that studied emergence in complex graphs with properties such as scale-free and small-world. Furthermore, the relationship between norm emergence and other graph parameters such as clustering factor and diameter are studied [10]. In recent work, the notion of role models has been studied and its effect in norm emergence in networks [11].

The scenario presented in this paper, is used to justify how agents can monitor other agents' activities, and how they can influence future interactions. A mix of techniques have been used to accomplish this; the spread of normative reputation through gossip, and sanctioning norm-violators by blocking their access to the network in order to achieve ostracism. Norm enforcement is studied using these techniques in societies with differing structures.

3   The Scenario

The multi-agent system in this paper is structured as a network. Thus, it is modelled as an undirected, irreflexive graph: $MAS = \langle Ag, Rel \rangle$, where Ag is the set of vertices and Rel the set of edges. Each vertex models an agent and each edge between two vertices denotes that the agents are linked. Agents can communicate through their links. Three kinds of graphs have been chosen for their significance: Tree, Random, and Small-World. A tree is a graph in which each node is linked to one parent and some number of children; only one node, the root node, has no parent, and the leave nodes have no children. A tree has a large average distance between nodes, and no clustering. A random graph, on the other hand, does not have any regular structure. The nodes in this type of graph can be linked to any other one with a given probability. Random graphs have a small average distance between nodes, but the clustering factor is very low. Small-world graphs reside half way between regular, structured graphs, and random ones. The average distance between nodes is as small as in a random graph with the same number of nodes and edges, but its clustering factor is orders of magnitude higher [14]. The small-world graphs in the simulations have been created by starting with a regular graph[2], and rewiring enough random edges to make the average distance between any two vertices significantly smaller. The different graph structures have been generated to have a similar average number of links per node.

A game-theoretic approach is used to model interactions between agents. Interactions are two-player prisoner's dilemma games. Agents ought to choose the abide action given by the norm (*i.e.*, an agent disobeys the norm by choosing the violate action). An agent is capable of interacting with another if there must be a path in the graph between the two. An *initiator agent* searches for a path that leads to a *partner agent* with which to interact. The *mediator agents* are those agents in the path between the initiator and the partner. The partner finding process is explained below, but first some terms need to be formally described.

An agent's a_i neighbours are the agents it is linked to directly in the graph: $N(a_i) = \{a_j \in Ag \mid (a_i, a_j) \in Rel\}$. Each agent maintains a set of agents it blocks (an agent cannot block itself): $B(a_i) \subseteq Ag \backslash \{a_i\}$. An agent a_i can search through the network by querying other agents a_j for a list of their neighbours. Since agents are autonomous, when queried for their neighbours agent a_j can respond with any subset of its real neighbours. $RN(a_i, a_j) \subseteq N(a_j)$ are the reported neighbours a_j will return queh queried by a_i. The set of reported neighbours depends on the blocking strategy of a_j. The strategies used in the simulations are explained below. A path is the route (without cycles) in the graph structure through which interaction messages are delivered. Paths are represented as finite sequences of agents $p = [a_1, a_2, \ldots, a_n]$ such that for all i with $1 \leq i \leq n-1$ and $n \geq 2$ it follows that $a_{i+1} \in N(a_i)$, and for all i, j with $1 \leq i, j \leq n$ and $i \neq j$ it follows that $a_i \neq a_j$. The initiator agent will always be the first element in the path, the partner agent will be the last, while the remaining ones are mediators.

---

[2] $C_{N,r}$ is a regular graph on N vertices such that vertex i is adjacent to vertices $(i + j) \bmod N$ and $(i - j) \bmod N$ for $1 \leq j \leq r$.

The process through which an initiator agent a_i finds a path to a partner agent a_n is as follows. First a_i creates the path $p = [a_i]$. Since an agent cannot interact with itself, the path with one agent is not valid. Then the initiator agent queries the last agent in the path (the first time it will be itself) to give it a list of its neighbours. It will choose one of the reported neighbours[3] (a_j) and add it to the end of the path $p = [a_i, ..., a_j]$. At this point the initiator can choose agent a_j as the partner, if a_j allows it. Otherwise, it can query agent a_j for its neighbours and continue searching for a partner. If the path's last element is an agent a_n that refuses to interact with the initiator agent, and a_n does not report any neighbours when queried, backtracking is applied. Agent a_n is removed and a different agent is chosen from the list of a_{n-1}'s reported neighbours and added to the end of the list.

A prisoner's dilemma game is played between the initiator and the partner, when the first has chosen the latter. Each interacting agent has complete knowledge of the game results and mediating path. Interacting agents may gossip the game results to all the mediators in the path. The information contained in *gossip* is a tuple with the agents' names and their strategy choices for the given game: $Gossip = \langle ag_i, choice_i, ag_j, choice_j \rangle$, where $choice_i$ and $choice_j$ are either *abide* or *violate*.

During the whole process agents can execute any of the following actions:

- Return a list of neighbouring agents when asked for its neighbours.
- Choose one of the agents of a list as a mediator.
- Request an agent to become the interaction partner.
- Accept or reject an invitation to interact.
- Choose a strategy to play in the PD game when interacting.
- Inform mediators of the outcome of the interaction.

The society of agents is composed of three types of agents, each one characterised by a different strategy for the actions it can execute. The *meek agent* will always abide by the norm, it will always report all its neighbours to any agent, and it will always accept an offer to interact from any agent. When searching for an interaction partner, a meek agent will request the last agent in the current path to become its partner with probability p, and with probability $1 - p$ it will ask for its neighbours[4], and it will choose an agent randomly from the list of reported neighbours. Finally, a meek agent will not gossip the results of its interactions. A *violator agent* follows the strategy of a meek agent, except that it never abides by the norm, therefore it is not a norm-abiding agent.

Finally, *enforcer agents* have the ability to block violators, which is essential to achieve their ostracism. Enforcer agents have the same strategy of meek agents with the following exceptions: They will add agents that they know to have

---

[3] To avoid loops, an agent that is already part of the path cannot be chosen again.

[4] The value of p is set to 0.3 in all simulations. Since the path length follows a geometric distribution $Pr(L = n) = (1 - p)^{n-1}p$, the path's length expected value is $E(L) = 1/p = 3.33$ and its variance $var(L) = (1 - p)/p^2 = 7.77$. In future work we plan to relax the constraints on partner searching.

violated the norm to the set of blocked agents, and when they interact with a violator they gossip the interaction results to all mediators. Enforcer agents will never choose an agent in their blocked set as a partner, and will reject requests to interact from agents in their blocked set. Therefore, enforcers never interact with a violator more than once. When an agent a_i queries an enforcer agent a_m for its neighbours, if a_i is in the enforcer's blocked set it will return an empty reply. On the other hand, if a_i is not on the blocked set, two different strategies are possible: The Uni-Directional Blockage (UDB) strategy, where all its neighbours will be returned $(RN(a_i, a_m) = N(a_m))$. And the Bi-Directional Blockage (BDB) strategy, where only those neighbours not in its blocked set are returned $(RN(a_i, a_m) = N(a_m) \setminus B(a_m))$.

Choosing one enforcement strategy over another entails a tradeoff. When the BDB strategy is chosen, violators will be more efficiently ostracized, the tradeoff is that initiator agents may also be blocked from reaching certain parts of the network, the cost is freedom. Intuitively, one can see that enforcer agents are better off with the UDB strategy. An enforcer will never interact with a violator, but it can use it as a mediator to reach other parts of the society. Meek agents, on the other hand, do not hold a memory of violating agents. Therefore, meek agents may choose violators unknowingly as their partner repeatedly. The BDB protects meek agents, by reducing the chances of them choosing violator agents.

In order to focus on the aspects such as network structure and simple blocking strategies, the following assumptions have been made to limit the number of variables:

- Agents cannot change their interaction strategy.
- Agents cannot lie when sending gossip.
- There are no corrupt enforcer agents.
- There is no noise (*i.e.*, an agent knows its opponent's chosen strategy).

These assumptions imply that modelling an agents' reputation is simple, and there is no redemption for violators. Since gossip is always truthful and there is no noise, the validity of information is permanent. Therefore, if there is any evidence that an agent has violated it must be a violator. Furthermore, since a violator will never change its strategy, sanctions must be indefinite. Relaxation of these assumptions will be studied in future work. Thus allowing for sophisticated violators which could trick enforcers into blocking other enforcers by giving them false information through gossip.

4   Simulations

This section shows the results of the simulations that have been run following the scenario specified in Section 3. In order to focus the experiments to see the effect of certain variables, the rest have been set with the same value for all simulations. Each simulation consists of a society of 100 agents, with 1000 rounds per simulation. In each round agents take turns to find a partner with which to interact. If an agent cannot find a partner its turn is skipped. As said before, interactions consist of the prisoner's dilemma game specified in Figure 1.

The parameters that change in each simulation can take up the following values:

- Percentage of Violators (V) - from 10% to 90% in 10% increments.
- Percentage of Enforcers (E) - from 0% to 100% in 10% increments[5].
- Type of Graph (G) - either tree, small world, or random.
- Enforcement Type (ET) - Uni-Directional Blockage (UDB), or Bi-Directional Blockage (BDB).

Exhaustive simulations have been run with all the possible combinations of parameter values. Each simulation is repeated 50 times in order to obtain an accurate average value. The following metrics have been extracted for each simulation: number of games played, violations received, and utility gained by an agent. The metrics have been calculated for the whole society and for each type of agent. Furthermore for each metric, both the mean and the variance have been calculated. The data gathered from the simulations support our hypotheses.

(H1) Norm violations are reduced by applying a simple local blocking rule. The different graphs in Figure 3 contain eight different lines, each one represents a different percentage of violating agents. The x-axis represent the enforcer to meek agent ratio, and the y-axis the average violations received by agents of each type. In all cases, the higher the percentage of violators, the higher the violations received by agents, which is intuitive. On the other hand, a higher ratio of enforcer to meek agents reduces the number of violations received by the society as a whole, and by norm abiding agents particularly. But this is not true for meek agents and violator agents, which receive more violations when the ratio of enforcers increases.

The improvement in the society as a whole is not significant, as seen in Figure 3(a). When just norm-abiding agent are taken into account, the reduction in violations received is much greater (see Figure 3(b)). This happens because when the enforcer ratio is high, most norm-abiding agents are enforcers. Enforcers will only interact with each violator at most one time, therefore violations received by norm-abiders are greatly reduced. Therefore, violators end up interacting with the few agents they have access to. This is the meek agents not being protected by enforcers and other violators. Both of which increase the number of violations received as the ratio of enforcers increases (see Figures 3(c) and 3(d)). Since meek agents are a small portion of the norm supporters, the fact that they receive more violations does not influence the total violations received by norm supporters as a whole. The number of games played by violator agents also supports this hypothesis. In average, violators play less games when the number of enforcer agents is high, because enforcers manage to ostracize some violators.

(H2) Network structure influences the enforcement capabilities. The simulations show that different multi-agent system organisational structures have different effects on norm enforcement. Figures 4(a) and 4(b) show the average norm violations (y-axes) for each of the different structures tested: Random,

---

[5] The percentage of meek agents is computed through the following formula: $M = 100\% - V - E$. Therefore, $V + E$ cannot be more than 100%.

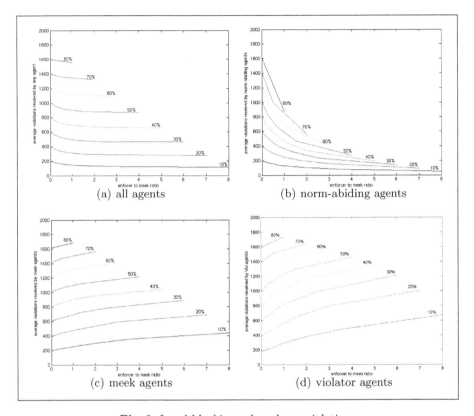

(a) all agents   (b) norm-abiding agents

(c) meek agents   (d) violator agents

Fig. 3. Local blocking rule reduces violations

Small World, and Tree. They represent the simulations where violator agents account for 10% and 20% of the population respectively. Therefore, at most there will be 90% or 80% of enforcers, respectively. The x-axes plots the percentage of enforcer agents. Both random and small world networks have an almost identical graph line. The tree structure has an altogether different graph line which greatly improves the enforcement capabilities. The fact that in a tree there is only one path between any two agents is the determining factor in making the society more secure to violations. In random and small world graphs, many paths can be usually found between any two agents. From the simulations it is deduced that the higher the number of paths that unite agents, the more vulnerable they are to non-normative attacks. On the other hand, the main difference between small world graphs and random graphs is their clustering coefficient. Since the two types of graphs have very similar results, the clustering coefficient can be ruled out from the variables that have an impact in norm-enforcement.

As an interesting side note, the tendency is that the more enforcer agents, the less violations. But in random and small world networks, when the percentage of enforcer agents reaches its maximum the percentage of violations received are increased (see Figure 4(b)). This happens because in both these networks violator agents manage to find paths that link them. Since at this point there are few meek agents for them to prey on, they are forced to interact with each other. Figures 3(c) and 3(d) show that the number of violations received by meek and violator agents increases with higher enforcer ratio. In an interaction between two violator agents, two violations are being accounted for and the average of violations is increased. A sub-society of violating agents is formed. This has been observed in all simulations with a ratio of violator agents of 20% and above, but it is not so acute in the simulations with 10% of violators (see Figure 4(a)). When the ratio of violator agents is low enough, enforcers manage to ostracise more of them, and they cannot interact with each other. When this happens, no sub-society of violating agents exists, they are completely blocked.

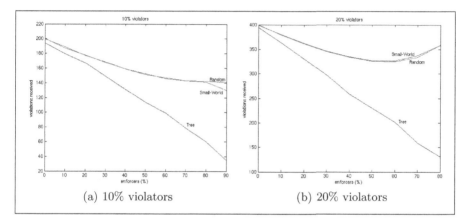

(a) 10% violators   (b) 20% violators

Fig. 4. Enforcement capabilities vary depending on structure

(H3) The enforcement strategy used by enforcer agents can reduce the number of violations received by meek agents. The x-axes in Figure 5(a) shows the enforcer to meek agent ratio. A higher ratio implies more enforcer agents. The y-axes measures the increment in efficiency at protecting meek agents from violations. Efficiency is calculated as the increment in percentage of the violations received by meek agents when enforcers use uni-directional blockage over bi-directional blockage (see Equation 1). A positive efficiency value means that BDB managed to stop more violations than UDB.

$$\Delta Efficiency = ((Violations_{UDB} / Violations_{BDB}) - 1) \times 100 \qquad (1)$$

In Figure 5(a) it can be observed that in random and small world networks the efficiency is always positive for any enforcer to meek agent ratio. It is also observed that for low ratio values the efficiency is increasing. But after a rate of 3 enforcers per meek agent the efficiency hits a ceiling. The results show that Bi-Directional Blockage has a higher efficiency at protecting meek agents from violator agents in these cases. The case of the tree network is different. The efficiency increment stays along the 0% line with some deviations. In networks organized as trees, the choice of enforcement strategy does not have a significant influence in the outcome. The reason being that the tree network ostracises violators quickly, independently of the blockage strategy used.

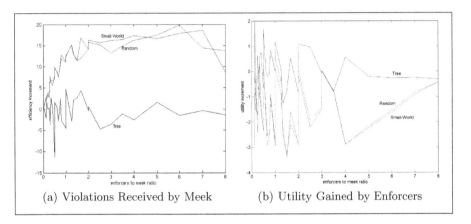

(a) Violations Received by Meek   (b) Utility Gained by Enforcers

Fig. 5. Enforcement strategy influences outcomes

The gain in efficiency at guarding meek agents comes at a cost. When enforcers use the BDB strategy they can ostracise themselves. This is the case when the enforcer is completely surrounded by violators. If the enforcer uses the UDB strategy it will use its neighbours as mediators, independently of their type. But when using the BDB strategy an enforcer is not be able to do this and therefore it cannot interact with anyone. Thus ostracising itself. This is a rare case but it can reduce the average utility gained by enforcer agents by up to 3% (see Figure 5(b)). The metric used to calculate the difference in utility can be seen in equation 2. A negative number means that the agent gains more utility when using a UDB strategy.

$$\Delta Utility = ((Utility_{UDB} / Utility_{BDB}) - 1) \times 100 \qquad (2)$$

(H4) Enforcement makes supporting the norm a rational strategy. The simulation data that refers to the utility gained by agents has been used to support this hypothesis. In the context of this paper, a strategy is said to be rational if the the agent will maximize the utility gained in the game. What has been tested is whether following the norm maximizes the agent's utility, and in which conditions. The simulation data has shown that when the ratio of enforcers passes a certain threshold, norm-abiding agents will gain more utility than norm-violating ones. This threshold depends on the amount of violating agents in the system. In a society with 10% of violator agents, five enforcers are needed for every four meek agents to make supporting the norm the rational strategy. For a society with 50% of violator agents, the ratio needs to be higher than 0.7 enforcers for each meek agent. The rest of simulations have inflection points between those two values. Strangely, societies with higher percentage of violators need a smaller ratio of enforcers to meek agents.

Finding a partner is a random walk through the network. Therefore, when a violator searches for a partner in a society where they are the majority, chances are that it will interact with another violator. When two violators interact the get a very low utility, thus a small number of enforcers interacting amongst themselves can easily win more utility than the violators, and even make up for the meek agents which are being preyed upon.

Figure 6(a) and 6(b) show the utility gained (y-axes) by norm supporting agents and violators respectively. Their x-axes show the enforcer to meek agent ratio. Each of the lines in the figures represent simulations with different percentage of agents in the society. As the number of enforcers increases norm supporters gain more utility. The opposite effect is observed for violator agents. When the enforcer to meek agent ratio is low, the utility gained by violator agents is much higher than the one gained by norm supporters. As the number of enforcer agents grows the values are reversed. The inflection point depends on the percentage of violator agents.

Interestingly, even though meek agents receive more violations as the rate of enforcer agents grows (see Figure 3(c)), the utility gained by them is not

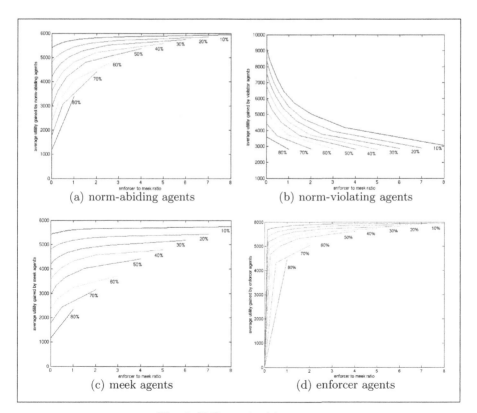

(a) norm-abiding agents   (b) norm-violating agents

(c) meek agents   (d) enforcer agents

Fig. 6. Utility gained by agents

decreased. It surprisingly increases (see Figure 6(c)). This is due to the fact that meek agents are still able to interact with other norm supporters. Enforcer agents will never interact with violators more than once, so they are restricted to interacting with other enforcers or meek agents. This helpful interaction from enforcers makes their utility increase despite receiving more violations. The ratio of violations to normative actions is lowered and the utility is increased.

5   Future Work

This paper is part of ongoing research on norm enforcement. Many other variants of this model will be simulated. In future work the set of assumptions about agents will be relaxed, by giving them the ability to change their strategies, to lie, and to allow enforcer agents to violate the norm (*i.e.*, corrupt enforcers). The perfect information assumption will be relaxed by adding uncertainty and noise. In these cases enforcer agents will need elaborate gossip techniques and reputation management to allow them to pick the right targets for enforcement. Futhermore, the agent's reputation can also be modelled by having the interaction mediators overhear the conversations they mediate. If overhearing is possible, there is no need to wait for interacting agents to report the interaction outcome. More so, other conservative blocking strategies can be studied; such as blocking off agents that mediate norm violators, or blocking agents until they are shown to be norm-abiders by interacting with the mediators.

The process an agent uses to search for a partner can influence the utility it gains. All agents in the simulations searched for partners using the same procedure, which was random. The process could be modified in many ways. For instance, the probability with which an agent is chosen as the interaction partner can be modified to make the average path length longer or shorter, this modification could be for all agents, or they could each have different average path length. Also, agents could choose an agent as the partner first and then try to find a path to it. These changes are not mutually exclusive, many combinations could be studied.

The impact of other network parameters (*e.g.*, clustering factor, diameter, number of links per agent, number of paths between agents) on norm enforcement should also be studied in future work. It has seen shown that a tree network is better from an enforcement perspective. Further studies could also relax the assumption of fixed networks. In order to find realistic models to be used in real networks, dynamic links must be allowed. Links could be added between agents dynamically and test how it affects norm enforcement. New enforcement techniques should be used that take advantage of the dynamic nature of the network.

Finally, related work has shown that when enforcement conveys a cost to the agent, the efficiency of enforcement diminishes [1,8]. The scenario in this paper does not consider such a cost associated to blocking violators. A cost could be associated to blocking interactions in order to test the enforcing efficacy in such a scenario. Enforcers would bear the cost of enforcement if they were not able to reject interactions. In such a case they would spare other agents from receiving norm-violations by receiving them themselves.

All these scenario modifications can also be used to research into the necessary conditions for norm emergence. Our goal is to find ways to apply this work to more realistic scenarios, such as security from malicious agents in open multi-agent systems over the internet.

Acknowledgements

This research has been partially supported by the Generalitat de Catalunya under the grant 2005-SGR-00093, and the OpenKnowledge[6] Specific Targeted Research Project (STREP), which is funded by the European Commission under contract number FP6-027253. The OpenKnowledge STREP comprises the Universities of Edinburgh, Southampton, and Trento, the Open University, the Free University of Amsterdam, and the Spanish National Research Council (CSIC).

A. Perreau de Pinninck is supported by a CSIC predoctoral fellowship under the I3P program, which is partially funded by the European Social Fund. M. Schorlemmer is supported by a *Ramón y Cajal* research fellowship from Spain's Ministry of Education and Science, which is also partially funded by the European Social Fund.

References

1. Axelrod, R.: An evolutionary approach to norms. The American Political Science Review 80, 1095–1111 (1986)
2. Boella, G., van der Torre, L.: Enforceable social laws. In: AAMAS 2005: Proceedings of the Fourth International Joint Conference on Autonomous Agents and Multiagent Systems, pp. 682–689 (2005)
3. Carpenter, J., Matthews, P., Ong'ong'a, O.: Why punish: Social reciprocity and the enforcement of prosocial norms. Journal of Evolutionary Economics 14(4), 407–429 (2004)
4. Castelfranchi, C., Conte, R., Paoluccci, M.: Normative reputation and the costs of compliance. Journal of Artificial Societies and Social Simulation 1(3) (1998)
5. Delgado, J.: Emergence of social conventions in complex networks. Artificial Intelligence 141(1), 171–185 (2002)
6. Grizard, A., Vercouter, L., Stratulat, T., Muller, G.: A peer-to-peer normative system to achieve social order. In: AAMAS 2006 Workshop on Coordination, Organization, Institutions and Norms in agent systems (COIN) (2006)
7. Hales, D.: Group reputation supports beneficent norms. Journal of Artificial Societies and Social Simulation 5(4) (2002)
8. Heckathorn, D.D.: Collective sanctions and compliance norms: a formal theory of group-mediated social control. American Sociological Review 55(3), 366–384 (1990)
9. Kittock, J.E.: The impact of locality and authority on emergent conventions: initial observations. In: AAAI 1994: Proceedings of the Twelfth National Conference on Artificial Intelligence, vol. 1, pp. 420–425. American Association for Artificial Intelligence, Menlo Park (1994)

---

[6] http://www.openk.org

10. Pujol, J.M., Delgado, J., Sangüesa, R., Flache, A.: The role of clustering on the emergence of efficient social conventions. In: IJCAI 2005: Proceedings of the Nineteenth International Joint Conference on Artificial Intelligence, pp. 965–970 (2005)
11. Savarimuthu, B.T.R., Purvis, M., Cranefield, S., Purvis, M.: Role model based mechanism for norm emergence in artificial agent societies. In: Proceedings of the International Workshop on Coordination, Organization, Institutions, and Norms (COIN), Honolulu, Hawai'i, USA (2007)
12. Taylor, M.: Community, Anarchy & Liberty. Cambridge University Press, Cambridge (1982)
13. Walker, A., Wooldridge, M.: Understanding the emergence of conventions in multi-agent systems. In: Lesser, V. (ed.) Proceedings of the First International Conference on Multi–Agent Systems, San Francisco, CA, pp. 384–389. MIT Press, Cambridge (1995)
14. Watts, D.J., Strogatz, S.H.: Collective dynamics of small-world networks. Nature 393, 440–442 (1998)
15. López, F.L., Luck, M., d'Inverno, M.: Constraining autonomy through norms. In: AAMAS 2000 and AAMAS 2002, pp. 674–681. ACM Press, New York (2002)
16. Younger, S.: Reciprocity, sanctions, and the development of mutual obligation in egalitarian societies. Journal of Artificial Societies and Social Simulation 8(2) (2005)

Retrieval of Case Law to Provide Layman with Information about Liability: Preliminary Results of the BEST-Project

Elisabeth M. Uijttenbroek, Arno R. Lodder, Michel C.A. Klein, Gwen R. Wildeboer, Wouter Van Steenbergen, Rory L.L. Sie, Paul E.M. Huygen, and Frank van Harmelen

Centre of Electronic Dispute Resolution – CEDIRE.ORG, VU University Amsterdam
http://best-project.nl

Abstract. This paper describes the experiments carried out in the context of the BEST-project, an interdisciplinary project with researchers from the Law faculty and the AI department of the VU University Amsterdam. The aim of the project is to provide laymen with information about their legal position in a liability case, based on retrieved case law. The process basically comes down to (1) analyzing the input of a layman in terms of a layman ontology, (2) mapping this ontology to a legal ontology, (3) retrieve relevant case law based, and finally (4) present the results in a comprehensible way to the layman. This paper describes the experiments undertaken regarding step 4, and in particular step 3.

Keywords: concept-based search, case law, information retrieval.

1 Introduction

[5, 9, 12] show that popular and influential applications in most countries are case-management systems. These systems helped to reshape the organization of courts and contributed to the reduction of case loads. Still, the judiciary is faced with more cases than they can handle.

Litigation is the traditional and public dispute resolution process, but several other so-called private dispute resolution processes exist of which the most prominent ones are negotiation, mediation, and arbitration. Arbitration and litigation are adversarial procedures, in which a third decides the case. Mediation and negotiation are consensual procedures in which the disputants aim at reaching agreement, either on their own or helped by a third called the mediator or facilitator. This third does not impose a decision upon the parties, but merely guides the procedure.

A decision to either go to court or to mediate (or negotiate, arbitrate) should be based on a well-informed choice. Currently the necessary information to make such a decision is often lacking. One of the aims of the BEST-project[1] is to provide litigants with information about the expected outcome of a court proceedings.

In literature as well as practice of Alternative Dispute Resolution the Harvard method is influential. It is based on work carried out in the setting of the so-called

---

[1] BATNA Establishment using Semantic web Technology, http://best-project.nl.

P. Casanovas et al. (Eds.): Computable Models of the Law, LNAI 4884, pp. 291–311, 2008.
© Springer-Verlag Berlin Heidelberg 2008

PON: the Project on Negotiation. This Harvard Negotiation Project introduced the concept of principled negotiation, which advocates separating the problem from the people. Fundamental to the concept of principled negotiation is the notion of *Know your best alternative to a negotiated agreement* (BATNA).

In the BEST-project we are developing a system that supports users by retrieving relevant case law on liability. In this way parties are given the opportunity to form a judgment about whether they could hold another party liable for certain caused damage or if they could be held liable themselves. Also, parties can determine how much room for negotiation is available

We develop a system for intelligent disclosure of case-law in which the retrieval is based on search terms provided by laymen. The main challenge we face is to match the different terminology used in case law and by laymen. Laymen describe cases in their own words, which differs from the vocabulary used by legal experts and in legal texts. We therefore decoupled the task of giving a meaningful description of the legal case at hand from the task of retrieving similar case law from the public available case law database www.rechtspraak.nl.

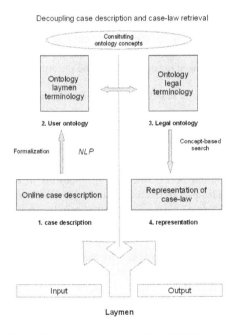

Fig. 1. Schematic overview of the BEST-project

Figure 1 shows a schematic overview of the retrieval process. First a case description is entered by the layman user. An ontology with layman concepts is used to structure the input and guides the user by entering relevant aspects of the case at hand. The laymen ontology is mapped to a second, legal ontology that is used for indexing case law. The retrieved case-law is then presented in a way comprehensible to the user and provides information relevant for his legal position.

In this contribution we first describe the thesaurus-based retrieval technique that we use and present the experimental results. Although the retrieval results were satisfactory, we considered this approach too static for relating the retrieval results to the laymen input. Therefore, in the following section 3 we propose a method to define search documents on the level of the concepts in a code section. This approach allowed us to use our thesaurus-based statistical retrieval techniques together with a visualization technique novel to the legal domain, which is described in section 4.

In section 5 a method is proposed to represent case-law to the users in a clear and comprehensible manner, based on recommender techniques.

In the last section we look ahead to our future work, that encompasses the first and second stage of the model.

2 Concept-Based Search

2.1 Applied Technique

We need to obtain retrieval results that:

- are relevant to the case description of the laymen, and
- show the conditions necessary to establish liability.

This information should contribute to better insight by the layman about his legal position. The retrieval experiments we conducted with a statistical indexing technique are described below.

We used a thesaurus-based statistical indexing technique [13]. A thesaurus is used to create a vector representation of each document. Documents are compared by their vector representations. For searching a "query document" is created, and the vector representation of this query document is compared with the vector representations of the other documents [7].

This technique has been implemented in a commercially available software tool[2]. The main advantage of this technique over standard information retrieval techniques based on the *vector space model* [14], is that the indexing is guided by a thesaurus. This means that only terms relevant to a specific domain are taken into account. The indexing method works roughly as follows [11]. The indexing algorithm first detects sentences in documents and removes stop-words. After this it normalizes the remaining words, which means that nouns are reduced to the singular form and verbs to the first person singular form. In our experiments, we have used a specialized normalization engine for the Dutch language. From these normalized terms or phrases, the relevant ones are then identified using a domain-specific thesaurus.

A list of the relevant concepts identified in a document is called a *concept fingerprint* of that document. For each identified concept a unique concept identifier is added to the fingerprint. This concept identifier is assigned a relevance score, based on term frequency and the specificity of the term in the thesaurus (which is the depth in the hierarchy), and the lexical similarity of the term with the textual contents. A fingerprint can be seen as a vector in a high dimensional space. The dimensions of

---

[2] by Collexis BV, http://www.collexis.nl

this space are formed by the concepts of the thesaurus. The weight (or value) in each dimension is the relevance score for the concept in the document. The search is performed by a matching engine in the software, which matches a search vector with the vectors of the indexed documents. The vector for the search query is calculated in a similar way as described above. The matching engine will compute the distance between the query vector and the vectors of the documents. The result of the matching engine will be a set of document vectors sorted on their distance to the query vector. This is presented to the user as a ranking on relevance of the indexed documents.

2.2 Experimental Set-Up

2.2.1 Data Source

Although the case law database used to disclose similar cases, the public website www.rechtspraak.nl, can be accessed online, for processing purposes we have locally stored all available cases. The approximately 100.000 cases is a low number, given the over 1 million legal verdicts annually. Nonetheless, this database contains almost all digitally available newer case law (1999-) in the Netherlands. The verdicts have some meta-data attached to them, e.g. the location of the court, the date of the verdict, a unique identifier (LJN)., and for around 50% of the verdicts (the newest) a summary of a few lines. Internally, the documents have no computer parsable structure, but are plain text.

2.2.2 Research Questions

First of all, we wanted to know whether a concept-based search technique as described above is suitable for the retrieval of case law in which a prototypical legal case is described. To obtain relevant retrieval results - that is a prototypical legal case similar to the case described by the layman - an effective search document has to be created. We conducted different experiments to find out what the best method is to create a search document.

In our first experiment we distilled the relevant terms for a specific legal case category from Code sections. So the search documents consisted of the terminology used in the text of the Code (we call this: *code-based fingerprints*). In our second experiment, we did the same for case law as we did for the Code, so now created case-based fingerprints. The search documents in this experiment consisted of terminology used in case law. Since we did not use automated techniques we labelled this method *case-based manually created fingerprint*. Finally, we selected a number of relevant cases and used these together as one search query (*case-based automatically generated fingerprints*).

Because terminology in the code differs from terminology in case law, we expected that the search with case-based fingerprints would be better suited to identify relevant cases. Second, we expected that the indexing process for the *generated fingerprints* would automatically distinguish the most important terms and therefore perform better than the manually created fingerprints.

2.2.3 Procedure

We started with a selection of specific legal categories for which we wanted to identify relevant case law. We chose three fairly different types of liability and a fourth one that is related to one of the other situations:

- liability for misleading advertisements;
- liability for non-subordinates;
- liability for real estate;
- liability for subordinates.

The reason for this choice is twofold. First, diverse legal cases allow us to check whether our set up is suited for distinguishing legal cases at all. Second, the broad selection prevents us from drawing conclusions based on a not representative subset of liability. The idea behind the two similar situations (liability for subordinates / nonsubordinates) is that this will learn us whether the technique is also able to distinguish different legal cases that are quite similar to each other but are based on different Code sections. For each of the four specific legal categories we did the following three things:

1. we distilled the relevant terms from the Code;
2. from the court decisions database we selected a number of relevant cases;
3. we analyzed relevant cases and made a list of important terms used.

In addition, we created a thesaurus with legal terms. The thesaurus is manually created from the terms identified in task 1 and 3 in the list above and the terms identified in the Code for other types of liability. This resulted in a thesaurus with 360 concepts. The structure of this thesaurus is imposed by the structure of the law itself, i.e. "liability" is the root concept with more specific types of liability below it, e.g. "liability for persons", which in turn has "liability for subordinates" below it. The relevant terms are placed below the types of liability for which they hold, including some synonyms. The thesaurus is used to index the data set (i.e. creating fingerprints for each document in the repository) and the other material is used to create different search documents for which fingerprints are calculated. We then used the search documents' fingerprints to search for relevant cases. The top of the highest ranked results were evaluated on relevance.

2.3 Results

2.3.1 Code-Based Fingerprints

In a first set of experiments, we evaluated the code-based fingerprints, i.e. the fingerprints with terms distilled from the sections of the code. We did not expect very good results here, as we assumed that the vocabulary used in the cases is different from the vocabulary in the code text. Nevertheless, Dutch law is build on the Code (in contrast with the Common Law tradition), so we could ignore this in our retrieval process. As can be seen in Table 1, the correctness figures for the 10 highest ranked are indeed quite low. For two fingerprints, this set did not contain any relevant result at all. In the other one, we only found 3 relevant cases, but also two cases in which the article searched for was only casually mentioned. Note that for section 6:170, we found 8 slightly relevant cases.

Table 1. Correctness figures for code-based fingerprints

Code sections	Correctness
6:170 subordinates (including slightly relevant 90%)	10 %
6171 non-subordinates	0 %
6:174 real estate (including slightly relevant 50%)	30 %
6:194 misleading advertisements	0 %
6:162 unlawful act (including slightly relevant 48%)	40 %

2.3.2 Case-Based Manually Created Fingerprints

We did the same experiment for case-based manually created fingerprints—fingerprints based on important terms identified by the expert in a selection of the case law. This resulted in the figures printed in Table 2. For two of the three sections the results are fairly good. For one article, the results are not so good; interestingly, this is a section for which a good result was obtained for the code-based fingerprints.

Table 2. Correctness figures case-based manually created fingerprints

Code sections	Correctness
6:171 non-subordinates (incl. slightly relevant 70%)	50 %
6:174 Real estate	10 %
6:194 Misleading advertisements (incl. slightly relevant 92%)	83 %

2.3.3 Case-Based Automatically Generated Fingerprints

Thirdly, we evaluated the performance of automatically generated fingerprints — fingerprints based on the full text of a set of pre-selected relevant cases. We started with fingerprints based on 5 case descriptions for 3 different legal cases. The results vary for the different Code sections (see Table 3). The table lists the number of cases used to create the fingerprint, the number of relevant cases as fraction of the total number of evaluated cases, and this fraction represented as a percentage. We evaluated the relevance of the first 15 returned documents, but we did not count the documents that were used to *create* the fingerprint. This explains the difference in the totals in the column with the correctness.

Table 3. Corectness figures of automatically generated case-based fingerprints

Code section	Corectness
6:170 subordinates	70 %
6:171 non-subordinates	0 %
6 :174 real estate	37 %
6 :194 misleading advertisements	77 %
6:162 unlawful act	32 %

A hypothetical explanation for the diverse results is that the sets of documents from which the fingerprint are generated are too small. To check this, we generated a fingerprint from a larger set of documents (20 cases) for the worst performing legal case, i.e. "real estate". Because the total number of cases in the data set for real estate are between 40 and 100, it is not realistic to base fingerprints on much more than 20 documents. As can be seen in the table, the results were still not good: only 2 out of the 10 cases highest ranked were relevant.

Finally, we have generated a fingerprint from a very large set of documents. We used the general "unlawful act" section 6:162 for this, as this is the only section for which we had enough cases (around 500) to do this experiment. The fingerprint for section 6:162 is based on 249 cases. To our surprise, the results were still disappointing: only 8 from the 30 highest ranked cases were relevant and not yet used to create the fingerprint. Even when the cases about the related section 6:174 were considered as relevant, we only count 13 cases. Moreover, the first case which was not part of the fingerprint appeared to be irrelevant.

2.4 Additional Experiments

While looking for an explanation for the results of the previous experiments, especially the under-performance of the fingerprint for "liability for real estate", we considered that the wide variety of the factual situations underlying a specific legal category probably blurred the legal similarity. For example, "liability for real estate" copes with all kinds of real estate, including roads, and accidents with all kinds of vehicles because of shortcomings in the road. However, we also found out that there are *typical phrases* that are used to prove a specific type of liability. Therefore, we extended the *case-based manually created fingerprints* with such phrases. We distinguished the different argumentation lines used to prove something and typical phrases used in the judges' argumentation. On average, we added around eight phrases per legal category. Translated examples of such phrases for "real estate" are:

- "causing danger for persons or objects",
- "owner of a property", and
- "requirements that in a given situation".

We have added these phrases also to the thesaurus and re-indexed the complete repository. The results of this experiment are listed in table 4. The figures indicate that there are more relevant cases returned than in previous experiments. What is also interesting, but not visible in the figures, is that the ordering seems to be better than in previous experiments: the relevant and irrelevant cases were less intermixed than before.

In this experiment we also counted the number of relevant cases that did not explicitly mention the section number. These are interesting cases, because they can be found by relevant wording only, and not because the section number is mentioned. As can be seen, there are at least some cases that are relevant, but do not literally contain the section number.

Table 4. Correctness figures for case-based manually created fingerprints after adding legal phrases

Code section	Correctness
6:170 subordinates	88 %
6:171 subordinates (including slightly different 53%)	47 %
6:174 Real estate (including slightly different 87%)	80 %
6:194 Misleading advertisements	80 %

Finally, we redid the last experiments with no other concepts in the thesaurus, i.e. we reduced the thesaurus to the four different types of liability and their relevant phrases. This resulted in a thesaurus of 25 concepts expressed in 50 terms (i.e., 25 synonyms). When using this thesaurus to index the complete data set, around 7000 documents (out of 68.000) were ignored because none of the terms in the documents were similar to terms in the thesaurus. The remaining documents were indexed with only 1.08 terms on average. This suggests that sensible results are unlikely, because it almost means that for each document a single keyword is attached. The correctness figures for this method were still quite high, 70% for the first 10 hits in for "sub-ordinates liability". However, all of them literally contained the section number. As we have seen in one of the previous experiments, there are also relevant cases in which the article number is not literally mentioned.

2.5 Discussion

Several observations can be made.

First, we noted that only for one legal category ("liability for misleading advertisements", Section 6:194) the results for the *automatic case-based generated fingerprints* were notably better than the *code-based fingerprints*. A possible explanation is that the specific code text uses very abstract formulations, which have only a few terms in common with actual cases. We noted that the code specifies a non-exhaustive list of possible misleading statements ("about the contents", "about the amount", etc.). The terms used in this list of typical misleading statements will not frequently occur, as they describe statements at an abstract level. These abstract terms are different from the concrete terms that are used in case law. Thus, even although case law contains the term 'misleading advertisement' very often, the resulting fingerprint will be quite different. The automatically generated fingerprints from the cases do contain the concrete terms from the cases, of course.

A second interesting observation is that when using code-based search, we found for some of the legal categories (e.g., sections 6:162, 6:170 and 6:194) many *indirectly relevant* cases, i.e. cases in which the article was only casually mentioned. This finding can possibly be explained by the *interpretive* character of the legal concepts mentioned in the code for these articles. When such concepts are not precisely defined the legislator intentionally left room for interpretation by judges. Legal reasoning that involves interpretation is a manifestation of the application of a vague concept. An example of such a vague concept is 'the reasonable man' or 'an act or omission violating a rule of unwritten law pertaining to proper social conduct'. In situations where vague concepts are used case law determines the meaning of these concepts. Court decisions often refer to a concept with an interpretive character, which causes a lot of indirectly relevant retrieval results. Therefore, a high number of *indirectly relevant* cases would be a sign of code text that is characterized by interpretive concepts.

Another observation, which is not directly visible in the figures, is that the analysis of the results showed that the type of cases returned for the *automatic case-based fingerprints* and the *code-based fingerprints* are very different for sections 6:171, 6:174, and 6:162, although the percentages of correctness are comparable. *Code-based fingerprints* resulted in cases that literally contained some non-interpretable concepts in code sections, while *case-based fingerprints* resulted in cases that define the meaning of interpretive concepts in the code. This suggest that *code-based fingerprints* are useful for finding *non-interpretive concepts*, e.g. concepts that have a precise meaning in the law, while *case-based fingerprints* are more useful to find *interpretive concepts*. This is in agreement with the intuition that the meaning of interpretive concepts is defined by case law.

Another interesting finding is that manually created fingerprints in general perform better than automatically generated fingerprints (except for one of our examples, i.e. "real estate"). This is contrary to what we expected. This might have to do with the large number of real world situations in which some legal concepts can be relevant. To describe these situations different (ambiguous) terms can be used. It is therefore more difficult to distinguish them only by looking at the terms used. This is in particular a problem for the automatic method, as it uses the number of occurrences of the terms as the measure to calculate the relevance. When manually creating fingerprints the most irrelevant terms are probably left out.

Finally, we have seen that by adding typical legal phrases the results improve. There are more relevant cases returned and the distinction between relevant and irrelevant seems to be crisper. However, the phrases alone are not sufficient. It seems that the phrases help to eliminate irrelevant cases in the top of the ranking (improve precision), but that additional concepts in the thesaurus are required for finding relevant documents that do not contain the literal article number (improve recall). A hypothesis is that the phrases are especially helpful for retrieving the concepts that need additional interpretation, i.e. the vague concepts.

3   Search Documents

3.1   Concept-Based Search Documents: Technique Enabling Visualization in a Later Stage

In the experiments described above we created search documents for each section of the code. The conditions to establish liability can be found in the relevant code section. To provide laymen with relevant information about his legal position, it is necessary to make at least clear which conditions need to be fulfilled to establish liability. For this reason we conducted a following series of experiments. We created search documents for each condition necessary to establish a specific type of liability. For example, in Dutch tort law liability based on the general section 6:162 BW can only be established if the following conditions are fulfilled:

- the presence of an unlawful act (that is: an infringement of a right, a violation of a statutory duty, and an act or omission violating proper social conduct);
- damage;
- a causal relation between the act and the damage;
- accountability.

For each of these conditions search documents were created. We did this for 15 different sections of Dutch tort law. Tort law doctrine has been used to determine the necessary conditions. However, doctrine was not always decisive. For the retrieval of case law also other factors should be taken into account, such as the relevance of a concept in the light of the case law to be retrieved or the different contexts in which the same concept is used. The following criteria have been used to divide a section into legal concepts:

a. The legal concept should have a certain level of broadness to make it applicable to a large category of case law;
b. The legal concept should be precise enough to be relevant in a particular factual context;
c. Tort law doctrine is the leading guideline;
d. Coherence between the different concepts distinguished.

3.2 Open Textured and Clear Concepts

Legal reasoning is indeterminate due to its open, procedural nature [8]. Bench-Capon & Sergot [1] share the view that indeterminacy of law is a consequence of open texture. They define an open textured term as one whose extension or use cannot be determined in advance of its application. This means that the application of an open textured concept in code sections cannot be derived from the code itself. Open texture is the main reason to treat the legal domain as a specific domain of retrieval. We used the following indicators [cf. 15] to determine the open textured character of a concept.

1. *Ambiguity* - A term is ambiguous if there are more definitions for one concept. Dutch Tort Law terminology is characterized by ambiguity. For example, the term 'accountability' could relate to the establishment of liability but it is also used to determine the amount of compensation that has to be paid.
2. *Granularity* - The degree to which a concept is abstract in its nature. Such as "amount" or "duration".
3. *Discretionary statutes* - Only the framework for discretionary room can be given, but discretion can be described in the form of a "shopping list". For example in section 6:194 different circumstances under which an advertisement will be judged misleading are enumerated.
4. *Jurisprudence* - Judges often give an interpretation of relevant, vague concepts. An example from section 6:162 is 'an act or omission violating proper social conduct'.
5. *Socio-political environment* - A changed socio-political environment could indicate that a certain term is subject to interpretation. In section 6:175 regarding the liability for waste products it is determined that a product will under any circumstances qualify as a waste product if a legally binding decision said so. New waste products come and others disappear, and the legally binding decision can be adapted to the newly identified (dangerous) waste products.
6. *Completeness of knowledge*- The last indicator of open texture is the completeness of knowledge in a specific domain or field. If there are two or even more definitions for the same term, classification ambiguity comes into play. To obtain relevant retrieval results an ambiguous concept should be characterized as an open textured concept and treated as such. The term "work" is an example of a term that leads to classification ambiguities. Work can relate to labor law issues but also to the object of copyright infringements (the created work). Search documents need to be defined in such a way that retrieval results are restricted to the right interpretation of a specific term.

Clear concepts do not have to be interpreted. An example of a clear concept is an act violating a statutory duty. All the possible violations can be found in the Dutch code. In case law the reason for unlawfulness of the act, such as acting in conflict with the obligation to identify, or the relevant section, can be mentioned.

3.3 Creation of Search Documents

Distinguishing between clear concepts and open textured concepts is relevant for retrieval, for it indicates a difference in the way natural language is used [6]. If code text is used literally for the creation of a search document, the retrieval results will be poor for open textured concepts because these concepts are interpreted or complemented by the judge.

Case-based fingerprints are search documents created for open textured concepts. These fingerprints are based on the terminology used in case law. The open texture necessitates that concepts are interpreted. Although it is not possible to determine the full scope of interpretation in advance it is possible to give an estimation about the room left for interpretation. Court decisions were manually analyzed to distil relevant terms for an open textured legal concept. For clear concepts code-based search documents were created. In case of clear concepts the code text alone suffices to obtain relevant retrieval results. The following two decisions have to be made for each search documents:

A. *Code-based or case-based* - The search document should be either based on code text or on case law terminology;
B. *Level of abstraction* - The search document should be abstract enough to retrieve as much relevant court decision as possible. Different legal categories are distinguished in case law for the concept "an act or omission violating an unwritten law pertaining to proper social conduct". These include situations of sports & play, negligence, creation of danger, etc. The search document therefore has to comprise all these categories. However, it is not necessary to define every sports & play situation there is. It is unnecessary to comprise terms as "tennis", "football", etc.

3.4 Experimental Set-Up

3.4.1 Data Sources
For these experiments we also used the case law database of the public website www.rechtspraak.nl. See for more information section 2.2.1.

3.4.2 Procedure
For each of the legal concepts a search document is built as described in the previous section. The retrieval software is used to query the database for documents (case law) similar to the search document, which results in a ranking of all documents. Subsequently, the 30 most similar documents for each of the search documents are analyzed manually on their relevance. To determine the relevance of retrieval results for code-based or case-based fingerprints we set the following criteria. The retrieved court decisions should interpret or mention the norms relevant to the legal concept for which the fingerprint had been created. The legal concept can be mentioned literally,

but a description of the relevant concept is also sufficient. The concept has to be mentioned at least indirectly.

3.4.3 Results

In table 5 an overview is given of the retrieval results for 5 essential concepts. The results showed a relevance of approximately 70% (see Table 5). The relevance score for concepts created for sections of the code that are not applied regularly were below average, while the relevance score for concepts of often applied sections were above average. Obviously, less court decisions are available in the database for the sections that are less regularly invoked.

Table 5. Relevance scores for individual concept queries

Section	Concept	Type	Relevance
6:162 BW	Act or omission violating an unwritten rule pertaining of proper social conduct	Case-based	100%
6:170 BW	Say over subordinates	Code-based	69%
6:174 BW	Danger for persons and objects	Code-based	90%
6:174 BW	Realized danger	Code-based	58%
6:174 BW	Requirements under certain conditions	Case-based	100%

4 Visualizing Overlap between Concepts

4.1 Motivation

Each search document of the conceptual retrieval technique just elaborates upon a single concept. The retrieval software calculates a similarity value between the search document and all documents in the database. This results in a ranking of all case law according to its similarity with the search document. We assume that a similarity above some threshold value implies relevance of these retrieved cases for the concept queried for. The threshold value is pragmatically chosen such that it provides a good balance between precision and recall for all query concepts. Because each code section is split into several concepts and hence search documents, an intuitive assumption is the following:

The relevance of a retrieved case for a specific code section increases with the number of concepts of that code section for which this case is relevant. Therefore, the intersection between the sets of retrieved cases for concepts of the same code section are probably the most relevant cases.

4.2 Procedure

For the visualization of the clustering of cases we use the clustermap viewer from Aduna2[3]. This software creates Venn-like diagrams of objects and show if they

---

[3] http://www.aduna.biz

belong to one or more sets. It allows for dynamically adding and removing of set specification, which can be helpful to see the effect of using different sets on the grouping of the objects.

Each object, in our implementation a court decision, is represented as a sphere. All retrieved cases that contain a specific concept are clustered and visualized as amoebalike shapes (blob shapes). If an object belongs to multiple clusters (which means that a court decision is relevant for more than one concept), the blob shapes overlap and the object is displayed in the overlap. The software can be configured in such a way that the darkness of the areas reflects the amounts of overlap. Therefore, one can immediately see which objects are in the highest number of clusters. For each object links to e.g. webpages can be added. In our implementation, we created direct links to the online version of the verdicts. This link points directly to the verdict at the website of rechtspraak.nl. Thus, our local database is only used to calculate the similarity between the cases and the search documents, but is not used to display the case to the user. An interface has been written that connects the Collexis search software to the Aduna clustermap viewer. This interface allows formulating queries for sets of concepts. We use this interface to specify sets of legal concepts that together represent a section of the code.

4.3 Visualization Experiments

We did some experiments with different combinations of the concepts for which we defined search documents. We chose the sets of concepts in such a way that we were able to visualize overlap between the cases for concepts that together establish a certain kind of liability.

We defined 28 combinations of legal concepts for 15 Code sections. We obtained approximately 900 different court decisions for the different combinations of legal concepts. The retrieved court decisions were sometimes partly overlapping for different combinations of legal concepts. Searches for some concepts resulted in a relative small number of cases (e.g. around 5), others in a much higher number (around 200).

For the evaluation of the results we set the following criteria. Court decisions are relevant if they deal with the type of liability, for which we created a specific set of clustered concepts, resembling the conditions that need to be met to establish liability based on a specific section of the code. For example, for the code section about "wrongful acts" a set of legal concepts is created, comprising the concepts "causality", "damage" and "wrongful act". The court decisions showed in the overlap between these concepts, handle about wrongful acts, and contain al three constituting concepts.

4.4 Examples

In this paragraph the search for cases about specific code sections is illustrated with three examples.

4.4.1 Real Estate

The first example (see Figure 2) shows the cluster map for concepts that constitute liability for real estate" (section 6:174). The concepts we considered are "real estate"(fp25), "possessor of real estate" (fp2), "danger for persons and objects" (fp12), "requirements under certain conditions" (fp6) and "realization of danger" (fp35). There are quite some cases in which "danger for objects and persons" (fp12) and "requirements under certain conditions" (fp6) play a role. There is also a reasonable number of cases in which both concepts are present. The picture shows that there is only one case in which all concepts are important. Inspection learned us that we obtain more relevant court decisions if the concepts "damage" and "possessor of real estate" were not included.

The relevance score is 100% if these concepts are excluded. These concepts are too broad ("damage") respectively to precise ("possessor of real estate") in formulation.

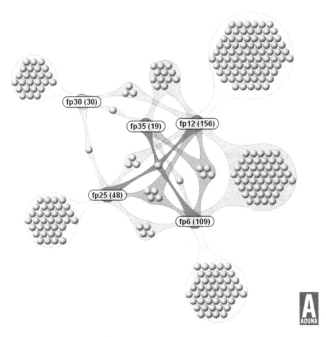

Fig. 2. The visualization of the grouping of relevant cases (yellow spheres) by the essential concepts of "liability for real estate"

4.4.2 Liability for Subordinates

A second example (see Figure 3) illustrates the clustering of cases for "liability for subordinates" (Section 6:172 Civil Code). The essential concepts are "fault of a subordinate" (fp9), "probability of a fault" (fp17), "say over subordinates" (fp36) and "damage to others" (fp28). In this example, it is immediately clear that there is no overlap between the documents returned for "damage to others" and the other returned documents. It also shows that there are eight cases for which three of the essential concepts are relevant. Those are included in the "darkest" part of the diagram. It is also interesting to see that it almost doesn't happen that a "fault of a subordinate" is relevant in a case without "say over subordinates" being relevant. Based on this we could hypothesis that the requirement "fault of a subordinate" is not very important when retrieving case law, as all that these documents are already retrieved when searching for "say over subordinates".

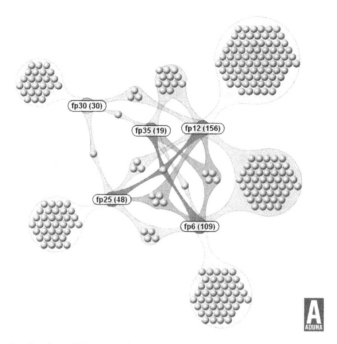

Fig. 3. The visualization of the grouping of relevant cases (yellow spheres) by the essential concepts of "liability for subordinates"

4.5   Results

The results (see overview in Table 6) showed relatively high scores on precision for the sections that are often applied to establish liability, such as the general tort law section. Poorer results in overlap were found for sections that are not often applied, such as the liability for representatives. The results showed an average of 60% relevance. The results for often applied sections show results up to 100%, while sections of the code that are rarely applied resulted in a relevance score of less than 40%.

The precision in general is good for some of the concepts. These results were in most cases better than the straightforward approach as described in section 2. We hypothesize that the poor results for the clusters of concepts that resembled less applied sections of the code is possibly also due to the fact that www.rechtspraak.nl exists since 1999 and that few court decisions about certain types of liability are available. To validate the recall, we used standard court decisions that contain the basic interpretation and argumentation for a liability section of the code. All these court decisions were from before the launching of rechtspraak.nl (1999), and therefore added to our database. We hypothesized that these basic court decisions would be displayed by the clustermap viewer. Poor results for the recall were obtained. The court decisions relevant for a specific category of liability were not displayed by the clustermap viewer. These poor results on the recall could possibly be explained by the use of different terminology in older court decisions that is not used in the fingerprints. Another possible explanation for the poor recall results is the limited manually composed thesaurus or the limited use of terminology for the manually created fingerprints.

Table 6. Overview of the results

Section	Essential concepts	Relevance score
6:162 BW unlawful act	Causality, damage and an act or omission violating unwritten law pertaining to proper social conduct	72% if the concept "damage" is not included. This concept proved to be irrelevant
6:170 BW subordinates	Fault of a subordinate, probability of fault, say over subordinates and damage to third	100% if the concept "damage to third" is not included. This concept proved to be irrelevant
6:171 BW non-sub-ordinates	Fault during work activities, non-subordinate, damage	38% if the concept "damage" is not included. This concept proved to be irrelevant
6:174 BW real estate	danger for persons or objects, requirements under certain conditions, realized danger (damage and possessor of real estate were irrelevant and not included in the evaluation of the results)	100% if the concepts "damage" and "possessor of real estate" were not included. These legal concepts proved to be irrelevant

The clustering results show that some concepts can be omitted. An example of a redundant concept for the retrieval of case law is "damage". The redundancy of this concept could be explained through the neutral character of the terminology related to the concept of damage. The concepts that combine possession and an object, for example "owner of real estate" seem to be too detailed and exclude a lot of relevant court decisions. If we observed the clustering results of concepts that only relate to the object, such as "real estate", the results for a set of concepts improved tremendously. Only 28 court decisions for the concepts "possessor of real estate" were retrieved, from which only one was part of an overlap, while for the concept of "real estate" 38 decisions were obtained, from which 9 were part of an overlap.

5 Presenting Relevant Court Decision

5.1 Motivation

Besides retrieval of relevant case law, the comprehensible presentation of the retrieval results is an important part of a successful system to provide laymen with information about their legal position. We assume laymen will have a problem reading the verdicts and understanding the different legal concepts, i.e. the conditions to establish liability. To present an understandable explanation of the relevant verdicts, we take two steps. First, we localize in the verdicts the legal concepts that are relevant for the user's case. With techniques from recommender systems we then decide which paragraphs are relevant for which concepts and we present the user the verdict based on these relevant paragraphs and apply also other recommender techniques. We also carried out a small user satisfaction research to find out whether the proposed presentation is indeed useful to prospective users.

5.2 Technical Implementation

Recommender Systems are usually divided into two approaches: Collaborative Filtering (CF) and Content Based Filtering (CBF). In Collaborative Filtering, the preferences of communities of similar users are used to decide on recommendations for the current user [4]. With Content Based Filtering the content of certain items is processed and based thereon a decision is made about whether the user will probably be interested in the item or not, based on some predefined user characteristics and a history of interest in earlier items [3].

Since we want to process the court decisions on content to explain their relevance Content Based Filtering might be helpful. The content of a paragraph decides whether the user will be interested in that paragraph or not. Of course, this is not based on the preferences of the user, but on the relevance of the legal content.

Another interesting prospect is that Recommender Systems sometimes provide a *reason* for the recommendation. An 'explanation mechanism' tries to explain why the program believes that the user will be interested in the prospective item [10]. We investigated whether the techniques used to establish the reason for recommendation are also feasible for explaining to the user why those specific verdicts are presented to him. However, in Recommender Systems the search for recommendable items is tied to the reasoning about why a certain recommendation was made while in our research, the search is conducted separately. Only afterwards we aim to re-establish the reasons behind the selection of the final set of verdicts. Also, history information about earlier recommendations is not available. As follows, the technique can only be applied on the content of the court decisions under scrutiny at the moment.

To test the effectiveness of the explanation system a small satisfactory research is conducted. Our basic assumption is that the explanation should convince the user that the presented verdicts are relevant for his own case. This relevance can exist in more in-depth information about the similarities and dissimilarities between his own case and the court decision represented by the system. Explanation systems that concentrate on this aspect are Keyword Style Explanation and Influence Style Explanation [2]. In Keyword Style Explanation the user is given a table explaining which words in his profile and in the content of the item had the most influence on the rank of the item. This can possibly be applied in our project to the occurrence of fingerprint terms. In Influence Style Explanation, the system tells the user how their interactions with the recommender system influenced the recommendation. In our project it might be possible to use this with the original description of the user case.

For the application of these techniques, we need to localize the legal concepts in the verdict, since they determine whether the content is relevant. This localization is described in the next section.

Since experiments showed us that it is impossible to localize the legal concepts that are extracted from the user's case in a direct manner (e.g. by keyword search), we decided to use the fingerprints from the search part of the project for localization. In GATE (General Architecture for Text Engineering) we first tokenized the relevant verdicts, then stemmed them, used a gazetteer to annotate words and phrases belonging to a concept, based on their fingerprint and finally used a transducer to be able to visualize the concepts belonging to the various annotations. We used the Snowball stemmer, a flexible gazetteer in combination with the OFAI gazetteer and the JAPE transducer. The fingerprints of each concept were provided to GATE in lists of all corresponding terms and phrases, also in stemmed version. In the next section we will describe how we processed these annotations with Recommender techniques discussed earlier to arrive at the final presentation of the verdict to the user.

5.3 Results

With the annotation of the terms from the fingerprints, we can now determine which paragraphs are relevant for which legal concepts. We used Keyword Style Explanation and designed a number of rules that state how many times a term, or multiple terms from the same fingerprint, must occur in a paragraph to deem that paragraph relevant for the particular concept that corresponds with that fingerprint. When a paragraph is relevant, we highlight it entirely (so the highlighting of the separate terms disappears) and provide the paragraph with a comment that explains the legal concept for which the paragraph is relevant. All legal concepts found in the verdict (corresponding to those extracted from the user case in another part of the program) are in general wording explained at the top of the verdict.

Besides this Keyword Style Explanation, we also used Influence Style Explanation. Certain terms or concepts were used to link the verdict to the user case. If for example the verdict was about a 'traffic accident', then the user would be notified whether this is a similarity or difference with respect to their case. This linking was done for multiple concepts in order to help the user apply certain aspects from the verdict to the user case.

5.4 User Satisfaction Research

As we were interested in the usefulness of this representation technique for court decisions a small scale user satisfaction research was conducted. The research group consisted of 21 participants and was divided into three groups of seven. Each group received a fictitious, but realistic, description of a case, a general explanation of the research, 4 verdicts and three different types of questions. The difference between the groups was the extra information given with the court decisions. Group 1 just received the verdicts, without any explanation. For Group 2 the court decisions were processed according to the Keyword Style, as explained in the previous section. Group 3 got the verdicts processed with Keyword Style *and* Influence Style Explanation.

Three different types of questions were formulated. The first category consisted of 'subjective' questions: propositions with an answering scale from 1 (I don't agree at all) to 5 (I agree completely). These were designed to measure the confidence the users have in the program and extent to which they feel the program is useful to obtain information about their legal position. The second category of 'objective' questions are in exam style. Those questions were designed to test the knowledge of the user about the provided case, the content of the legal concepts, and information about their legal position based on what they learned from the presented court decisions. The third and last category of questions had an open character in which the users could express what they liked about the program, what they missed and anything else they wanted to share. Some personal information was obtained to account for differences in age, education and legal knowledge.

Our overall hypothesis was that the groups would perform in increasing order. We hypothesized that group 1 would have the lowest scores for the subjective questions (meaning the highest confidence in and satisfaction with the program) and perform worst on the objective questions compared to the other groups. For group 2 these scores would improve, while group 3 would perform best on the subjective questions as well as on the objective questions. This hypothesis is based on the expectation that the extra information provided to group 2 and group 3 will contribute to an improved understanding of the presented court decisions relevant to gain more information about their legal position. The extra information provided to group 2 and 3 can help to enhance confidence in information provided by an online information system and also improve knowledge about their legal position. We hypothesize that the participants of group 1 need more time to complete the whole survey, since they will have to read the verdict on their own to find out what is relevant, whereas the other groups have the relevant paragraphs highlighted already. Out of the 21 surveys sent, we got 15 back; 5 in each group coincidently.

Table 7. Time needed to compleet the survey

Respondent	1	2	3	4	5	Average
Group 1	100 min	55 min	60 min	100 min	50 min	60,8 min
Group 2	60 min	40 min	55 min	45 min	60 min	43,3 min
Group 3	30 min	35 min	20 min	40 min	35 min	26,7 min

Table 8. Average scores on the subject questions (scale 1-5)

Question	1	2	3	4	5	6	Average
Group 1	2,4	2,8	2,6	3	3,8	-[4]	2,9
Group 2	3	3,2	3	3,8	3,2	3,4	3,3
Group 3	3,4	3,8	3,4	4	3,8	3,6	3,7

In relation to the objective questions, answers were given in free text, which makes it impossible to analyze them with average numbers. However, interesting differences between the groups were observed. None of the respondents in group 1 mentioned the legal concepts 'damage', 'causality' and 'an act or omission violating unwritten law pertaining to proper social conduct', where most of those in group 2 and 3 did. Further, all respondents believed that a judge would grant the victim full compensation of the medical expenses for his foot. The majority of those in group 1 and 2 believed that the judge would not grant expenses made because of the depression. Reason given for this belief was that the victim had had depressions before, so the causal relationship could not be established in their eyes. In group 3 there were remarkably more respondents believing that the depression-related expenses *would* be granted. Answers to the

---

[4] This question was about the extra information provided. Group 1 did not get any extra information, hence this question wasn't relevant to that group.

question whether a compensation for not being able to play sports anymore would be granted were rather varying. Main reason for this was the need for more detailed information about sports history of the victim and alternative career prospects. Almost none of the respondents believe that a judge will grant all claimed damages: from the reactions it seems they just assume that a judge will never give you exactly what you ask for. Finally, group 3 is more reluctant to accept the offer in respect of a settlement than the other groups (4 of the 6 would not accept the offer, whereas in groups 1 and 2 only 2 of the 6 would not accept the offer).

The responses to the open questions might even have been the most useful for our research, the participants considered the task very difficult. However, apart from group 1, the average score was above 'neutral' towards the positive side of the scale. This indicates that they *did* learn something from the program (as could also be seen with the open questions), although they thought it was too difficult for them.

Taking all the results together, we think we can be cautiously optimistic. The participants of group 3 were positive about their gained understanding of their case, and most of them did answer the objective questions in the way we envisioned beforehand. However, the verdicts are still very hard to read because of the legal jargon.

6 Future Work

In our future work we will concentrate upon stage 1 and 2 of the system as described in section 1. We will collect case descriptions entered by laymen to analyse the terminology they use to describe legal liability cases. We already launched a website, staikinmijnrecht.nl (freely translated: Am I legally right?), and will analyse the input we collect from this site. This will help us in developing a layman ontology. Right now we are beginning to develop the legal ontology, based on the analysis of the legal domain already undertaken, and the search concepts as described in section 3. This legal ontology is used to index case law.

In the end both ontologies are mapped to enable the retrieval of case law based upon a case description given by the laymen in his own wording. Only then we will know how successful the combination of the two parts of the project described in this contribution, viz. retrieval of case law and presenting the results, turns out. This will not be an simple enterprise, but the insights we gained so far makes us feel confident towards the future.

References

1. Bench-Capon, T.J.M., Sergot, M.J.: Towards a rulebased presentation of open texture in law. In: Walter, C. (ed.) Computer power and legal language, pp. 39–61. Qourum Books, New York (1988)
2. Bilgic, M.: Explanation for Recommender Systems: Satisfaction vs. Promotion. Computer Sciences Austin, University of Texas. Undergraduate Honors: 27 (2004)
3. Bing, J.: Designing text retrieval systems for "conceptual searching". In: International Conference on Artificial Intelligence and Law, Boston (1987)
4. O'Donovan, J., Smyth, B.: Trust in recommender systems. In: International Conference on Intelligent User Interfaces, ACM Press, San Diego (2005)

5. Fabri, M., Contini, F. (eds.): Justice and technology in Europe: How ICT is changing the judicial business. Kluwer Law International, The Hague (2001)
6. Fluit, C., van Harmelen, F., Sabou, M.: Ontology-based Information Visualization: Towards Semantic Web Applications. In: Visualising the Semantic Web, 2nd edn. Springer, Heidelberg (2005)
7. Klein, M.C.A., van Steeenbergen, W., Uijttenbroek, E.M., Lodder, A.R., van Harmelen, F.: Thesaurus-based retrieval of case-law. In: Proceedings JURIX 2006, pp. 61–70 (2006)
8. Lodder, A.R.: Law, Logic, Rhetoric: a Procedural Model of Legal Argumentation. In: Rahman, S., Symons, J. (eds.) Logic, Epistemology, and the Unity of Science, ch.26. Logic, Epistemology, and the Unity of Science Series, vol. 1. Kluwer Academic Publishers, Dordrecht (2004)
9. Lodder, A.R., Oskamp, A., Schmidt, A.H.J. (eds.): IT support of the Judiciary in Europe (ITeR deel 43), Den Haag: SDU 2001 (2001)
10. McSherry, D.: Explanation in Recommender Systems. Artificial Intelligence Review 24(2), 179–197 (2005)
11. van Mulligen, E.M., van der Eijk, C., Kors, J.A., Schijvenaars, B.J., Mons, B.: Research for research: tools for knowledge discovery and visualization. In: Proceedings of the AMIA Symposium, pp. 835–839 (2002)
12. Oskamp, A., Lodder, A.R., Apistola, M. (eds.): IT support of the judiciary in Australia, Singapore, Venezuela, Norway, The Netherlands and Italy. IT & Law series no. 4. Cambridge University Press, TMC Asser Press (2004)
13. Salton, G.: Automatic text processing: the transformation, analysis, and retrieval of information bycomputer. Addison-Wesley Longman Publishing Co., Inc., Boston (1989)
14. Salton, G., Buckley, C.: Term weighting approaches in automatic text retrieval. Technical report, Ithaca, NY, USA (1987)
15. Stranieri, A., Zeleznikow, J.: Knowledge Discovery from Legal Databases, Law and Philosophy Library, vol. 69. Springer, Heidelberg (2005)
16. Stuckenschmidt, H., van Harmelen, F., de Waard, A., Scerri, T., Bhogal, R., van Buel, J., Crowlesmith, I., Fluit, C., Kampman, A., Broekstra, J., van Mulligen, E.: Exploring Large Document Repositories with RDF Technology: The DOPE Project. IEEE Intelligent Expert 19(3), 34–40
17. Uijttenbroek, E.M., Klein, M.C.A., Lodder, A.R., van Harmelen, F., Huygen, P.: Semantic Case Law Retrieval – Findings and Challenges. In: Proceedings SW4Law workshop 2007 (2007)
18. Wildeboer, G.R., Klein, M.C.A., Uijttenbroek, E.M.: Explaining the Relevance of Court Decisions to Laymen. In: Lodder, A.R., Mommers, L. (eds.) Proceedings of JURIX 2007, Amsterdam, Berlin, etc, pp. 129–138. IOS Press, Amsterdam (2007)

ICT-Supported Dispute Resolution

Claudia Cevenini and Gianluigi Fioriglio[*]

CIRSFID – University of Bologna,
Via Galliera 3, Bologna, Italy
{claudia.cevenini,gianluigi.fioriglio}@unibo.it

Abstract. This paper aims at describing how the use of Information and Communication Technologies can positively contribute to the resolution of disputes. Once a conflict arises, the parties have on the one side the possibility to resort to Courts (judicial dispute resolution); on the other side, they can agree to submit the issue to an arbitrator or mediator (alternative dispute resolution). While in judicial dispute resolution and partly in arbitration the introduction of ICT necessarily has to comply with the rules of procedural law, mediation allows for a higher freedom and possibly for entirely on-line procedures. Both cases are examined below.

Keywords: ICT, information and communication technologies, judicial dispute resolution, alternative dispute resolution, arbitration, mediation, on-line dispute resolution, ADR, ODR, ALIS project.

1 Introduction

In today's Information Society the resolution of disputes is a crucial problem, as the laws get more and more complex, while the possibilities of illegal behaviour and the negative effects thereof are multiplied by the new technologies. On the other side, the same technologies can effectively support litigating parties, lawyers, judges, arbitrators and mediators.

This paper will examine the current state-of-the-art of the use of ICT in dispute resolution, both in judicial and alternative proceedings and will introduce how research is exploring new perspectives of evolution.

2 ICT Support for Judicial Dispute Resolution

Today the courts have to deal with a growing number of cases: this trend seems common to many countries. For instance, in 2006 the Italian Supreme Court of Cassation ("Suprema Corte di Cassazione") had 100.609 civil trials unsettled; the percentage of unsettled trials is constantly growing (7,34 % in 2006) and the average duration of a trial is 902 days (from initial filing until final judgement), as explicitly stated by the

---

[*] This paper is partly based on research conducted for the EC project ALIS (FP6-027968). Claudia Cevenini wrote par. 1, 4, 5 and 7. Gianluigi Fioriglio wrote par. 2, 3 and 6.

P. Casanovas et al. (Eds.): Computable Models of the Law, LNAI 4884, pp. 312–322, 2008.

same Court [3]. It should be underlined that the Supreme Court is the third instance of a trial and it judges on matter of laws, while first and second instance courts judge on matters of fact. The trials decided by these latters are longer since they need to acquire evidences (see [1] for a quick analysis of the Italian judicial system) and their staff is inadequate. However, this situation constitutes a violation of art. 111 of the Italian Constitution, which states that trials should be of reasonable length.

The growing number of cases is maybe due to the risks caused by a society that is getting more and more global and complex; however the different factors are too many and too composite to be analyzed in this contribution (please see [6] for an overview of these problems). Thus, it appears more useful to underline which are the biggest problems that the different judiciary machines have to face: in particular, inefficiency of the courts and excessive length of trials. The use of ICT systems and tools can produce several benefits and partly solve these problems (see also [7]): the use of ICT "is considered one of the key elements to significantly improve the administration of justice" [2] and it can enhance "efficiency, access, timeliness, transparency and accountability" [2].

At the present time, "basic" ICT systems and tools are widely used by all the actors involved in civil and penal trials. For instance, the judges most frequently use computers to write decisions, interact with their staff, consult databases, etc.; court clerks can manage documents, interact with judges and lawyers, etc.; lawyers can write acts, consult databases, communicate and interact with Court offices, etc. Furthermore, 'end users' (the parties of a dispute) could benefit from improved, more transparent and more efficient court trials because proceedings would be faster and possibly cheaper.

It is clear that ICT tools and techniques can automate some of the Courts' activities, but the current perspective is to have an evolution of the system and not a revolution. In the near future, more sophisticated tools could be developed to support judges in making decisions (advanced expert systems, such as the one developed in the ALIS project that will be mentioned hereinafter). Maybe, someday in the future it could be possible to have an automatic judge: this may be the next step forward for Artificial Intelligence applied to Law. Many people think human affairs are so complex that a computer system will never be able to deal with them and substitute actual judges, however the impossibility to create such systems today does not imply that it will not possible in a future time (in fact, actual limitations can be surpassed in the future). However, is it desirable to have an automated judge? Maybe yes, maybe not, but for sure it could avoid discrepancies in judgements – and this is certainly desirable.

3   The Italian On-Line Civil Trial

The Italian On-Line Civil Trial has been instituted by the Decree of the President of the Republic (DPR) n. 123/2001 (see, among others, [8] and [18]) and involves seven courts throughout the Italian territory. It is a significant e-government project that aims at automating the information and documentation flows between the different actors of civil trials. It is not a 'new' type of trial; instead, it is a support to the current civil trial: in other words, it constitutes an improved way of communication between judges, lawyers and courts' offices.

The Italian On-Line Civil Trial is also regulated by the technical rules set in the Decree of the Minister of Justice (D.M 14.10.2004) and by another 'technical' Decree of the Minister of Justice (D.M. 15.12.2005) that specifies how Document Type Definitions (DTD) must be structured to be legally valid (these regulations can be downloaded from [5]; see also [15]).

In general terms, each lawyer can write, sign and deposit a legal deed (receiving the proof of transmission) without physically going to the courts; furthermore, he or she can receive communications from court clerks and have an on-line access to the filed documents related to his or her own legal cases. He or she can also ask and receive copies of the filed documents and legal deeds. The payments related to the procedure can be made on-line or off-line, but in the latter case the paper receipt must be shown in the next court hearing.

Judges can manage and plan tasks, activities and documents related to the proceedings assigned to them; they can also create, digitally sign and transmit decisions, building a local database of case-law. Court clerks can benefit from the automatic insertion and upgrading of proceedings and from automatic notifications to lawyers and other subjects (e.g. expert witnesses).

Administrative offices will have a longer operational time and so both external and internal users will have a benefit. It should be underlined that these offices are usually open only at the same time in which courts' hearings are held and thus lawyers can have serious time constraints, while a longer operational time will solve this problem. Furthermore, administrative staff and court clerks will benefit from a better distribution of the work to be carried out.

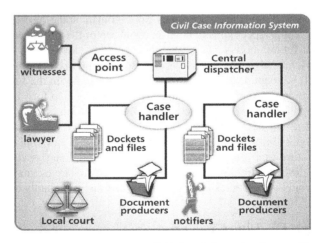

Fig. 1. Graphical representation of the Civil Case Information System (Source: [5])

Given these premises, we can take a quick look at the way the system works. A registered lawyer can write and sign a deed with a specific software. The Italian Ministry of Justice has realized a demo prototype to test the system (see [5], download section). This demo has no legal value, however it may be useful to test the features of the system. The first version was written as a plug-in for Microsoft Word, but now it is a Java application and so it can be used virtually on every system which supports Java.

The deed is stored in Acrobat Portable Document Format (PDF) and may include attachments in the foreseen formats (PDF, RTF, TXT, JPG, GIF, TIFF, XML files and/or ZIP, ARJ and RAR files that contain files in the above mentioned formats; it must include the scanned proxy statement that must be digitally signed). The lawyer must sign the deed with an electronic signature. He or she must have a smart card and his or her signature must be certified in compliance with the Italian rules and regulations on electronic signatures. A XML file that contains the signature is then attached to the document. At the time of writing, the size of the whole message is limited to 10 Megabytes. Another XML file contains the information needed by the system in order to execute the operations related to court clerks' activities.

Apart from lawyers, other external users, such as Court's experts, may be authorized. Each user must register and open a particular type of Certified Mail account ("casella di Posta Elettronica Certificata", PEC: it is regulated by the Legislative Decree n. 82/2005 "Codex of the Digital Administration") specific for the On-Line Civil Trial: it is called Certified Mail Account of the On-Line Civil Trial ("casella di Posta Elettronica Certificata per il Processo Civile Telematico", PECPT) and can receive messages only from other points of access and from the Central dispatcher.

External users (including lawyers) have no direct Internet access to the courts' archives. They connect to an Access Point that sends the data to the Central Dispatcher. Each Access Point authenticates the users (through the smart card and an electronic certificate) and then communicates with the Central Dispatcher via a secure channel.

The Central Dispatcher analyzes the received data and sends them to the appropriate Case Handler.

Each Case Handler processes received documents and sends acknowledgements and outbound notices to the Central Dispatcher. The receipts are sent to the Point of Access that will send them to the specific Certified Mail account of the registered user. Furthermore, the Case Handler maintains the status of the proceedings and their related documents, handling all the related workflow.

Internal users can have access to the information related to their activity through a specific network ("Rete Unitaria della Giustizia", RUG) or the point of access set by the Ministry of Justice. Judges, court clerks and administrative staff are internal users.

At the time of writing, the On-Line Civil Trial is operational only in the Tribunal of Milan and only for the deposit of petitions for injunction decrees, because this procedure goes through typical deeds and the related information flow is rather simple [4].

Traditionally, justice is not as fast as technology and the legal changes and evolutions are usually slower than the technological ones. The On-Line Civil Trial is emblematic of the difficulties that emerge in the realization of a convergence between these two worlds.

On the one hand, the real results are far from the foreseeable ones and in six years we still do not have a *real* and full-scale implementation of the On-Line Civil Trial. Another criticism is due to the choice of Acrobat PDF as the adopted standard. It is actually a proprietary format and it should be better to use or to develop a open format in order to make all the software developers and users free from every commercial obligation with a private company.

On the other hand, it should be pointed out that, as it has been said, the first results of the implementation in the Tribunal of Milan are very good and maybe in the near future the On-Line Civil Trial will be adopted throughout all the Italian courts. The hope is that the waiting time will not be too long.

4 The Use of Information and Communication Technologies in Arbitration

Instead of resorting to Courts, the parties of a dispute can agree to opt for a private judgment, such as arbitration or mediation. While the former shares several traits with Court procedures and is regulated by the law, the latter is a more informal procedure, which aims at assisting the parties in mutually agreeing on a satisfactory solution. In both cases, ICT can play a strategic role in terms of cost reduction, higher efficiency and speed.

As concerns arbitration[2], at present it is still impossible to carry out it entirely on-line, owing to legal constraints, and unlike with electronic court proceedings no specific discipline has been drafted. Despite this, some steps of the arbitration procedure can be dematerialised, in the light of the recognition of the legal validity of electronic documents[3].

The starting point is the arbitration clause, which has to be made in writing or otherwise it shall be deemed null and void. This requirement is deemed respected also in case the will of the parties is expressed with "telegraph, news ticker, facsimile or telematic message" in compliance with the rules on transmission and reception of electronically transmitted documents. The clause may be part of an electronic contract. In this latter case, should contract conditions be drafted and imposed by one of the parties, a double signature would be required as the clause would be deemed vexatious.

The communication by a party concerning the appointment of its arbitrator has to be notified in writing to the other party. This step can be substituted with a notification by certified electronic mail, which guarantees origin, non modification and time-stamping of the messages and is recognised as legally equivalent.

Also the acceptance by the arbitrators, which has to be made in writing, can be substituted with an electronic procedure and be performed with a digitally signed electronic document[4]. The digital signature can directly be affixed to the electronic version of the arbitration clause.

During the proceedings, the arbitrators can hear witnesses, either directly in person or by receiving written answers to questions within a given term. As happens with other requirements in writing, these papers can be substituted by digitally signed electronic documents. The introduction of ICT tools such as chat lines, forums or, more

---

[2] The Italian legal system will here be taken as a reference. The rules on arbitration are foreseen in articles 806 to 840 of the civil procedure code, as last modified by legislative decree n. 40 of 2 February 2006.

[3] An interesting analysis of the admissibility of on-line arbitration within the current legal framework is provided in [16].

[4] According to the Italian law, the requirement of the written form is respected by an electronic document with digital signature. On the contrary, the legal validity and relevance of an unsigned electronic document is addressed by the judge on a case-by-case basis.

simply, allowing communication through videoconferencing would positively contribute to this step. A witness finding itself in a distant location could be heard by the arbitrator and information could be exchanged in a bi-directional way, thus enabling clarifications that through the simple production of documents would not be possible. No provision, however, allows the on-line meeting between arbitrators and witnesses.

The arbitration procedure has as its outcome the issuing of a decision by the arbitrators. This is deliberated by majority and put in writing. It can be thus deemed that this can be substituted by the issuing of an electronic document, digitally signed by the arbitrators. Besides, each arbitrator has the right to ask that the decision, or part of it, is deliberated by the arbitrators in an in-person meeting. No reference is made to the possibility of a virtual meeting, by way, for example, of audio-video conferencing tools. The absence of such a provision, as in the case of witness hearings, appears quite anachronistic, also considering the level of sophistication presently achieved by communication systems.

The decision is drafted in one or more originals and signed at least by the majority of the arbitrators. An original - or a true copy thereof - is sent to each party. This passage, as the other ones which foresee a document drafting and transmission, can be accomplished with electronic means. In this case, besides, the law imposes the requirement of the date of each signature: each electronic document should therefore be affixed with a legally valid time-stamping.

If one of the parties intends to have the decision enforced at national level, it shall deposit the decision at the clerk's office of the Court where the seat of the arbitration is located. If the decision is formally regular, the Court declares its enforcement by decree. The virtualisation of this step is closely linked with the implementation level of the on-line civil trial. The same can be said of the challenge for nullity, the revocation of the decision and the third party opposition before the Court of Appeal, as well as of the recognition and enforcement of, and opposition against, foreign arbitration decisions.

In general, automating certain passages of the arbitration procedure should be performed paying maximum attention to the compliance with the rules and regulations which set the equivalence between paper documents and electronic documents. Should this equivalence not be ensured, there would arise the risk of nullity of the arbitration decision for lack of certain formal requirements.

It can be observed that at present the rules on arbitration allow for a rather limited support by information and communication technologies, as the dematerialisation mainly affects the passage from paper to electronic document exchange. Considering the state-of-the-art of technology, much more could be done to improve the quality of this procedure, taking advantage of further tools, as has been possible up until now only for less formal alternative procedures, as will be seen later.

5 On-Line Mediation

The development of a mediation actually poses less problems in terms of potentialities of the use of ICT compared with arbitration as, except for particular cases, it is not strictly regulated by the law.

Starting from traditional mediation, performed with the support of a human mediator, it should be said that its total virtualisation may not always be necessarily seen as an improvement. This is linked with the fact that, unlike the judge and the arbitrator, the mediator is usually called on not to make a decision in lieu of the parties but to support them in finding common points and reach an agreed decision by themselves. This means that the psychological element is as important as the knowledge and application of the rules. In this context, two different views can be expressed. On the one side, the inter-personal communication by way of electronic tools which do not enable the viewing and hearing of the other parties and the mediator (e.g. e-mail, chat lines, forums, etc.) has the advantage of letting the parties concentrate more on the real object of the dispute. In this case, the personal implications, such as the anger or desire of prevarication, etc. pass to a lesser level. This would constitute a positive element for the fast and effective resolution of a dispute; however, on the other side, it should be remembered that the persons having limited ICT skills or those who express themselves better in person may be disadvantaged.

At present mediation is still often performed by in-person meetings, where the parties directly explain the issue to be solved and, possibly, produce paper documents. Over the last ten years, however, several attempts – especially in the United States, first at academic then at commercial level - have been made to pass to on-line procedures, either assisted by a human mediator ("open" model) or totally automatic (the so-said "blind" model).

A significant example of the open model is Squaretrade[5], one of the most used on-line dispute resolution (ODR) systems at international level; it is used by eBay and e-commerce services like Verisign and PayPal [11]. The owner of a website exposing the Squaretrade sign in its web pages shows it is willing to resort to its services in case a dispute arises. At first, the system aims at letting the parties solve the case by themselves. Should the outcome be negative, a mediator intervenes, to assist them in finding an agreement. Should this again not lead to a solution, the mediator then asks whether the parties would like it to suggest them a solution or a settlement. It is up to the parties, in any case, to accept this or not. The procedure is performed with the support of an advanced messaging functionality, which makes it possible for both the parties and the mediator to be constantly updated.

Another example, operating at national level, is Risolvionline[6], a system to solve e-commerce disputes designed and managed by the Milan arbitration chamber.

A party wishing to promote a mediation attempt, can fill in an on-line form (Figure 2) and send it electronically. The scheme is very simple: the information required only includes the party's contacts and a brief description of the issue; if useful, files can also be attached.

Risolvionline then invites the counterparty by e-mail to take part in the mediation. If this latter agrees to participate, it only needs to fill in a form available through a link in the same e-mail. At this point, a mediator is assigned to the case and date and time of the mediation meeting are fixed. The meeting is performed through chat lines or, upon request of the parties, by e-mail. The chat line is accessible with user id and

---

[5] Squaretrade is accessible at http://www.squaretrade.com (last access: 26.02.2008).

[6] Risolvionline is accessible at http://www.risolvionline.com (last access: 26.02.2008). For a more detailed description of the system, see [13].

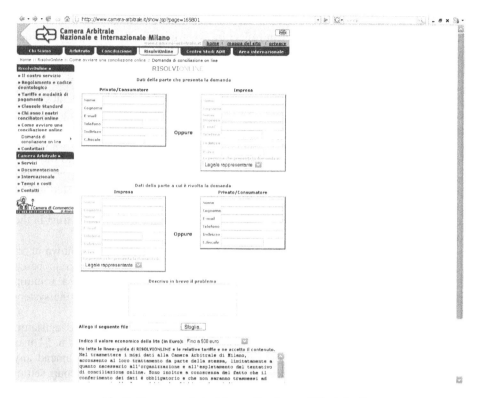

Fig. 2. The screenshot of the on-line form of Risolvionline

password and the procedure is governed by the mediator in a similar way as in a face-to-face meeting. If the outcome is positive and the parties manage to reach a decision, the mediator drafts a transcript, which shall be printed and signed by each party in double copy. Risolvionline will send each signed copy to the other party so that each one of them will have the document signed by the other.

This last step of the procedure could be more easily accomplished with the aid of digital signatures, and it may appear strange that a totally on-line ICT-assisted proce-dure necessarily has to end with a traditional exchange of paper documents. However, this decision is probably imputed to the still limited diffusion of digital signatures.

As concerns the blind model, an interesting example can be provided by Cyberset-tle[7]. Unlike open systems, it makes it possible to settle a case without the intervention of a human mediator, by using a completely automatic system which matches the parties' offers and demands.

The parties have access to the system through a secure user login and then are asked to select the type of claim (e.g. property damage, medical malpractice, etc.). Before offers can be made, the system only asks for basic information about the case. Users then submit three rounds of offers and can assign a limited time to respond. The offers by a party are not disclosed to the other party. The absence of contact between

---

[7] Cybersettle is accessible at http://www.cybersettle.com (last access: 26.02.2008).

the parties helps them concentrate more on the issue to be solved, and the limitation to three offers is deemed to make them reflect on what is really the sum they consider reasonable. If the parties do not reach an agreement on-line, they can solve the dispute through traditional negotiation and with the assistance of telephone facilitators.

This system is mainly used for disputes where the parties need to reach an agreement on a monetary amount, as for example in insurance claims. Using an automatic system makes it possible to reach a faster and cost-effective decision, and also lets the professionals who assist the parties concentrate on more complex cases.

Mediation systems, like the ones illustrated before, have the advantage, especially as regards disputes arising from e-commerce and on-line activities, to avoid the total disruption of the relationship between the parties. Besides, they can also generate trust in perspective users, who would buy from a web site as they know they would not be forced to go to Court in case of a dispute, a solution which appears impracticable in case of petty controversies. In most cases, consumers admit they still have trust in the vendor and would buy from it again in the future [10].

A more advanced perspective in constructing an ODR environment is shown in recent studies [14], which combine dialogical reasoning with game-theoretic based negotiation techniques: the environment facilitates the parties towards a solution through a dialogue support tool; should they not yet reach an agreement, the system proposes them a possible resolution.

The relevance of ICT tools in mediation has also been recognised by the legislator, as it is witnessed, for example, by the E-commerce directive[8]. In whereas n. 51 it is affirmed that Member States "should be required, where necessary, to amend any legislation which is liable to hamper the use of schemes for the out-of-court settlement of disputes through electronic channels; the result of this amendment must be to make the functioning of such schemes genuinely and effectively possible in law and in practice, even across borders".

6   Intelligent Technology in Dispute Resolution: The ALIS Project

ALIS (Automated Legal Intelligent System) is a European funded project[9]. The system, currently in development, will be able to analyze the parties' requests and tell which ones are compliant and which ones are not, before looking for an agreement that is fair to all parties.

Intellectual Property is the chosen domain, since it constitutes a perfect test bed for the system, owing to its relevance in the Information Society and its level of complexity. Once implemented, the ALIS system could be applied to potentially any other field of the law.

The project aims at developing a powerful system more effective and sophisticated than 'traditional' ODR systems thanks to its strong scientific basis. In fact, it will be based on the theoretical outcome from three scientific fields: game theory, computational logic

---

[8] Directive 2000/31/EC of the European Parliament and of the Council of 8 June 2000 on certain legal aspects of information society services, in particular electronic commerce, in the Internal Market (Directive on electronic commerce).

[9] Specific Targeted Research or Innovation Project financed within the VI Framework Programme of the European Commission, Priority 2, Information Society Technologies.

and legal reasoning. ALIS technology will use ontology and semantics web languages with a research approach aiming at combining them with artificial intelligence and game theory. At the time of writing, ALIS ontology is an advanced stage of development. It has been tailored on the specific requirements of the system, but the pre-existent work in this field (such as IPROnto – Intellectual Property Rights Ontology [12]) has been taken into account.

ALIS could be used (among others) by judges to quicken their decisions thanks to the automated legal analysis of the cases they have to decide and by lawyers to speed up their work thanks to the automated legal analysis of their own legal cases. Furthermore, citizens could benefit from the use of ALIS, because they should be able to query the system in order to know how law applies to 'real life' cases.

Owing to its advanced features, the system should be an extremely powerful tool for ADR. In fact, it should automate the reasoning and deciding process that today is adopted by legal experts, combining strictly legal questions with 'real life' problems thanks to the theoretical bases of Game Theory. In this way it should be possible to have an *ex ante* evaluation of the *ex post* consequences of a particular dispute.

The whole judiciary system may also substantially benefit from such system, because it could reduce the number of disputes that should be decided by the courts since the parties would be able to know the possible outcome of a trial without the need to start any legal proceeding.

7 Conclusions

The state-of-the-art in judicial and alternative proceedings shows how only part of the big potential of ICT has been positively implemented up until now. The future perspective as initiated by scientific research, at the same time, appears extremely ambitious, albeit promising. Making the most of ICT in solving cases can, however, derive only from effectively passing from theory to practice, through a close collaboration and the consolidation of synergies between the academia and the end-users (e.g. legal practitioners, judges, citizens, etc.), who in the near future may be called on to make use of newer, more advanced intelligent systems on a day-by-day basis.

References

1. Fabri, M.: Information and Communication Technology for Justice: the Italian Experience. In: Oskamp, A., Lodder, A.R., Apistola, M. (eds.) IT Support of the Judiciary. IT & Law, vol. 4, pp. 111–133. Asser Press, The Hague (2004)
2. Velicogna, M.: Justice Systems and ICT. What can be learned from Europe? Utrecht L. Rev. 3, 129–147 (2007)
3. Corte Suprema di Cassazione – Ufficio statistico: Statistiche anno 2006 (last access: 26.02.2008),
   http://www.cortedicassazione.it/Documenti/AGCivile2007.pdf
4. Pacchioli, P., Pappalardo, F.: Il decreto ingiuntivo telematico con valore legale: l'esperienza del tribunale di Milano. Diritto dell'internet 2, 203–210 (2007)
5. Ministero della Giustizia, Processo Civile Telematico (last access: 26.02.2008),
   http://www.processotelematico.giustizia.it

6. Fioriglio, G.: Temi di informatica giuridica. Aracne, Rome (2004)
7. Fabri, M., Langbroek, P.M. (eds.): The Challenge of Change for Judicial Systems. IOS Press, Amsterdam (2000)
8. Buonomo, G.: Processo telematico e firma digitale. Giuffrè, Milano (2004)
9. Camardi, C. (ed.): Metodi on line di risoluzione delle controversie. Arbitrato telematico e ODR, CEDAM, Milan (2006)
10. Maggipinto, A. (ed.): Sistemi alternativi di risoluzione delle controversie nella Società dell'Informazione, Nyberg Edizioni, Milan (2006)
11. Shah, A.: Using ADR to Resolve Online Disputes. Richmond Journal of Law & Technology 10 (2004)
12. IPROnto – Intellectual Property Rights Ontology (last access: 16.01.2008), http://dmag.upf.es/ontologies/ipronto/
13. Sali, R.: Risolvionline Experience: A New ODR Approach for Consumers and Companies. In: Proceedings of the UNECE Forum on ODR 2003 (2003)
14. Lodder, A., Zeleznikow, J.: Developing an Online Dispute Resolution Environment: Dialogue Tools and Negotiation Systems in a Three Step Model. The Harvard Negotiation Law Review 10, 287–338 (2005)
15. Celentano, F.: L'utilizzo del DTD nel processo civile telematico. Diritto dell'Internet 4, 415–418 (2006)
16. Morek, R.: Online Arbitration: Admissibility within the Current Legal Framework (last access: 26.02.2008), http://www.odr.info/papers.php
17. Katsh, E., Rifkin, J.: Online Dispute Resolution: Resolving Conflicts in Cyberspace. John Wiley & sons inc, Hoboken (2001)
18. Contaldo, A., Gorga, M.: Le regole del processo civile telematico anche alla luce della più recente disciplina del SICI. Diritto dell'Internet 1, 5–23 (2008)

Concepts and Fields of Relational Justice

Pompeu Casanovas[1] and Marta Poblet[2]

[1] UAB Institute of Law and Technology (IDT), Faculty of Law, Autonomous University of Barcelona, Bellaterra, Barcelona-08193, Spain
http://idt.uab.cat
[2] ICREA Researcher at the UAB Institute of Law and Technology
{pompeu.casanovas,marta.poblet}@uab.es

Abstract. This paper intends to introduce and explore the broad conceptual background of relational justice according to the current state of the art. *Relational Justice* (RJ) is defined as the justice produced through cooperative behavior, agreement, negotiation, or dialogue among actors in a post-conflict situation. We found concepts stemming from at least thirty different fields, going from behavioral sciences (neurology, brain sciences, primatology, social psychology, etc.) to criminology, jurisprudence, and philosophy. One of these contributing fields is Artificial Intelligence (AI), which uses several techniques to grasp the practical knowledge of negotiators and mediators and builds tools to support both negotiation and mediation processes. However, contrary to the legal ontologies field, there are no developed ontologies of Relational Justice yet representing the conceptual richness of the domain.

Keywords: legal concepts, legal ontologies, legal systems, dialogue, relational justice, restorative justice, ADR, ODR.

1 Introduction

Our focus is *Relational Justice* (RJ), which we define broadly as a bottom-up justice, or the justice produced through cooperative behavior, agreement, negotiation or dialogue among actors in a post-conflict situation (the aftermath of private or public, tacit or explicit, peaceful or violent conflicts). The RJ field includes Alternative Dispute Resolution (ADR) and Online Dispute resolution (ODR), mediation, Victim-Offender Mediation (VOM), restorative justice (dialogue justice in criminal issues, for juvenile or adults), transitional justice (negotiated justice in the aftermath of violent conflicts in fragile, collapsed or failed states), community justice, family conferencing, and peace processes.[1]

---

[1] Only in the field of restorative justice we may distinguish different separate processes and situations according to prevailing legal cultures and legal systems: community mediation programmes, victim offender reconciliation programs, victim offender mediation (VOM), conferencing, youth justice, family groups conferences in New Zealand, conferencing in Wagga Wagga (Australia), community groups, conferencing circles, Navajo justice, sentencing circles, healing circles [1]. In Europe, to consider one example, juvenile justice differs considerably as regards processes, procedures, environments, and relation with courts [2]. Differences among mediation forms, institutionalization policies and legislations are even broader [3].

P. Casanovas et al. (Eds.): Computable Models of the Law, LNAI 4884, pp. 323–339, 2008.
© Springer-Verlag Berlin Heidelberg 2008

The aim of this paper is to show the conceptual complexity of this kind of justice, which is not solely based on the application of fundamental legal concepts—norms, rules, normative systems, rights, duties, etc.—but on both *behavioral* concepts from different theoretical fields and the singular, non-homogeneous *experiences* and *practices* of negotiators, facilitators, and mediators. Focus, processes and goals are therefore combined in a *continuum* of approaches [4].

Before any attempt to represent knowledge in a computational system or in a platform of ODR services we first need to consider the epistemological problems of knowledge acquisition. How to represent the different aspects and dimensions of experiences and practices of RJ as knowledge? How to elaborate ontologies capturing RJ knowledge? One way to proceed is to have a look on all the theoretical, scientific, and practical fields involved in the generation of relevant concepts.[2] This task should be distinguished from *ontology mediation* (mapping, aligning and merging) [6], *knowledge engineer mediation* (among conflicting domain ontologies [7], or through wiki tools [8] [9]), and MAS *ontology negotiation* (among intelligent agents) [10].

It is worth mentioning that we are not identifying either the domains in which negotiation, mediation and ADR techniques may apply (i.e. family, real estate, environment, commerce, armed conflicts, etc.). We are focusing instead on concepts such as *empathy*, *reciprocity*, or *remorse*, which contribute to set up the structural frameworks to understand, explain and develop mediation and negotiation processes. We therefore propose a general overview of the theoretical and practical concepts that, emerging from both academic and professional fields, constitute conceptual kernels in the area of RJ.

2   Concepts and Fields

We found at least thirty academic fields focusing on conflict resolution and justice. We used four criteria of identification: (i) authoring (quotations and cross-discussions and fertilization in a stable community), (ii) focus (agreement on common problems, discussion on research approaches) (iii) object (agreement on definitions, common language, conflicting theories), and (iv) methodology (comparable data, experiments or outcomes).

This meta-analysis is not entirely satisfactory and results are not homogeneous, because there are no discrete criteria to satisfy a discriminatory function, either for individuals or for collectivities. Consider, for instance, a psychologist who is both a practitioner and an academic philosopher. Similarly, we may define AI & Law as a single academic field, or have it included into the broader field of Applied Artificial Intelligence. Choices do not go without theoretical discussions. (In the case of AI & Law, we preferred the second option because there are many authors focusing on mediation and AI who belong to different communities).

Moreover, for us [4], shifting from restorative justice to relational justice also means to adjust our lens to a wider scope, since new theoretical fields come into play.

---

[2] In this sense, this is a complementary paper to the micro-foundations for Restorative Justice that we set up in [4]. We realized that we could expand our arguments to a broader notion of justice.

For instance, economy and game theory (allocation of rights) play a more fundamental role in conflict resolution and management research than in VOM studies. In addition, recent developments in neuroeconomics have shown for the first time the neural foundation of social preferences, trust and social punishment [11]. As the NBIC [Nano-Bio-Info-Cogno] convergence shows [12], there is a growing interaction and synergy through scientific and technological fields.

However loose this taxonomic exercise may be, it draws the present complexity of thinking of a bottom-up justice. Micro-foundations of social behavior have already been incorporated to model agents' behavior in multi-agent systems (MAS) developments [13]. To understand social phenomena at the macro-level dimension (i.e. the functional violation of social norms or the emergence of collective properties) electronic or human agents must be conceived both with intentions, plans and goals *and* with the capacity to be affected by their own cognitive representations. In other words, they must incorporate an *emotional* dimension. At the micro-level, then, *rationality* and *emotion* cannot be conceived as opposed, but as intertwined. Fig. 1 below shows a general framework for the micro-foundations of RJ.

2.1 From Empirical to Philosophical Approaches

To organize the different conceptualizations, we split up micro-foundations of RJ into four macro-domains: (i) empirical research on mind, language, forgiveness, empathy, and emotions; (ii) social research on culture, language, apologies, and micro-situations; (iii) economic, social, political and philosophical research on conflict and dialogue; and (iv) social, political, jurisprudential and philosophical research on rights and legal systems.

In this way, we start from the most empirical and fundamental research on social neuroscience (including recent trends in neuroeconomics), cognitive science, primatology, and basic social psychology (see box 1 in Fig. 1) and we draw a large intellectual bow up to the more common and general legal language of jurisprudence and ethics—rights, duties, rules, principles and norms—(box 8 in Fig. 1). The last kind of reflections may be more or less empirically grounded, may have a more or less practical or fundamental orientation, may choose a more or less literary or artificial language, may have different degrees of consistency, coherence and soundness, but they do not intend to be evaluated through the methods of normal science.

In between, we have all the specific research on conflict, dialogue, negotiation, and mediation emerged from human and social sciences (linguistics, anthropology, sociology, psychology, political science, economics), philosophy (logic, epistemology, argumentation), and technology (computation and artificial intelligence, including MAS and virtual or electronic institutions).

2.2 Natural Conflict Resolution, Aggression and Conciliation Patterns

Natural Conflict Resolution [14] is the title of a well-known handbook for primatologists. The main idea is to substitute a conciliatory or cooperative pattern to the aggressive one that pioneers like Konrad Lorenz set up for natural life. From this point of

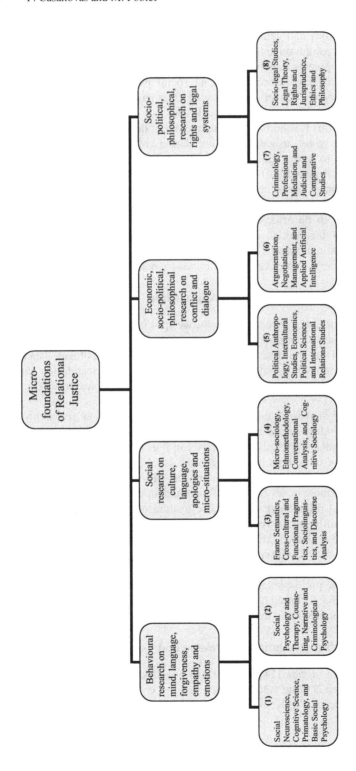

Fig. 1. A general framework for micro-foundations of Relational Justice

view, conciliatory behavior is as 'natural' as violence. Moreover, from an evolutionary point of view, a necessary condition for survival relies on the acquisition and management of knowledge on how to handle conflicts that could diminish the capabilities of the group. "Aggression as an antisocial instinct is being replaced by a framework that considers it a tool of competition and negotiation." [15]

Empathy [Einfühlung], isopraxis (produced by *mirror neurons*), *enaction, embodied cognition, consolation* and *reconciliation* are some of the concepts used within the framework of a *relational model* of aggression. Reconciliation is considered a heuristic concept, capable of generating testable predictions about stable relationships. Protection of cooperative bonds is crucial in non-human primates with social experience and triadic relations. Humans experience the same physiological changes participating in conversation and watching it later in a video. *Mind reading*, making attributions about the mental states (desires, beliefs, intentions) of others, may be conceptualized as a different cognitive process than empathy. *Empathy* means sharing feelings and emotions in absence of any direct stimulation to themselves. Aureli [in 16] considers it a kind of *intervening variable*, an epistemic construct used to explain complex webs of variables.

There is a strong debate on the ultimate bases of empathy and the theory of mind lying behind it [16].[3] Yet, neurological bases for shared pain between loved couples, e.g., have been detected by functional Magnetic Imaging Ressonance (fMIR) experiments[4], and social neuroeconomics takes advantage from it interpreting brain activations involved in *altruistic, fair* and *trusting behavior*. The *self-interest* hypothesis assumed by classical behavioral sciences is being replaced instead by the idea of *strong reciprocity* in cooperative behavior [19].

Empathy plays a fundamental role in empirical psychological studies on *forgiveness* (and *unforgiveness*) as well. *Pre-offence closeness, apology, sincerity, memory, rumination, anger, shame, avoidance, revenge, current closeness,* are some of the variables taken into account in experimental models. There are several models in the literature relating to individual, family and social behavior.[5] But all of them tend to emphasize the relational nature of variables and the importance of emotions in concepts such as *innocence, guilt* and *remorse*. Social meaning and concepts contribute to trigger feelings and emotions. However, there is no agreement yet on the composition of basic or primary emotions (*fear, joy, disgust, rage* and *surprise*) [22].

2.3 The Role of Culture and Language in Interaction Patterns

Micro-situations have been mainly analyzed by linguists and sociologists. Frame-semantics, cognitive linguistics, cross-cultural pragmatics, functional pragmatics,

---

[3] The *Perception-Action Model* (PAM), by de Waal and others, is grounded in the idea that perception and action share a common code of representation in the brain [16]. The *Somatic Markers Hypothesis* (SMH) by Damasio and others contends that bio-regulatory signals, including those that constitute feelings and emotion, provide the principal guide for decisions [17].

[4] Tania Singer experiments on wives observing their husband's pain show that there are strong anatomical connections between regions constituting the pain matrix, and this leads to the suggestion that these regions are highly interactive [18].

[5] Forgiveness is a well-trodden path in social and family psychology. There are relational models based in prototypes, narratives, interactions, flows and regression analysis [5] [20] [21].

sociolinguistics, corpus-based, and discourse analysis have contributed to have a better knowledge of the elements, structure, processes, and functions of linguistic interactions.

Some of the notions involved share a common tradition in linguistics and philosophy of language: i.e. *locutionary and illocutionary acts, speech acts, events, context, competence, indexing, reference, co-reference*. Others have been developed in parallel with cognitive science and AI: the notions of *script, schema, slot, prototype, frame, framing, reframing, mental space, semantic field, semantic space, mental model* [23] e.g. And, still, other concepts have been used along with new logical trends in philosophy: *inference, inferencing, entailment, presupposition, natural and non-natural meaning, conventional implicature, conversational implicature* [24]. Finally, a few of them have been developed through the empirical analysis of linguistic interactions or reflection on the phenomenology of speech: *sociolects, idiolects, contextual cues, diglossia, deixis, turn-taking, adjancy-pairs, switching codes, sociolinguistic competence, face-threatening acts [FTAs]* [25] [26].

More specifically, stemming from this tradition, cross-cultural pragmatic research has focused on the linguistic content and expression of *politeness, apologies* and *excuses* in different natural languages and cultures [27]. There are different existing frameworks to analyze them. Researchers have used three main paradigms to situate their analysis: (i) the 'maxims model' (Leech, Lakoff), (ii) the 'conversation contract model' (Fraser and Nolen) (iii) and the 'relevance theory model' (Sperber and Wilson) [5].

Ethnometodology, cognitive sociology, conversation analysis and micro-sociology have tried to grasp the way in which language, expression, and thought are combined in a *situated meaning* and in a *situated, shared, tacit and socially distributed knowledge*.[6] Some of their originally ideas, figured out in the reaction against functionalism in the fifties and sixties, have been useful to develop later more precise cognitive and computer science applications, e.g. the *Parallel Distributed Processing* model [28] or the *Situated Cognition* model [29].

It is worth saying too that pragmatic analyses sometimes offer non conclusive results. The notions of *gender language* and *gender speech*, for instance, remain controversial.[7] Nevertheless, pragmatic approaches show a good understanding of speakers, concrete issues at stake, and situations they describe.

2.4 Context, Negotiation and Dialogue Processes in Conflict, Violence and Reconciliation Patterns

Anthropologists and political scientists have stressed the importance of culture and language, especially when violence is involved, in markets, communities, societies, states and political organizations. Differences between *binary* (*negotiation*) and

---

[6] E. Gofmann, H.Garfinkel, A. Cicourel, R, T.Scheff and S. Retzinger are some of the names contributing to the qualitative analysis tradition in conflict and negotiation. In spite of the differences among them, they all share a detailed micro-analysis approach.

[7] Focusing on the apologies in British English, Deutschmann carried out a corpus-based analysis on about 3.000 excuses contained in the BCNweb. He could not find any significant differences between men and women style of apologizing [30].

triadic (*mediation, arbitration, adjudication*) models of conflict resolution have been discussed in the literature since the sixties, following the debate between functional and cultural anthropology within the American and European traditions [31].

Contemporary post-war situations in the late 20th c., in which *mobs, mafias* and *private armies* operate at a sub-state level, require new concepts to describe and explain them. Negotiation and peace processes in the absence of the state (in *collapsed states* and *failed states*) have fostered new refinements of the functions and types of mediators involved: *explorer, convener, decoupler, unifier, enskiller, envisioner, guarantor, facilitator, legitimazer, enhancer, monitor, enforcer, reconciler* [32].

Those functions are related to context and the *level of escalation of conflict*. Interestingly enough, in *transitional justice* —the complex aftermath of violent conflicts— it is not possible to face social justice in a simple way. Either in Peru, Argentina or the Czech Republic, in distant places with different times and actors, forms of relational justice are combined with a sort of community resilience, the requirement of public recognition of crimes, and punishment.

Taken from this point of view, negotiation and dialogue processes are not only produced through dialogical argumentation forms, but through complex social processes, in which even the most common categories —such as *court* and *trial*— have to be rebuild and implemented within a new political and economic environment. Kimberley Theidon, e.g., has been able to reconstruct recently one of such processes, carried out by peasant communities in the mountains of Peru after the defeat of Sendero Luminoso [33].

This kind of highly descriptive work, rooted in history and analysis of particular cases, has been taken into account, but left behind at the same time, in other fields with a different theoretical background. Problems such as the allocation of rights, the reckoning of the best strategic move, or the impact of conflict into the markets, lead to reduce the human and political complexity of possible scenarios to set forth theoretically manageable problems: reduction of actors to only two players, precise definition of cases, allocation of resources among competitive activities, difference between types of tactics and strategies in negotiation analysis according to hypothetic scenarios [34] [35] [36].

However, again, applied theory (either in international policy analysis, organization studies, management, or business analysis) introduces new levels of complexity to cope with real situations and explain the processes and outcomes of conflicts. Therefore, *distributive* (e.g. reckoning of compensation for a loss) and *procedural justice* (e.g. negotiation rules) are usually complemented with the so-called *interactive justice* (e.g. personal attitudes, emotional impact, communicative skills) [37], [38]. Some recent economic trends on intuition are following the old *motto* advanced by Herbert Simon: "Our task, you might say, is to discover the reason that underlies unreason" [39] [40].

2.5 Theory and Practice of Mediation in Law and Legal Systems

Studies on mediation identify four different types to perform mediation: (i) *facilitative* (neutrality of mediator), (ii) *evaluative* (assistance and help offered to the parties to structure their position), (iii) *transformative* (mediator helps the parties to transform or change the situation), (iv) *narrative* (storytelling to get a new common version of what happened). Two-party bargaining is, since Raiffa's book, divided into two parts: distributive and integrative [34]. Sometimes a *therapeutic style* is distinguished from the narrative or the transformative ones, and a *settlement-driven* style is generally distinguished from a *dialogue-driven* (or transformative) one.

However, from an empirical point of view, when modeling epistemic situations (in business, markets, organizations or political arenas), a *hybrid* position is usually taken, because understanding situated strategic moves requires combining elements stemming from different classifications too. E.g., elements of power (pressure, sanctions...) combined with justice types, or social combined with procedural justice. Power always matters.[8] To add complexity to the situation, a distinction may be made between regulatory and meta-regulatory strategies (regulation of regulation, regulation of law, regulation of access to justice initiatives) [41].

Mediators and negotiators use a particular professional language, and particular metaphors and folk concepts to handle cases and to refer to their own work. One of these most popular metaphors is *window of opportunity*. But there are more of them, related to situations where mediators intervene: *cold* or *hot* negotiation, *in the shadow of the courts, get the hamster off the treadmill, being under the covers...* [42].

In the eighties and nineties, this language and attitudes of mediators were challenged by legal scholars. Criticisms were thrown especially on the supposed 'neutrality' of the mediator. Maintaining such neutrality would lead to a paradox, because the intervention of non-intervention was viewed as untenable: a hidden agenda [43]. Actually, from this perspective, strategies like BATNA and practical books *Getting to Yes* can be easily seen as lawyers' manipulative intents to not loose control over the situation.[9]

In recent times, proactive attitudes are considered more acceptable as a part of the process: the outcome is viewed as a result of the tension between mediator pressure and party autonomy [46]. Mediators themselves are aware of what they call "micro-level paradoxes"[10], within a cooperative democratic framework (along with courts and the legal system) [47].

3   Discussion and Future Research Trends

It is our contention that the language of Relational Justice is being produced not only through practice of ADR and ODR, but through the theoretical discourse and expanding work of all the scholars and reflective practitioners who try to figure out institutions and legal values from their practices and procedures. In this sense, there are two interesting issues to be faced. The first one is related to institutionalization: how to

---

[8] Aquino et al. put it in this way: "(...) power and justice are intertwined: one cannot really understand justice dynamics without understanding power dynamics and vice versa, because the concern for justice acts check on the use of power" [38].

[9] BATNA: Best Alternative To a Negotiated Agreement [44] [45].

[10] "Mediators sometimes use what are known among psychotherapists as 'paradoxical interventions' to move the process along; that is, suggesting one thing while meaning another. For example, when we talk with a party who is hell-bent on proving her case in court, we might discuss all the advantages of a trial because the disadvantages would simply deepen her resistance to settlement." [44]

map the dialogic concepts, terms and techniques used in ADR into legal concepts and procedures so as to add value to the outcome of the mediation process. The second issue is how to grasp and preserve this type of RJ procedures and practices through the formal languages of the Semantic Web and W3.0. The use of technology clearly influences the use of mediation [48]. We agree with that.

In online mediation, capital letters are the online equivalent of shouting.[11] Users sometimes mimic real face-to-face dialogues. But when the entire process is online, without the mediation of a real person (at least at the first steps), perhaps they don't. What should therefore be taken into account by electronic agents?

We have seen several types of variables —empathy, emotion, culture and professional practice, to summarize them— that cannot be ignored while taking a users-centered approach to ODR.

Perhaps the structure of online communication may alter the transformative side of mediation. But analysts of negotiation processes have noticed the power of *reciprocation*, the strong tendency to match another's person behavior [50]. This tendency acts in human-machine interfaces as well. As far as it has been researched, the Internet is not producing new kinds of emotions, but intensifies the existing ones [51]. The particular position of the machine may facilitate the empowerment of users and the balancing of emotions *within* rational communication (not out from it).

Quite recently, Walton and Lodder have proposed the use of a Rational Rule (RR) to act as a sort of cooperative conversational maxim between opponents.[12] [52].

Especially in difficult interactions, RR could be a helpful device if users decide to adopt an additional control over their own dialogue. The enactment of such a rule could be shared by both parties as well, and in this case we would not see negotiation and argumentation paradigms as mutually exclusive, but mutually inclusive. Argumentation devices and schemas could be modeled precisely to reach "coherent dialogues across incommensurable worlds".[13]

A pluralist approach to ODR implies respecting cultural constraints that users may have in their understanding of what are they doing through dialogue[14], *and* giving them the opportunity to gain control over their own moves.

The idea of *collaborative design* is interesting too, and not incompatible with adding some rules to the argumentation process. Stemming from normative argumentation pragmatics, Aakhus describes the work of mediators as "communication by design", as they *redirect*, *temporize* and *relativize* the dialogue between disputants [56]. A pragmatic reconstruction of this "disagreement space", could help to build up useful tools for ODR purposes.

---

[11] "I JUST WANT TO BE DONE WITH HER AND NEVER DEAL WITH HER AGAIN! LET'S JUST STOP ALL THE HASSLE AND RETURN MY MONEY! MANY, MANY THANKS!" [49]

[12] "(RR) When a proponent puts forward a valid argument with premises $P=(P_1, P_2...Pn)$ that are all commitments of the respondent and conclusion C, the respondent must, at the next move, either accept C or retract commitment to at least one of the premises P."

[13] Littlejohn and Domenici (2001), quoted in [53]. See also the ODR environment proposed by Lodder and Zeleznikow within a three-step mode [54].

[14] See [55] on the difference between Arabic and Hebrew concepts of what negotiation is. "To negotiate peace, rivals must agree on what is 'to negotiate' and what 'peace' is".

The last issue we would like to address deals with ontology construction. On the one hand, dialogue and mediation have been already linked [57]. Several dialogue typologies have been identified [58]. Walton and Godden have reflected on the ways to model and embed persuasion dialogues into negotiation dialogues [59]. On the other hand, there are some works on ontology already done within the ecommerce field [60], collaborative tasks [61], negotiation [62] and negotiation agents [63]. There are some attempts to apply XML to mediation (the so-called ODR XML).[15] And, of course, this book (LNAI 4884) has shown some interesting work on ODR ontological proposals (OPENKNOWLEDGE, BEST, ALIS).

However, the fundamental concepts of relational justice have not been captured by any ODR core ontology yet. In this paper, we have tried to explore their richness. ODR is an open wide growing and promising field. We think that this is to be understood as a reason to incentivize future trends in this direction, because we believe that ODR ontologies are legal ontologies as well.

Acknowledgments

This work has been developed within the EU COST Action A21 *Restorative Justice Developments in Europe*. We did a great bulk of work during our stay at the CIRSFID and the ITTIG, in Bologna, in the summer of 2006. We warmly thank Giovanni Sartor, Giuseppe di Federico and Anna Mestitz. We thank Arno R. Lodder, John Zeleznikow and Pablo Noriega for their help and useful comments in writing this paper.

References

1. McCold, P.: Primary Restorative Justice Practices. In: Morris, A., Maxwell, G. (eds.) Restorative Justice for Juveniles Conferencing, Mediation and Circles, pp. 41–58. Hart Publishing, Oxford-Portalnd (2001)
2. Mestitz, A., Ghetti, S. (eds.): Victim-offender Mediation with Youth Offenders in Europe. An overview and comparison of 15 countries. Springer, Dordrecht (2005)
3. Singer, J., Makie, K., Hardy, T, Massie, G. (Eds.): The EU Mediation Atlas: Practice and Regulation. CEDR (2004); Aertsen, I., Daems, T., Robert, L. (Eds.): Institutionalizing Restorative Justice, Devon. Willan Publishing (2006)
4. Dignan, J., Marsh, P.: Restorative Justice and Family Group Conferences in England: Current State and Future Prospects. In: Morris, A., Maxwell, G. (eds.) Restorative Justice for Juveniles Conferencing, Mediation and Circles, pp. 85–101. Hart Publishing, Oxford-Portland (2001)
5. Casanovas, P., Poblet, M.: Micro-foundations of restorative justice: a general framework. In: Mackay, R., Bošnjak, D.J., Pelikan, C., Stokkom, B., Wright, M. (eds.) Images of Restorative Justice Theory, p. 258. Verlag für Polizeiwissenschaft, Frankfurt am Main (2007)
6. de Bruijn, J., Ehring, M., Feier, C., Martín-Recuerda, F., Scharffe, F., Weiten, M.: Ontology Mediation, merging, and Aligning. In: Davies, J., Studer, R., Warren, P. (eds.) Semantic Web Technologies. Trends and Research in Ontology-based Systems, pp. 95–113. John Wiley & Sons, Ltd., Chichester (2006)

---

15 We thank Arno Lodder and John Zeleznikow for this information. See http://www.oasis-open.org/committees/download.php/133/OdrXML%20Charter%202002.09.15%201jk.doc

7. Aschoff, F.R., Schmalhofer, F., van Elst, L.: Knowledge Mediation: A Procedure for the Cooperative Construction of Domain Ontologies. In: Proceedings of the ECAI-2004 Workshop on Agent-mediated Knowledge Management (AMKM 2004), pp. 29–38 (2004)

8. Vrandečić, D., Pinto, S., Tempich, C., Sure, Y.: The DILIGENT knowledge process. Journal of Knowledge Management 9(5), 85–96 (2005)

9. Casanovas, P., Casellas, N., Tempich, C., Vrandečič, D., Benjamins, V.R.: OPJK and DILIGENT: ontology modelling in a distributed environment. Artificial Intelligence and Law 15, 171–186 (2007)

10. Bailin, S.C., Truszowski, W.: Ontology Negotiation: How Agents Can Really Get to Know Each Other. In: Truszkowski, W., Hinchey, M., Rouff, C.A. (eds.) WRAC 2002. LNCS, vol. 2564, pp. 320–334. Springer, Heidelberg (2003)

11. Fehr, E., Camerer, C.F.: Social neuroeconomics: the neural circuitry of social preferences. TRENDS in cognitive science 11(10), 227–419 (2007)

12. Bainbridge, W.S., Roco, M.C.: Manging Nano-Bio-Info-Cogno Innovations. Converging Technologies in Society. Springer, Heidelberg (2006)

13. Castelfranchi, C., Giardini, F., Marzo, M.: Relationships between rationality, human motives, and emotions. Mind & Society 5, 173–197 (2006)

14. Aureli, F., de Waal, F.M.B. (eds.): Natural Conflict Resolution. University of California Press (2000)

15. de Waal, F.M.B.: Primates – A Natural Heritage of Conflict Resolution. Science 289, 586–590 (2000)

16. Preston, S., de Waal, F.M.B.: Empathy: Its ultimate and proximal bases. Behavioral and Brain Sciences 25, 1–72 (2002)

17. Damasio, A.R.: The somatic marker hypothesis and the possible functions of the prefrontal cortex. Philosophical Transactions: Biological Sciences, 351 1346, 1420–1513 (1996)

18. Singer, T., Frith, C.: The painful side of empathy. Nature Neuroscience 8(7), 845–846 (2005)

19. Fehr, E., Fischbacher, U., Gätcher, S.: Strong Reciprocity, Human Cooperation and the Enforcement of Social Norms. Human Nature 13, 1–25 (2002)

20. Murray, R.J.: Forgiveness as a Therapeutic Option. The Family Journal: Counseling and Therapy for Couples and Families 10(3), 315–321 (2002)

21. McCullough, M.E., Rachal, K.C., Worthington Jr., E., Brown, S.W., Hight, T.L.: Interpersonal Forgiving in Close Relationships: II, Theoretical Elaboration and Measurement. Journal of Personality and Social Psychology 75(6), 1586–1603 (1997)

22. Ekman, P.: Emotions Revealed. Understanding Faces and Feelings. Widenfield & Nicholson, London (2003)

23. Nerlich, B., Clarke, D.D.: Semantic fields and frames: Historical explorations of the interface between language, action, and cognition. Journal of Pragmatics 32, 125–150 (2000)

24. Levinson, S.: Pragmatics. Cambridge University Press, Cambridge (1983)

25. Foley, W.A.: Anthropological Linguistics. An Introduction. Blackwell Publ., Oxford (1997)

26. Saville-Troike, M.: The Ethnography of Communication. An Introduction. Blackwell Publ., Oxford (2003)

27. Blum-Kulka, S., House, J., Kasper, G.: Cross-cultural pragmatics: Requests and Apologies, Ablex, Norwood, NJ (1989)

28. Rumelhart, D.E., McClelland, J.L., the PDP Research Group: Parallel Distributed Processing. Explorations in the Microstructure of Cognition. vol. 1. Foundations. vol. 2. Psychological and Biological Models. The MIT Press, Cambridge (1986)

29. Clancey, W.J., Sachs, P., Sierhus, M., Hoof, R.V.: Brahms: simulating practice for work systems design. International Journal of Human-Computer Studies 49, 831–865 (1998)
30. Deutschmann, M.: Apologising in British English, Doctoral Dissertation. Skrifter från moderna språk 10. Institutionen för moderna språk, Umeå Universitet (2003)
31. Nader, L. (ed.): Law in Culture and Society. Aldine Publ., Chicago (1969)
32. Lederach, P.: Building Peace. Sustainable Reconciliation in Divided Societies. United States Institute of Peace Press, Washington (1997)
33. Theidon, K.: Justice in Transition. The Micro-politics of Reconciliation in Postwar Peru, Journal of Conflict Resolution 50(3), 433–457 (2006)
34. Raiffa, H.: The Art and Science of Negotiation. Harvard University Press, Cambridge (1982)
35. Garfinkel, M.R., Skarpedas, S.: Economics of Conflict: An Overview. University of California (Irvine) (2006),
    http://ideas.repec.org/p/irv/wpaper/050623.html
36. Yiu, K.T.W., Cheng, C.O.: A Study of Construction Mediator Tactics. Part II: The Contiongent Use of Tactics, Building and Environment 42(I2), 752–761 (2007)
37. Chebat, J.C., Slusarczyk, W.: How emotions mediate the effects of perceived justice on loyalty in service recovery situations. Journal of Business Research 58, 664–673 (2005)
38. Aquino, K., Tripp, T., Bies, R.J.: Getting Even or Moving On? Power, procedural Justice, and Types of Offenses as Predictors of Revenge, Forgiveness, Reconciliation, and Avoidance in Organizations. Journal of Applied Psychology 91(3), 653–668 (2006)
39. Simon, H.A.: Making Management Decisions:The Role of Intuition and Emotion, Academy of Management Executive, February, pp. 57–64 (1987)
40. Kahnemann, Daniel: Maps of Bounded Rationality: Psychology for Behavioral Economics. The American Economic Review 5, 1449–1475 (2003)
41. Braithwaite, J.: Meta-regulation for Access to Justice: Presentation to General Aspects of Law (GALA). Seminar series, University of California, Berkeley, November 13 (2003),
    http://www.law.berkeley.edu/centers/kadish/gala03/Braithwaite%20Kent.pdf
42. Jameson, J.K., Bodtker, A.M., Jone, T.: Like Talking in a Brick Wall: Implications of Emotion Metaphors for Mediation Practice. Negotiation Journal 22(2), 199–207 (2006)
43. Cobb, S., Rifkin, J.: Practice and Paradox: Deconstructing neutrality in Mediation. Law and Social Inquiry 16(1), 36–62 (1991)
44. Fisher, R., Ury, W.: Getting to Yes. Negotiating Agreement Without Giving. Houghton Mifflin Company (1981)
45. Fisher, R., Shapiro, D.: Beyond Reason. Using Emotions as You Negotiate. Random House Business Books (2006)
46. Hoffman, D.: Paradoxes of Mediation, American Association Dispute Resolution Magazine, Fall/Winter (2002) (2005), http://bostonlawcollaborative.com/documents/2005-07-paradoxes-of-mediation.pdf
47. Olson, S.M., Dzur, A.W.: Revisiting Informal Justice: Restorative Justice and Democratic Professionalism. Law & Society Review 38(1), 139–176 (2004)
48. Uijjttenbroek, E.M.: The influence of motives and styles in mediation online dispute resolution. In: Lodder, A., Rule, C., Zeleznikow, J. (eds.) Proceedings of 4th International Workshop on ODR, Palo Alto, June 8, pp. 31–35 (2007)
49. Raines, S.S.: Can Online Mediation Be Transformative? Tales From the Front. Conflict Resolution Quarterly 22(4), 437–451 (2005)
50. De Dreu, C.K.W., Carnevale, P.J.: Disparate Methods and Common Findings in the Study of Negotiation. International negotiation 10, 193–203 (2005)

51. Ben-Ze'ev, A.: Privacy, emotional closeness, and openness in cyberspace. Computers in Human Behavior 19, 451–467 (2003)
52. Walton, D., Lodder, A.: What Role can Rational Argument Play in ADR and Online Dispute Resolution. In: Zeleznikow, J., Lodder, A. (eds.) Second International ODR Workshop. Wolf Legal Publishers, Tilburg (2005)
53. Putnam, L.: Transformations and Critical Moments in Negotiations. Negotiation Journal 20(2), 275–295 (2004)
54. Lodder, A.R., Zeleznikow, J.: Developing an Online Dispute Resolution Environment: Dialogue Tools and Negotiation Support Systems in a Three-Step Model. Harvard Negotiation Law Review 10, 237–288 (2005)
55. Cohen, R.: Negotiating Across Cultures, 2nd edn. Institute for Peace, Washington (1997)
56. Aakhus, M.: Neither Naïve nor Critical Reconstruction: Dispute Mediators, Impasse, and the Design of Argumentation. Argumentation 17, 265–290 (2003)
57. Gordon, T., Märker, O.: Mediation Systems, Online mediation. In: Märker, O., Trénel (eds.) Neue Medien in Der Konfliktvermittung-Mit Bespielen Aus Politik Und Wirtschaft, Sigma edn., Berlin, pp. 61–84 (2002); Thiessen, E., Zeleznikow, J.: Technical Aspects of Online Dispute Resolution—Challenges and Opportunities, http://www.odr.info/unforum2004/thiessen_zeleznikow.htm
58. Katsh, E.: Online Dispute Resolution: Some Implications for the Emergence of Law in Cyberspace, International Review of Law Computers & Technology, 21(2), 97–107 (2007)
59. Walton, D.: The place of Dialogue Theory in Logic, Computer Science and Communication Studies. Synthese 123, 327–346 (2000); Sartor, G.: A Teleological Approach to Legal Discourses, EUI WP LAW n. 28 (2006)
60. Walton, D., Godden, D.M.: Persuasion dialogue in online dispute resolution. AI and Law 13, 273–295 (2005)
61. Tamma, V., Phelps, S., Dickinson, I., Wooldridge, M.: Ontologies for supporting negotiation in e-commerce. Engineering Applications of Artificial Intelligence 18, 223–236 (2005)
62. Ermolayev, V., Keberle, N., Tolok, V.: OIL Ontologies for Collaborative Task Performance in Coalitions of Self-Interested Actors. In: Arisawa, H., Kambayashi, Y., Kumar, V., Mayr, H.C., Hunt, I. (eds.) ER Workshops 2001. LNCS, vol. 2465, pp. 390–402. Springer, Heidelberg (2002)
63. Anumba, C.J., Ren, Z., Thorpe, A., Ugwu, O.O., Newnham, L.: Negotiation within a multiagent system for the collaborative design of light industrial buildings. Advances in Engineering Software 34, 389–401 (2003)
64. Bailin, S.C., Truszkowski, W.: Ontology Negotiation Between Intelligent Information Systems. The Knowledge Engineering Review 17(1), 7–19 (2002)

Appendix: Tables

Table 1. Basic empirical research on mind, language, empathy and emotions

Fields	Authors	Focus	Object	Methodology
Social Neuroscience	Farrow, Clark Lawrence Singer, Saxe Damasio LeDoux , Harris, Iacoboni, Preston	Brain and neural functioning	Empathy and emotions (forgiveness) in brain areas. Somatic markers hypothesis (SMH).	Controlled lab experiments, Functional MRI (fMulti Image Ressonance)
Cognitive Science	Gardner Minsky Rumelhart McClelland Hollan, Hutchins, Thagard	Intelligence and social behavior	ToM (Theory of Mind). Scripts and cognitive patterns in social cognition	Neural networks, scripts, cognitive modeling
Primatology	De Waal, Aureli Flack, Cords Schaffner	Aggression and conciliatory behavior	Empathy and cooperative interaction patterns in monkeys, apes and humans. Relational model of aggression.	Behavioral observation, social networks, distance analysis, kinship, social niches reconstruction
Basic Social Psychology Research	Ekman, Rolls Lazarus Bandura , Ortony, Gallup	Categorization, basic emotions and communicative behavior	Conceptual representation, empathy and universal (basic) emotions in human expression (bodily and linguistic)	Neural networks, controlled lab experiments, genetics, regression analysis, prototype and semantic analysis

Table 2. Applied social psychology on empathy, forgiveness, apologies and evaluation

Fields	Authors	Focus	Object	Methodology
Social Psychology, Therapy and Counseling	Enright, McCullough Barber, Allan Worthington Wade, Rye Kearns, Franz Jolliffe, Petrucci Mcpherson Finchman Mullet, Murray	Forgiveness, guilt, shame and basic emotions in interactions, family situations and collective behavior	Empathy, faceworking, forgiveness, apology, revenge and avoidance patterns in conflict, post-conflict situations and mediation.	Controlled lab experiments, tests, descriptive statistics, regression analysis, evaluation scales (TRIM, PTRIM), prototype analysis
Social Psychology and Narrative Analysis	Baumeister Zeichmeister Stillwell Schütz Leith Romero	Narratives by victims and offenders	Identity and roles. Speech and writing patterns expressing emotions (shame and guilt)	Textual and narrative analysis, rol games, controlled lab experiments.
Social Psychology and Criminology	Strang Maxwell Morris Sherman Karremans Bayley Robbins Arrigo Darley, Harris Hay, Kressel van den Boss	VOM/RJ evaluation, recidivism, victim-offender attitudes. Juveniles. Divorce.	Personality factors, family disorders, violence, restoration effects.	Evaluation analysis. Indexes, scales. Regression analysis. Meta-analysis.

Table 3. Applied linguistic research on politeness, apologies, excuses and cultural contexts

Fields	Authors	Focus	Object	Methodology
Frame Semantics and Cognitive Linguistics	Fillmore, Lakoff, Johnson	Structure of meaning encoding pragmatic and encyclopedic knowledge	Frames, scripts, prototypes behind lexical groups of words. Rhetorical devices esp. metaphor.	Linguistic phenomenology, prototype analysis
Linguistics and Cross-cultural Pragmatics	Tannen, Wierzwicka, Blum-Kulka, Nelson, Rojo Risako, Liebersohn	Cultural competence and linguistic assumptions within social behavior	Discourse, and event and speech acts in apologies and excuses in diverse cultural contexts and natural languages	Linguistic phenomenology, comparative data analysis,
Linguistics and Functional Pragmatics	Tannen, Fraser, Jacobs Heisterkamp, Glover Holmes Stewart	Pragmatic and linguistic competence in verbal and non-verbal interactions	Frames, code-switching, distal deixis, contextual-cues, in apologies, negotiation and mediation sessions	Linguistic phenomenology, transcript and videotape analysis
Sociolinguistics, Discourse Analysis and Corpus-based analysis	Brown, Levinson, Deutchsmann, Davies, Márquez Meier	Politeness, social and linguistic and pragmatic rules in social behavior	Idiolects. Sociolects. Structure, types and functions of excuses and apologies in verbal sequences in natural language (English, Spanish...)	Linguistic phenomenology, descriptive statistics, data-bases organization methods, prototype analysis

Table 4. Sociological research on micro-situations, cognition, emotions and discourse

Fields	Authors	Focus	Object	Methodology
Micro or Interactional Sociology	Simmel, Goffman, Scheff, Retzinger	Social emotions in interactions. Violence cycles and social structure.	Face-working and human interaction patterns: shame as a social bond	Participant observation, ethnography, interviews, conceptual modeling.
Ethnomethodology, Discourse and Conversational Analysis	Garfinkel, Goodwin, Antaki Garcia Greatbach Dingwall Presser Lowerkamp	Discourse sequences, turn talking, adjancy pairs, code-switching	Argumentative talking patterns and situated cognition	Audio and videotaping, transcript data analysis, discourse and cognitive analysis, conceptual modeling
Cognitive Sociology	Cicourel Engeström Middleton Edwards	Social distributed cognition in organizations, institutions and workplaces	Situated meaning: explicit and implicit assumptions in interaction patterns	Audio and videotaping, transcript data analysis, discourse and cognitive analysis

Table 5. Research on social and political violence, conflict resolution, reconciliation, allocation of resources and rights, and neural bases of preferences

Fields	Authors	Focus	Object	Methodology
Anthropology and Conflict Resolution	Bateson, Nader Greenhouse Lederach Roberts Theidon Davidheiser	Violence and reconciliation in communities and societies	Transitional justice, forgiveness and reconciliation patterns	Ethnography, history, narrative analysis, conceptual modeling
Communication and Intercultural Conflict Studies	Hammer, Ting-Toomey Oetzel Kurogi Van Ginkel Goto Chan Trubisnky	Cultural competence in inter-ethnic conflicts.	Conflict resolution styles in individualist and collectivist cultures.	Factor analysis, regression, focus groups, conceptual modeling
Political Science, Conflict Resolution and International Relations Studies	Berkowitz, David, Choi Didier, Marret Lefranc, Minow	Inter, intra and infra-state conflicts	Policy, transitional justice and peace processes	Game theory, social networks, history, regression and path analysis, conceptual modeling
Economics and Conflict Resolution	Arrow, Raiffa, Axelrod, Harsanyi, Garfinkel, Fehr	Allocation of resources, effects of conflict on economic outcomes, property rights, neurological bases	Policy, competitive decision making, cooperative behavior, theory of complexity, neuroeconomics	Game theory, choice theory, optimization techniques statistics,probabilistic calculus, axiomatic models

Table 6. Empirical and theoretical research on dialogue, argumentation, negotiation, and mediation

Fields	Authors	Focus	Object	Methodology
Argumentation and Dialogue	Walton, van Eemeren, Jacobs, Grootendors, Jackson, Prakken Sartor, Lodder	Practical and dialectical reasoning	Inferences, chains and arguments. Induction, deduction and abduction.	First order, modal and non-monotonic logic, observation, conceptual modeling (dialectic systems)
Negotiation Studies	Carnevale De Moor Cobb, Ross Putnam Kolb, Botker Jameson,Dewulf, Karsten, Curhan, van Merode	Communication and strategic behavior in mediation and dispute resolution	Agreement, settlement-driven processes. Strategic and tactic moves	Game theory, descriptive statistics, focus groups, participant observation, narrative analysis
Management and Organization Studies	Simon, Chen, Yiu Cheung, Bradfield, Chebat Aquino, Benoit Tomlison Dörrenbächer	Decision making in workplaces and organizations. Communication inter and intra-organizations.	Trust repair, image repair, organizational design, conflict management and tactic behavior	AI design, regression analysis, game and bounded rationality analysis, conceptual modeling
Applied Artificial Intelligence	Zeleznikow, Katsh, Bench-Capon, Wyner Bellucci, Nitta Aakhus, Aldrich Tzeng, Gordon Ben-Ze'ev Castelfranchi	Human-machine interaction and design	Internet. Computer/user interface. On-line Dispute Resolution (ODR). Multiple-Agent Systems (MAS)	Program design, game theory, program implementation cycles, conceptual modeling

Table 7. Criminological and judiciary research on practices and outcomes of mediation and VOM

Fields	Authors	Focus	Object	Methodology
Criminology	Christie, Wall, van Stokkom, Daly, Pelikan, Cohen, Dignan, Mika, Young, Bazemore, de Haan, Strang, Cohen, Morris	Practices of VOM, Mediation, Juvenile justice. Conferencing	RJ outcomes and processes. Low and severe crimes. Juvenile and gendered violence/ restoration.	Participant observation, surveys, descriptive statistics, ethnography, meta-analysis
Social Work and Professional Mediators' Studies	Umbreit, Vos Coates, McCold Démaret, MacKay, Schroeder Hofman, Raines Coppola, Tyler Crossland ,Sousa	Professional counseling, policing, and practice	Mediation processes and outcomes. Tactics, strategies, cases.	Reflective practice, case analysis, professional experience accounts, descriptive statistics.
Comparative RJ and Judicial Studies	Miers, Martin, Willensems, Mestitz,Getti, Aertsen, Deklerck, Vanfraenchem	RJ organization and practice. Linkage to judicial settings and administrative settings.	Functional anchorage of RJ (juveniles, VOM, mediation) into legal and judicial systems in national countries.	Semi-structured and in-deep interviews, surveys, descriptive statistics, comparative organizational data analysis

Table 8. Legal, social, political and philosophical framework (Rule of Law) for rights and values

Fields	Authors	Focus	Object	Methodology
Socio-legal studies	Olson, Dzur Roche, Faget Bonnafé-Schmitt Lemley, Hudson, Hoffmann, Morrill, Morrison, Resta	Regulation, transformation of law, policy, legal anchored mediation and professional work	Institutionalization of RJ practices in a globalized world. Deliberative democracy, governance and law	Legal, political and socio-legal modeling. Meta-analysis.
Legal Theory, Rights and Jurisprudence	Zehr, Braithwaite Walgrave, van Ness, von Hirsh, Hartmann, Bosnjak, Ashworth Johnstone Benstain Reimund	RJ as a general transforming model	Reflection on the RJ political and legal paradigm. Role and function of RJ within changing Rule of Law models.	Legal design, rights' discourse, legal, criminal and political argumentation. Normative analysis
Philosophy and Ethics	Wright, Elster, Nussbaum, Bennett Smith	Distributive justice, restorative justice, and moral values	Justice within socioeconomic and political models	Normative analysis

Author Index

Lecture Notes in Artificial Intelligence (LNAI)

Vol. 4944: Z.W. Raś, S. Tsumoto, D.A. Zighed (Eds.), Mining Complex Data. X, 265 pages. 2008.

Vol. 4938: T. Tokunaga, A. Ortega (Eds.), Large-Scale Knowledge Resources. IX, 367 pages. 2008.

Vol. 4933: R. Medina, S. Obiedkov (Eds.), Formal Concept Analysis. XII, 325 pages. 2008.

Vol. 4930: I. Wachsmuth, G. Knoblich (Eds.), Modeling Communication with Robots and Virtual Humans. X, 337 pages. 2008.

Vol. 4929: M. Helmert, Understanding Planning Tasks. XIV, 270 pages. 2008.

Vol. 4924: D. Riaño (Ed.), Knowledge Management for Health Care Procedures. X, 161 pages. 2008.

Vol. 4923: S.B. Yahia, E.M. Nguifo, R. Belohlavek (Eds.), Concept Lattices and Their Applications. XII, 283 pages. 2008.

Vol. 4914: K. Satoh, A. Inokuchi, K. Nagao, T. Kawamura (Eds.), New Frontiers in Artificial Intelligence. X, 404 pages. 2008.

Vol. 4911: L. De Raedt, P. Frasconi, K. Kersting, S. Muggleton (Eds.), Probabilistic Inductive Logic Programming. VIII, 341 pages. 2008.

Vol. 4908: M. Dastani, A. El Fallah Seghrouchni, A. Ricci, M. Winikoff (Eds.), Programming Multi-Agent Systems. XII, 267 pages. 2008.

Vol. 4898: M. Kolp, B. Henderson-Sellers, H. Mouratidis, A. Garcia, A.K. Ghose, P. Bresciani (Eds.), Agent-Oriented Information Systems IV. X, 292 pages. 2008.

Vol. 4897: M. Baldoni, T.C. Son, M.B. van Riemsdijk, M. Winikoff (Eds.), Declarative Agent Languages and Technologies V. X, 245 pages. 2008.

Vol. 4894: H. Blockeel, J. Ramon, J. Shavlik, P. Tadepalli (Eds.), Inductive Logic Programming. XI, 307 pages. 2008.

Vol. 4885: M. Chetouani, A. Hussain, B. Gas, M. Milgram, J.-L. Zarader (Eds.), Advances in Nonlinear Speech Processing. XI, 284 pages. 2007.

Vol. 4884: P. Casanovas, G. Sartor, N. Casellas, R. Rubino (Eds.), Computable Models of the Law. XI, 341 pages. 2008.

Vol. 4874: J. Neves, M.F. Santos, J.M. Machado (Eds.), Progress in Artificial Intelligence. XVIII, 704 pages. 2007.

Vol. 4870: J.S. Sichman, J. Padget, S. Ossowski, P. Noriega (Eds.), Coordination, Organizations, Institutions, and Norms in Agent Systems III. XII, 331 pages. 2008.

Vol. 4869: F. Botana, T. Recio (Eds.), Automated Deduction in Geometry. X, 213 pages. 2007.

Vol. 4865: K. Tuyls, A. Nowe, Z. Guessoum, D. Kudenko (Eds.), Adaptive Agents and Multi-Agent Systems III. VIII, 255 pages. 2008.

Vol. 4850: M. Lungarella, F. Iida, J.C. Bongard, R. Pfeifer (Eds.), 50 Years of Artificial Intelligence. X, 399 pages. 2007.

Vol. 4845: N. Zhong, J. Liu, Y. Yao, J. Wu, S. Lu, K. Li (Eds.), Web Intelligence Meets Brain Informatics. XI, 516 pages. 2007.

Vol. 4840: L. Paletta, E. Rome (Eds.), Attention in Cognitive Systems. XI, 497 pages. 2007.

Vol. 4830: M.A. Orgun, J. Thornton (Eds.), AI 2007: Advances in Artificial Intelligence. XIX, 841 pages. 2007.

Vol. 4828: M. Randall, H.A. Abbass, J. Wiles (Eds.), Progress in Artificial Life. XII, 402 pages. 2007.

Vol. 4827: A. Gelbukh, Á.F. Kuri Morales (Eds.), MICAI 2007: Advances in Artificial Intelligence. XXIV, 1234 pages. 2007.

Vol. 4826: P. Perner, O. Salvetti (Eds.), Advances in Mass Data Analysis of Signals and Images in Medicine, Biotechnology and Chemistry. X, 183 pages. 2007.

Vol. 4819: T. Washio, Z.-H. Zhou, J.Z. Huang, X. Hu, J. Li, C. Xie, J. He, D. Zou, K.-C. Li, M.M. Freire (Eds.), Emerging Technologies in Knowledge Discovery and Data Mining. XIV, 675 pages. 2007.

Vol. 4811: O. Nasraoui, M. Spiliopoulou, J. Srivastava, B. Mobasher, B. Masand (Eds.), Advances in Web Mining and Web Usage Analysis. XII, 247 pages. 2007.

Vol. 4798: Z. Zhang, J.H. Siekmann (Eds.), Knowledge Science, Engineering and Management. XVI, 669 pages. 2007.

Vol. 4795: F. Schilder, G. Katz, J. Pustejovsky (Eds.), Annotating, Extracting and Reasoning about Time and Events. VII, 141 pages. 2007.

Vol. 4790: N. Dershowitz, A. Voronkov (Eds.), Logic for Programming, Artificial Intelligence, and Reasoning. XIII, 562 pages. 2007.

Vol. 4788: D. Borrajo, L. Castillo, J.M. Corchado (Eds.), Current Topics in Artificial Intelligence. XI, 280 pages. 2007.

Vol. 4775: A. Esposito, M. Faundez-Zanuy, E. Keller, M. Marinaro (Eds.), Verbal and Nonverbal Communication Behaviours. XII, 325 pages. 2007.

Vol. 4772: H. Prade, V.S. Subrahmanian (Eds.), Scalable Uncertainty Management. X, 277 pages. 2007.

Vol. 4766: N. Maudet, S. Parsons, I. Rahwan (Eds.), Argumentation in Multi-Agent Systems. XII, 211 pages. 2007.

Vol. 4760: E. Rome, J. Hertzberg, G. Dorffner (Eds.), Towards Affordance-Based Robot Control. IX, 211 pages. 2008.

Vol. 4755: V. Corruble, M. Takeda, E. Suzuki (Eds.), Discovery Science. XI, 298 pages. 2007.

Vol. 4754: M. Hutter, R.A. Servedio, E. Takimoto (Eds.), Algorithmic Learning Theory. XI, 403 pages. 2007.

Vol. 4737: B. Berendt, A. Hotho, D. Mladenic, G. Semeraro (Eds.), From Web to Social Web: Discovering and Deploying User and Content Profiles. XI, 161 pages. 2007.

Vol. 4733: R. Basili, M.T. Pazienza (Eds.), AI*IA 2007: Artificial Intelligence and Human-Oriented Computing. XVII, 858 pages. 2007.

Vol. 4724: K. Mellouli (Ed.), Symbolic and Quantitative Approaches to Reasoning with Uncertainty. XV, 914 pages. 2007.

Vol. 4722: C. Pelachaud, J.-C. Martin, E. André, G. Chollet, K. Karpouzis, D. Pelé (Eds.), Intelligent Virtual Agents. XV, 425 pages. 2007.